Pest and Disease Managemer Farmers, Growers and Sn

A COMPLETE GU

Pest and Disease Management for Organic Farmers, Growers and Smallholders

A COMPLETE GUIDE

Gareth Davies, Phil Sumption
and Anton Rosenfeld

In association with Garden Organic

THE CROWOOD PRESS

First published in 2010 by
The Crowood Press Ltd
Ramsbury, Marlborough
Wiltshire SN8 2HR

www.crowood.com

British Library Cataloguing-in-Publication Data
A catalogue record for this book is available from the British Library.

ISBN 978 1 84797 150 0

Typeset by SR Nova Pvt Ltd, Bangalore, India

Printed and bound in Singapore by Craft Print International Ltd

Contents

Conventions Used in This Book

PLANT, PEST AND DISEASE NAMES

The first time a crop, weed, pest or disease is mentioned we give the commonly accepted UK name followed by the standard biologically accepted (*Latin*) name to provide an absolute reference. This is because common names vary widely between regions and countries. Thereafter the same common name is used although the full Latin name is repeated where this increases understanding or provides an important reference point. The tables (9–17) on common pests and diseases of crops also provide a reference point for common and Latin names.

ABBREVIATIONS

°C	(degrees) centigrade
cm	centimetres
g	grams
ha	hectare
m(2)	metre (squared)
mm	millimetre
£	UK pound (sterling)

ACRONYMS

AM (F)	Arbuscular mycorrhizal (fungi)
ACOS	(UK) Advisory Committee on Organic Standards
CRD	(UK) Chemical Regulation Directorate
Defra	(UK) Department of Environment, Food and Rural Affairs
EU	European Union
FWAG	Farming and Wildlife Advisory Group
GM	Genetic modification
GMO	Genetically modified organism

GO	Garden Organic (formerly The Henry Doubleday Research Association)
IFOAM	International Federation of Organic Agriculture Movements
IPM	Integrated Pest (and Disease) Management
NIAB	(formerly) National Institute of Agricultural Botany
OELS	Organic Entry Level Scheme (UK)
PSD	Pesticide Safety Directorate (now superseded by CRD)
RSPB	Royal Society for the Protection of Birds
UK	United Kingdom
VAM	Vesicular arbuscular mycorrhizal fungi

Preface and Acknowledgements

PREFACE

All farmers and growers are conversant with pests and diseases, and none more so than organic farmers and growers. The incidence of pests and/or disease is in some ways the most obvious manifestation of the state of a crop, and often provides the most striking physical signs that the crop is not healthy. Perhaps because of this it is also one area that has received a great deal of research attention over the years, and countless technological solutions have been proposed for 'controlling' pests and diseases, especially during the last thirty or so years.

In this book we have tried to present an alternative viewpoint, one that puts the ecological reality of crop protection at the centre of the farm view and at the centre of any attempt to control pests and diseases. In fact one of the consequences of this view is to question the belief that pests and diseases can be controlled in any strict sense of the word, but to promote the idea that the farm system should be managed to suppress those things that we don't want, such as pests and diseases, and promote those that we do, namely healthy crops producing a sustainable yield.

Members of the organic movement are strong advocates of this world view. Organic agriculture holds that crop 'health' is central to farming, and that this arises from both soil, and wider ecosystem, health. This in turn depends on promoting diversity, and in promoting strong links between the diverse elements that make up the farm system – its soil, its water, its organisms, and the people who interact with it. A great deal of ecological research has also begun to back this position. A teaspoon full of soil contains a myriad of organisms, the vast majority microscopic. A cursory examination of any crop canopy or field margin will reveal a host of (mainly) small creatures. The fact is that most of them are kept in bounds most of the time, and in fact most of them are highly beneficial. The relationships between these organisms that make up the soil and terrestrial food webs can be incredibly complex. It is these relationships and processes that serve to ensure that nutrients and energy cycle through the ecosystem, and are thus in effect vital for plant health. Pest and disease organisms are just a small part of this complexity.

Taking this perspective it becomes pertinent to ask what will be achieved by attempting to sweep away and drastically simplify this system: to remove the complexity? What are the chances that removing large segments of this diversity, say by applying an insecticide or fungicide, will actually improve crop health? Modern ecological farmers have asked this question and found the answer, which is to reformulate it to ask how we can work *with* this system rather than knock it down. The truth is that pests and diseases are thus just a small facet of agroecosystems, albeit at times a frustrating and costly one, but that even they have an ecological place or niche.

The central tenet of organic agriculture is that feeding the soil will feed the crop. To this we could add that feeding the soil will feed the ecosystem. In this book we have attempted to promote pest and disease management from the viewpoint of ecosystem management. Pest and disease management on organic farms is about working with and emulating the natural cycling of nutrients and resources through the farm, and the experience of countless organic farmers and growers is that this works. It is in this spirit that this book is presented. We hope that it will be a sound basis for giving farmers and growers an understanding of how pests and diseases fit into their farm systems and in giving them ideas on how to build pest and disease management into their crop enterprises.

Lastly we would like to say that we hope this book is not an end in itself, but that farmers and growers will improve on the knowledge in it by sharing their ideas, successes and enthusiasm with each other. In doing this they will help develop the organic movement in all its diversity, at the same time as playing a vital role in the ongoing development of organic pest and disease management practice.

ACKNOWLEDGEMENTS

The authors wish to thank the numerous people involved in bringing this book to publication, and in particular the various people who have taken part in, and contributed to the organic research programmes at Garden Organic. We particularly wish to thank colleagues in other organizations who have greatly contributed to our knowledge about pests and diseases, and without whose contributions to our research projects we would have been much the poorer. Special mention must go to Peter Gladders of ADAS, who ran the DOVE (Diseases of Organic Vegetables) project, financed by Defra and aimed at improving advice to organic growers on disease control; and to Rosemary Collier of the University of Warwick (HRI), who has collaborated on many organic pest management programmes, also in the main funded by Defra. Much of the information in this book stems from these projects.

We would also like to thank the numerous farmers and growers who have contributed their knowledge, experience and information to our research programmes, and whose own research and stories have greatly improved our understanding over the years. It goes without saying that any mistakes or omissions are ours and not theirs.

1 Principles for Organic Pest and Disease Management

Managing pests and diseases on crop plants is one of the principal concerns for farmers or growers converting to organic production, but experience has shown that if ecological practices are incorporated into the production system they are not normally as damaging as at first feared. Obviously a large number of pest and disease species attack crops, but contrary to initial impressions, they actually only encompass a small proportion of the total number of species that can be found in a typical farming landscape. An ecologist as opposed to a farmer might ask why there are not more pests and diseases, and might note that farmers and growers are fortunate in that they only actually have to deal with a limited number of organisms intent on damaging their crops, and that in fact the hidden majority of the animals and micro-organisms on the farm actually aid the production of crops. Indeed, one of the central tenets of organic farming practice is that farming is embedded in ecology, and that only a deep understanding of agro-ecosystems, which embraces all organisms on the farm, will lead to sustainable, profitable and equitable farming approaches. Pests and diseases of crop plants need to be understood in this context.

The aim of this book is therefore twofold. Firstly, and straightforwardly, it aims to describe the causes of pest and disease problems that organic farmers and growers face, and also the principal problems encountered in temperate organic vegetable and cereal production systems. Secondly, and perhaps more importantly, it aims to provide an underlying framework that will help organic farmers and growers take decisions about managing pests and diseases on their farms, and so limit the damage they cause. This is necessary because pest and disease management is dynamic and the decisions that need to be made will vary depending on the farm, the season and the year. What is required in this situation is, above all, a flexible approach that plans to prevent problems but is also capable of reacting appropriately when problems arise. We concentrate on those methods suited to small, medium and large scale commercial production in temperate agriculture and horticulture, although a lot of the information is also applicable in a wider range of crop and vegetable production situations.

In this chapter we develop a basic framework for approaching pest and disease management in organic farming systems. In subsequent chapters we expand upon both the implications of these ideas and the methods available to organic farmers and growers to manage pests and diseases by describing in more detail the methods available for preventing pest and disease populations building up, and the measures available to more directly control them. It is important to understand the likely costs of pest management techniques in order to offset this against any potential benefits of leaving them, and we present ideas on how to assess these costs and benefits. We go on to provide a summary digest of the pests and diseases that are more likely to be encountered, together with a range of techniques or approaches that might be adopted as part of any strategies for their management. We also illustrate how they might be incorporated into integrated cropping strategies in organic vegetable and cereal production systems.

GENERAL PRINCIPLES FOR PEST AND DISEASE MANAGEMENT IN ORGANIC SYSTEMS

Overview

Pests and diseases can be classified in many different ways, but for the purposes of organic pest and disease management it makes sense to classify them according to their functional role in ecological systems. Many books also classify them according to their notional place in the 'tree of life', where plants, animals and microorganisms are grouped by their degree of affinity and/or relatedness. In practice most classification systems compromise between the two methods as they are both useful depending on the circumstance.

In a functional and ecological context pests are generally herbivorous organisms that attack, eat, damage, and sometimes kill agricultural crop plants. Characteristically they are animals that seek out and devour plant tissues, so causing damage. The familiar 'holes' seen in leaves are a familiar example of herbivore damage, but herbivores also cause a wide range of other symptoms (*see* Chapters 7 and 8). In many cases they are limited to being able to exploit and eat a few different plant species or a group of plant species, and in this case they have numerous adaptations for locating their 'scarce' host plants and digesting them. Others have an ability to eat a wide range of plant species, so-called generalists, although they still often have adaptations for locating, processing and digesting a limited (though broad) range of plant types (*see* Chapter 2). Invertebrate pest species include

Typical pest herbivore: caterpillars on brassicas.

flatworms, nematodes, molluscs, arachnids and especially insects. Vertebrate pests are mainly birds and mammals such as rabbits, deer and rats.

A disease is broadly defined as an impairment of the normal functioning of an organism, and diseases are generally caused by pathogens. Pathogens are normally microscopic organisms and are characteristically found within plant tissues or cells which they find more or less by chance, although some are carried to plants by herbivores or other living vectors. Plant pathogens have a wide range of adaptations for gaining entry to, and taking over, a plant's normal metabolism and diverting it to their own ends, which is to grow and produce more pathogens. Because they have limited mobility, they often survive from one generation to the next by producing large numbers of infective bodies that are released into the environment and are more or less resistant to decay. This life strategy can lead to long-term and persistent disease problems once pathogens become established on a farm.

Diseases manifest themselves through a large range of symptoms, from wilting to spotting to blotching to death (*see* Chapter 9). Pathogens include fungi, protozoans, bacteria, mycoplasmas and viruses. In some cases nutritional disorders are sometimes also referred to as diseases, but they are not discussed in this book as a discussion of their management more properly belongs in a book on soil fertility.

Nematodes and other small invertebrate organisms can be referred to as either pests or diseases in different books. This is perhaps because they are normally found within plant tissues but also have a limited ability to seek out plants, and are sometimes also transmitted by herbivores or other vectors so demonstrating a bridge between classic pest and pathogen types. In this book we consider them to share more of the characteristics of pests and refer to them as such.

Typical disease: grey mould on basil.

DAMAGE DUE TO PESTS AND DISEASES

Pests and diseases are normally only remarked upon when they are having a detrimental effect on people, or impact negatively on their aesthetic expectations. In the widest sense, pests and pathogens compete with humans for natural resources, or directly feed on people, their crops, and/or their livestock thus causing some form of 'damage'. It is a truism that a pest or pathogen in one situation (say, a crop) can be an interesting herbivore or parasite in another (in a nature reserve). In many cases it is also important to distinguish between the injury that pests and diseases cause, which is their direct deleterious effect on the host plant, and the damage that they cause, which is the measurable loss caused by the injury. Damage is normally measured as a loss of yield or a reduction in quality of the crop.

Pests and diseases are thus defined by the damage they do, and this is by no means trivial and can be considerable. Depending on the circumstances and source of the information, they are normally credited with damaging anywhere between 30 and 50 per cent of crops grown worldwide. This represents a significant proportion of world harvests. This is obviously an average, and any farmer or grower can testify that crop losses in any specific situation can vary anywhere between 0 and 100 per cent. Organic farmers and growers would hope to manage pests and diseases to limit damage to the lower end of this range, although they should realize that in an open ecological system there will always be some damage present, and that in some way this represents the presence of 'nature' in their farming systems.

In this book we will also define pests and diseases in a more restricted sense: that is, pests and diseases are defined as organisms that cause 'economic damage' to crops. This implies that the damage to crops is quantifiable in some way (normally as lost harvest or reduced quality expressed in monetary terms), and that a judgement can be made as to whether the cost of implementing any management methods will be covered by any increased production and/or quality, leading to improved sales. This concept is explored in more detail in Chapter 6, but implies that pest and disease management decisions and practices are rooted in an economic or business framework, and that pest management will result in some economic benefit (normally monetary) to the farmer or grower.

Despite this, such decisions are often by no means clear cut, as judgements of this nature will often involve a great deal of guesswork about future outcomes, both ecologically and socio-economically. In addition, much economic decision making does not take into account costs external to the farm system. These costs are, in a real sense, absorbed and paid for by society as a whole but are not generally incorporated in the price of the produce. It is one of the aims of this book to help provide a focus for, and a sound basis to, decision making for pest and disease management, and to reflect on some of these external costs which organic farmers and growers have, in some sense, chosen to try and internalize in their production systems, thereby putting themselves at a competitive disadvantage in an open free market system.

TYPES OF DAMAGE DUE TO PESTS AND DISEASES

There are no plant parts or tissues that are not subject to some form of exploitation or parasitism by pests and diseases. Examples of pests or diseases that attack and cause injury to roots, shoots, leaves, flowers, fruit and seed can easily be found. The specific details of the types of injury likely to be caused are described in more detail in the relevant chapters (7–9) and not described fully here. Suffice to say that it is evident that not all plants suffer simultaneous attack to all parts, and the severity (or negative consequences) of pest and disease attack characteristically varies between crops and seasons. The likely severity of pest and disease attack will be a function of various factors which have been summarized below, but which are explored more fully in Chapters 2–5.

The severity or degree of damage caused by a pest or disease is often related to the type of damage done. Pests or diseases that directly consume or attack the harvestable part of a plant, be it roots, shoots, leaves, fruit or seed, will usually cause more severe losses than those that attack non-harvested parts. This is because they are directly removing the yielding component of the harvest. In contrast those pests or diseases that attack other plant parts only have an indirect effect on yield, for example by removing foliage that reduces photosynthetic capability, or tunnelling in roots that interferes with root development and nutrient take-up.

Direct damage to cabbage by caterpillars.

Although indirect damage can obviously be severe when whole plants are removed, as in stem-cutting caterpillars, or foliage destroyed as in leaf blights, a surprising amount of indirect damage can be tolerated by plants without, or with only minor, yield loss to the harvestable parts.

Severity of damage is also a function of the pattern of attack of the pest or disease and their interaction with other organisms. This is discussed more fully in Chapter 2, but it is recognized that crop plants can potentially be attacked by a range (or spectrum) of different pests and diseases, and that the interaction between these can have an important bearing on the severity of damage. These relationships are sometimes referred to as the 'pest and disease complex' of a crop. Within this complex it is likely that only a few pests and/or diseases will be key – that is, play a major role. Others will be secondary or minor pests that are nevertheless capable of becoming more important if the key pests are removed or other conditions changed or perturbed in some way. For example, the blanket application of biocides is known to disrupt pest complexes and cause relatively minor pests to assume key roles as their competitors and/or natural enemies are removed.

Key pests are usually those that cause the most damage, and are consequently those that dominate pest or disease management decisions about a crop (*see* Chapter 6). They are often perennial in the sense that they are recognizably present in most, if not all, seasons and often cause direct damage to the crop. They are often capable of building up to high numbers in any one season. In contrast, minor pests are only likely to be important in some seasons, and in others may not be noticed. They are also more likely to be those species that cause indirect damage to the crop. However, some pests and diseases, although minor in their own right, are capable of causing serious damage in their role as vectors or transmission agents for pathogens, especially viruses.

Powdery mildew on carrot causes indirect damage.

PEST AND DISEASE MANAGEMENT

Pests and diseases that cause damage are not normally tolerated, and most farmers seek to mitigate or reduce potential or actual damage in different ways. Crop protection and pest management certainly has a long cultural history, and in many ways is central to the historical development of agricultural methods and techniques. For example, many cultural techniques regarded as best agricultural practice, such as rotation, are effective, at least partially, because they also manage pest and disease complexes on crops.

Crop, pest and disease management for most of recorded human history has been a slow evolution of technological approaches (often without any scientific understanding of why things worked) embedded within specific cultural settings which have given rise to many 'traditional' farming systems worldwide. These farming systems served to help adapt societies (at least temporarily) to the ecological conditions that surrounded them. Within these traditions and systems there is a wide range of pest and disease management methods, many based on cultural rituals or habits rather than scientific understanding, which nevertheless work to manage pest- and disease-causing organisms.

More recently pest and disease management practice has changed as a better understanding of the relationship between crops and their pest and disease complexes has been gained through the application of increasingly sophisticated scientific and ecological methods. It has also been increasingly changed by the application of social science methods as regards the interaction between culture, society and farming, which helps to define what is acceptable and desirable. Arguably though, crop protection practice has been most radically changed with the rise of modern industrial agricultural (or commodity) approaches, and it is no coincidence that the widespread use of 'pesticides' to kill pests and diseases has come at the same time as adoption of 'economically efficient' agricultural technologies. Whilst this has been a period of rapid innovation in production methods, it has come at a large environmental and social cost as both the social and ecological scientists have been crowded out by the voices of the 'harder' technological sciences, which take economic efficiency as their cue.

In particular, defining the use of pesticides with regard to economic thresholds and economic injury levels has become a standard research approach for agricultural scientists working in pest and disease management. This approach takes the view that crop protection decisions should be based on an economic evaluation of the likely cash loss of yield to any pest or disease, which should then be actively managed as long as the costs of management do not exceed the cost of the yield loss. The time at which management action should be taken is the point at which the pest or disease has developed to the extent that economic damage will result if no action is taken – the so-called economic threshold. The population level at which a pest or disease will go on to cause economic damage is known as the 'economic injury level', and the aim of defining a threshold is to manage the pest so that this population level is not attained within the crop (Chapter 6).

This model has suffered from many shortcomings, not least the difficulty (or even impossibility) of predicting the various thresholds and levels for each pest and/or disease, and then combining them to account for the pest and disease complex on any crop in any season. Whilst the development of ever more sophisticated pesticides has been a logical outcome of this model of research and development, it has long been recognized by agricultural scientists that a more harmonious model that takes into account both environmental and social factors, as well as technical ones, is likely to result in the development of more truly sustainable pest and disease management approaches. Indeed the organic approach to pest and disease management has been well informed, and has made a valuable contribution to these so-called integrated pest management approaches.

INTEGRATED PEST MANAGEMENT

Over the last thirty or so years, as the implications of both 'green revolution' and 'industrial or commodity' agricultural practices have become evident, a series of integrated pest and disease management approaches (also known as IPM) has arisen, which strive to minimize negative ecological or health side effects whilst maintaining positive economic returns to farmers. In this book we consider organic pest and disease management to be the logical end-point of all such IPM programmes because of

Natural biological control is central to organic pest management.

the far-reaching underlying principles and assumptions behind organic agriculture (*see* below). In any case, both approaches share much in common, and because of this, we describe below in some detail the rationale behind integrated pest management approaches as a basis for organic pest and disease management programmes.

IPM is defined as a pest management system that takes into account the life cycles of the pests (and/or diseases), their ecological characteristics, and the situation in which they are to be managed. It identifies the most appropriate techniques and methods and, in as compatible a manner as possible, uses them to maintain pest and disease populations at levels below those at which they are likely to do any economic damage. In order to achieve this, it attempts to identify the points at which the pests and diseases are most vulnerable to management and control in the whole farm system. Pre-eminence is given to proactive techniques such as cultural controls, cropping methods, plant breeding and biological control methods followed by direct control methods. This is also true of organic management programmes, except that some methods used in IPM approaches are not acceptable for use in organic systems as they contradict some of the other underlying organic principles. Technologies prohibited in this case most obviously include synthetic pesticides (insecticides, fungicides) and genetic modification (GM), which all contradict one or more of the basic organic principles outlined below.

ORGANIC PRINCIPLES

In addition to general IPM and ecological principles (*see* above and Chapter 2), organic farmers and growers subscribe to another set of principles: those of the organic movement. These 'organic principles' are rooted in experiential (or traditional) agricultural knowledge, but build on this by incorporating the strong insights provided by modern scientific inquiry, especially in the fields of ecology, agricultural technology and sociology. Organic principles are especially beholden to agroecological knowledge, which incorporates a socio-economic analysis of agricultural systems and a rich tradition of developmental and participatory methodologies which has arisen over the last forty or so years in community development programmes.

IFOAM Principles

Organic principles have been defined in many ways depending on the context and audience, but there are four main guiding principles behind organic farming practice as defined by IFOAM (the International Federation of Organic Agriculture Movements), to which the majority of organic organizations look for guidance and inspiration. Each principle is articulated through a statement followed by a detailed explanation (excluded here for brevity, but available on the IFOAM web site), and they are intended to inspire action, to encourage the organic movement in its full diversity, and to serve as a guide to the development of positions, programmes and standards.

These principles are:

1. Health – organic agriculture should sustain and enhance the health of soil, plant, animal, human and planet as one and indivisible.
2. Ecology – organic agriculture should be based on living ecological systems and cycles, work with them, emulate them and help sustain them.
3. Fairness – organic agriculture should build on relationships that ensure fairness with regard to the common environment and life opportunities.
4. Care – organic agriculture should be managed in a precautionary and responsible manner to protect the health and well-being of current and future generations and the environment.

Pest and disease management is not separate from these principles, and all of them can be seen to have some bearing on pest and disease management decisions and techniques. Arguably the primary principle is that of ecology, and organic pest and disease management practices should be, above all, based in a recognition of the farm system as part of a living ecological system with its own rhythms and cycles. Pest and disease management practices should seek to emulate, strengthen and sustain these cycles so that pests and diseases, though part of the system, are suppressed at levels below those at which they do economic damage. Likewise the principle of health implies a holistic approach to pest and disease management problems, so that they are perceived as part of the farm system that needs to be managed; this principle especially implies a careful consideration of soil management practices on pest and disease problems within the farm system (*see* Chapter 3). The principle of care urges a precautionary approach to management practices, and is particularly pertinent in the case of direct pest and disease management techniques (*see* Chapter 4); it will preclude the use of any inherently 'risky' or environmentally damaging technologies. These three principles should all emerge from a careful reading of this book, rooted as it is in the principles of ecological and rational pest and disease management.

The final principle, that of fairness, argues for a more ethical approach to pest and disease management programmes, and is perhaps the principle that needs further discussion in relation to pest and disease management. IFOAM characterizes fairness 'by equity, respect, justice and stewardship of the shared world, both among people and in their relations to other living beings'. This principle emphasizes that, aside from ensuring adequate conditions for those involved in implementing pest and disease management programmes, the natural and environmental resources that are used for production and consumption should be managed in a way that is socially and ecologically just, and should be held in trust for future generations. This requires systems of production, distribution and trade that are open and equitable and account for real environmental and social costs. It is arguably this aspect of pest and disease management where organic farming has a potentially unique and positive contribution to make, as compared to purely ecological or IPM approaches that are more rooted in technology than ethics.

ORGANIC CERTIFICATION

Organic principles have become encoded as organic standards. The most popular certifying or control bodies in the UK are the Soil Association and Organic Farmers and Growers, although they have contemporaries in all countries. These bodies set their own standards to which farmers and growers have to adhere in order to be certified, and they inspect farms, usually on an annual basis, to verify that the farming system conforms to their 'organic standards'. The resultant products can be sold as originating from certified organic holdings.

All organic standards in Europe (including those of the control bodies in the UK) have to conform to a minimum organic standard laid down under EU legislation on organic standards which came into force in 2009 under Council Regulation (EC) 834/2007. This replaces and builds on the original Council Regulation (EEC) 2092/1991. These regulations and legislation are the basis for national, including UK, organic standards, which are implemented under the Organic Products Regulations 2004, and through the Compendium of UK Organic Standards. This is overseen by Defra in the UK, which is the competent authority for the purposes of council regulations on organic farming. The Advisory Committee on Organic Standards (ACOS), a body made up of actors in the organic sphere (such as farmers, researchers and marketers), assists Defra by providing advice on the development and implementation of organic standards, and by assisting with the supervision of the control system including the approval and control of the organic inspection bodies which, in turn, licence the individual organic operators.

In the UK, all pest and disease management programmes and techniques need to conform to the minimum legal EU requirements, and also to meet standards laid down by a relevant control body. In this book we mainly defer to this legislation in reporting on allowable organic pest and disease management practice, whilst understanding that actual practice in any country will depend on particular circumstances and/or control body decisions.

STATUTORY REGULATIONS

Apart from organic standards, a large and complex body of laws covers many, if not most, aspects of farm production in most countries with developed economies. This is understandable given that agriculture accounts for the overwhelming proportion of rural land use, especially in Europe. Pest and disease management is one of the most heavily regulated areas within agriculture, and pest and disease management in organic systems is not exempt to this regulation. Although much of the legislation in this area relates to the use of pesticides, which organic farmers will not generally be using, they still have to be aware of the large number of other regulations and keep themselves updated so as not to inadvertently transgress them. Legislation can include health and safety laws, quarantine and seed laws, plant breeders' rights, animal cruelty legislation and general environmental or wildlife laws. Organic farming practices are not exempt from national laws even though they might clash with, or even contradict organic principles, and a certain amount of pragmatism might be needed in this respect.

As an example, in the UK there is a series of laws and acts that affect plant protection practices; these include the following:

Plant Quarantine and Phytosanitary Acts: These acts regulate the movement and sale of plant and seed material. They are aimed at preventing pests and diseases that could cause serious damage to crops, and at preventing plants that could become weeds establishing in the UK. There are many of these laws, and Defra's website should be consulted for further details. The latest include the Council Directive 2000/29/EC, which aims to protect plant health within the EC, and the Plant Health (England) Order 2005 arising from this, the implementation of which Defra is responsible for in England and Wales, the Scottish Executive in Scotland, and other bodies in Northern Ireland.

Seed Purity Acts: Also administered by Defra through various agencies, these acts aim to provide good quality plant varieties and seeds, deemed essential for safe, reliable and efficient crop and food production. Areas of legislation include Plant Breeders Rights, National Listing and Seed Certification Services.

Health and Safety at Work Legislation: This legislation is based on the Health and Safety at Work *etc.* Act 1974, which places general duties on both employers and the self-employed, and also seeks to protect anyone else who may be affected by work activity such as visitors and members of the general public 'so far as is reasonably practicable'. In legal terms this means that an employer has to make a cost-risk analysis, and assess on the one hand the risk of the work and on the other hand the difficulty, time, trouble and expense of steps needed to avoid the risks; this applies to pest management operations as much as any other work activity.

Food and Environment Protection Act 1985 (FEPA) and associated **Pesticide Use Acts** and **Pesticide Safety Directorate (PSD)** and/or **Chemical Regulation Directorate Legislation:** These all govern the application of crop protection chemicals under a legally binding Code of Practice for all professional users of plant protection products in England and Wales in respect of Part III of the Food and Environment Protection Act 1985 (FEPA) and the regulations controlling pesticides, particularly plant protection products, under that part of the act. Organic farmers and growers should be familiar with the COSHH regulations (Control of Substances Hazardous to Health) and the Control of Pesticides Regulations (COPR), and should adhere to the code of practice when applying any organic crop protection products, and should realize that it is illegal to apply products (including home-made ones) to control pests and diseases which have not been registered and tested for safety and efficacy under this legislation.

The Plant Protection Products Regulations (PPPR) also have a bearing on the registration and use of pesticides under European Directive 91/414/EC. This latter regulation aims to harmonize the registration of plant protection products in the EU, and all existing pesticides used in the EU are currently being re-evaluated under the directive. This is likely to result in many products, including some currently permitted for use in organic systems, being withdrawn from use from 2010 onwards.

Wildlife and Countryside Act: This act and various amendments consolidate and amend existing national legislation to implement the Convention on the Conservation of European Wildlife and Natural Habitats (Bern Convention) and Council Directive 79/409/EEC on the Conservation of Wild Birds (Birds Directive) in Great Britain. It is complimented by the Wildlife and Countryside (Service of Notices) Act 1985, which relates to notices served under the 1981 Act, and the Conservation (Natural Habitats, &c.) Regulations 1994 (as amended), which implement Council Directive 92/43/EEC on the conservation of natural habitats and of wild fauna and flora (EC Habitats Directive).

Many of the provisions of these acts limit the types of method that can be used for crop protection purposes, especially where vertebrates such as deer and birds are concerned, and the Defra website should be consulted on this.

Hunting Acts: These acts also limit the types of activity that can be used for 'pest control'. Some of these are included in the Wildlife and Countryside Acts, as described above.

Animal Welfare Acts: These acts aim to protect animals from mistreatment; the most recent is the 2006 Act. The act contains a 'duty of care' to animals, which means that anyone responsible for an animal must take reasonable steps to ensure that its needs are met. This means that a person must look after an animal's welfare as well as ensure that it does not suffer. The welfare of farmed animals is in addition protected by the Welfare of Farmed Animals (England) Regulations 2007 (S.I. 2007 No.2078), which are made under the Animal Welfare Act. These regulations replaced the Welfare of Farmed Animals (England) Regulations 2000 (as amended) on 1 October 2007. The new regulations ensure that farmed animal welfare legislation is aligned with the provisions of the Animal Welfare Act.

Water Framework Directive: This directive came into force in 2000 and established a legal framework for the protection, improvement and sustainable use of all water bodies in the EU; it is being gradually implemented at national level. In the UK a programme of measures is intended to be operational by 2012, and it has been suggested that it is likely to have an impact on pesticide usage, as water quality is an important consideration under the directive.

A glance at this by no means exhaustive list will give an indication of the complex legislative framework that can potentially impact on crop protection decisions, even in organic farming systems, and organic farmers and growers would be well advised to keep up with the latest legislative developments and to take legal advice in areas where they are not sure of the acceptable course of action.

ORGANIC PEST AND DISEASE MANAGEMENT

All the principles and approaches discussed to this point need to be combined into practices on organic farms, and it is the aim of this book to show how this can be done, especially in temperate and organic vegetable and cereal production systems. This section indicates how the subsequent chapters can be consulted to develop strategies on pest and disease management on organic farms, based on the underlying principles and rationale. The subsequent chapters of this book therefore provide a framework for pest and disease management practices in temperate organic vegetable and cereal production. They draw on practical examples from (mainly) UK and European organic farms, but we believe the principles and practices outlined will serve to illustrate the potential for organic farm production in all situations.

Developing Pest and Disease Management Strategies on Organic Farms

Combining the general principles for pest and disease management and applying them to organic systems gives a unique 'organic' approach. Organic pest and disease management programmes and strategies will obviously be integral to the farm system and likely to be more or less complex and situation specific. Indeed, developing organic pest and disease management approaches (in line with the principles outlined above) requires the following:

- An understanding of the ecological relationships within a farm system, specifically that between the crop plant, the pest or disease organism, and the environmental factors that influence their development at all levels from the plant to the field to the landscape. Modern ecological understanding indicates that some of these relationships can be quite complex, and these are discussed in more detail in Chapter 2.
- An understanding of the economic factors underlying a farm production system. Factors include not only the measures available to the farmer to manage pests and diseases (*see* Chapters 3, 4 and 5) and the strategies available (Chapters 7, 8 and 9), but also the likely levels of damage caused and the resultant monetary loss to the farmer and the impact on profitability (*see* Chapter 6).
- An understanding of the socio-economic decision-making behaviour of the various actors in the farm system, be they farmer owners, farm managers, workers or marketers (*see* Chapters 6 and 10).
- An understanding of the policy framework in which the direct actors (above) act, which will largely be set by national legislature (for example pesticide regulations, plant quarantine legislation, health and safety at work) (*see* Chapter 1).

Knowledge and communication are central to organic pest and disease management.

- An involvement of the various actors (workers, farmer-owners, farm managers, marketers) in the decision-making processes as regards an analysis of problems and the definition of solutions within the farm system. Decisions made at all these levels (1–4) will impact on pest and disease management practices, and ultimately they may or may not reflect the actual 'economic value' of a crop or crop loss due to pests or diseases (Chapters 6 and 10).

Implementing Pest and Disease Management on Organic Farms

Following the principles outlined above, it follows that organic pest and disease management strategies should be based, in the first instance, in working with the natural ecological cycles on a farm, and should build on the premise of as minimal an intervention as possible to get the job done. Farmers and growers should work to design systems with pest and disease management approaches built in and only begin intervening where problems arise or adjustments are needed. Adding elements to a management strategy should always work from the least disruptive upwards in order to try and prevent untoward perturbations in the farm agroecosystem.

Incidentally, wider ecological considerations will also be important when adding new practices, with those that use the least resources, or use scarce resources more efficiently, being preferred over those that require excessive resource use or have other wider external costs to the farm system. On this basis pest and disease management in organic systems can be ordered into a hierarchy of strategies, sometimes referred to by some authors as 'strategy phases':

First phase (fundamental) strategies: These strategies can be defined as indirect preventative methods of pest and disease management. Such practices include crop production strategies as part of a long-term farm plan, and are broadly defined as cultural practices that are largely compatible with natural and ecological processes, of which the farm is a part. The ecology of organic farms is discussed in more detail in Chapter 2. The practices arising at this level include soil management, crop rotation, plant resistance, and preventing the establishment and spread of pathogens (hygiene). They are the types of practice most often associated with 'traditional' agricultural knowledge, and indeed often require detailed site-specific knowledge to implement, but have been improved by the enormous increase in scientific knowledge centred on agroecology. They are discussed at length in Chapter 3.

Second phase strategies: These include cropping and vegetation management practices that enhance the effects of natural enemies on pest and pathogen populations, and also exert other direct effects on these same populations. Many of these practices have also arisen from traditional knowledge, but others have been designed or improved upon using the most up-to-date scientific knowledge. Apart from adaptive cropping strategies, such as spacing or specific tillage practices, these strategies also include conservation biological control, intercropping and trap cropping. Also included in this category, despite their direct mortality effect on pests and diseases, is the development of an increasing array of barrier or trapping methods that either directly kill or prevent pests and pathogens completing some part of their life cycle. These are all discussed in Chapter 4.

Third phase strategies: These include inundative and inoculative releases of biological control agents against pests and pathogens. Although their development and use is based on a sound knowledge of ecological science, the number of different factors involved makes their effective use difficult in practice. In general they are most effective, and have found the most use in protected cropping, although there are examples of their use on a field scale. These are discussed in Chapter 4.

Fourth phase strategies: These pest and disease management strategies include the direct interference with pest and disease populations using pesticides or fungicides of biological and/or mineral origin. There is also an increasing use of materials that disrupt the normal life processes, such as mating, of pests and pathogens. All these methods are briefly summarized in Chapter 5.

2

The Ecology of Pest and Disease Management

As emphasized in the previous chapter, organic pest and disease management is, above all, rooted in a systemic, biological and ecological approach rather than a technological one *per se*. Organic farmers and growers should use the most appropriate interventions and technologies compatible with organic principles to manage pests and diseases in any specific situation. Biology and ecology are fundamental to understanding which interventions are likely to be effective, and farmers will need to have a basic knowledge of the underlying ecological factors at play when pests or diseases invade a crop and cause damage. This will allow them to take a proactive informed approach to their management and the goal of keeping them at levels below which they cause economic damage, both to the farm business and, by implication, to wider society as a whole.

In this chapter we describe the biological and ecological factors that need to be taken into account in any organic pest and disease management programme. Pest and disease attack needs to be understood both from a broad ecological perspective, and as a narrower, more intimate relationship between the pest or pathogen and host (crop) plant. The former ecological perspective seeks to understand the causes of pest and disease outbreaks on a landscape or farm scale, whilst the latter seeks to explain, then describe, the individual relationship between the pest or pathogen and crop plant. Both have a bearing on the types of management method that farmers and growers can use to manage pests and pathogenic organisms, leading to management methods on different scales ranging from the individual crop plant or crop, up to the farmed landscape as a whole. The consequences, and also the pest management techniques arising from this understanding, are described more fully in subsequent chapters (especially Chapters 3, 4 and 5).

ECOLOGICAL CAUSES OF PEST AND DISEASE OUTBREAKS

Pests and diseases arise as a natural consequence of our intervention to manage ecological systems to our benefit, and this is especially true in agriculture, which depends fundamentally on ecosystem resources (for example soil, water, air, sunlight) to deliver its primary produce. We will naturally be in competition for these resources with other organisms which also use them, including pests and pathogens. We should, however, also recognize that the majority of the organisms in the ecosystem are either neutral or positively beneficial to the farm system. For instance, without detritivores and decomposers, plant material would not be broken down and recycled as nutrients through the soil ecosystem. Such a holistic perspective recognizes that pest or disease problems are part of the process of farming, and that in solving any such problem it is better, at least initially, to stand back and take a 'whole farm' point of view before deciding on specific management methods in any particular situation. There is rarely a single factor that is entirely responsible for a pest and/or disease outbreak, and consequently there is likely to be a range of appropriate management methods that can be chosen to address any particular problem. Selecting the best management methods and approaches is discussed in subsequent chapters (but especially Chapter 10).

Agricultural and Ecological Systems

At a basic level, farms represent an overlap between human defined social and economic systems and biologically defined ecological systems. These are not different systems, but rather two sides of the same coin. The types of crop grown in any situation reflect not only the underlying biological

Many organisms are neutral or even beneficial to farming.

and ecological conditions, but are also the result of a historical process rooted in culture and circumstance. Social and economic systems often work to change underlying ecological systems but are obviously limited by them. We can work to modify the environment by draining a marsh, irrigating a crop or applying a fertilizer, but ultimately, the more we do this, the more external energy and resources we are likely to need in order to practise agriculture, and arguably, the less likely such systems are to be sustainable in the long run.

The sum total of social, economic and ecological interactions in a farm system are often referred to as the 'agroecosystem' (*see* below), although some people restrict this definition to the ecological relationships within a farm or farmed landscape. We will use the word in the former sense. Agroecosystems arise when people modify natural ecosystems to their own ends. So the interactions between crops, livestock, physical environment and other organisms, including pests and diseases, are all part of the agroecosystem.

Abiotic and Biotic Factors

At the lowest level, ecological interactions in the agroecosystem are mediated by abiotic and biotic factors, which combine to influence the types of biological communities that arise in any area and the interactions that occur within them.

Abiotic factors: These are the non-living and physical aspects of the ecosystem. Those of most concern to farmers and growers include climate and weather (temperature, rainfall, wind, humidity), soil (physics, chemistry), light (day-length, insolation) and land (height, topography). These obviously all interact to some extent and, in fact, are responsible for the characteristic landscapes we see around us, and delineate the ways in which we can farm. For instance, temperature tends to fall with altitude, and rain tends to be deposited on the windward rather than the leeward side of elevated land. Similarly the quantity of rain is likely to determine the availability of water in streams and rivers and in the soil, and may exhibit a periodicity that limits the growing season.

Biotic factors: These are the sum of the interactions that occur between the living individuals in any particular community or ecosystem. Populations of animals, plants or microorganisms use resources to grow and reproduce, and will tend to increase until resources become limiting in some way. In all ecosystems one of the principal limiting factors is competition for resources with other individuals. Predation and parasitism of individuals also occurs in all ecological systems. These are both examples of biotic interactions.

In a wider sense, biotic interactions can be divided into trophic (feeding) relationships between species in an ecosystem, often represented by food webs, and non-trophic relationships, which are other interactions. These are numerous, but include not only competitive associations (for example, for light between plants, for egg-laying sites between insects, for shelter between animals) but also mutualistic associations by which organisms cooperate for mutual benefits (for example mycorrhizal associations between fungi and plants, pollination associations between bees and flowering plants).

The Characteristics of Agroecosystems

Biotic and abiotic factors combine in any one particular area to give a characteristic ecosystem. A crucial difference between agroecosystems and 'natural' ecosystems is that the former are actively created and managed. Farmers and growers are the managers of these systems and can influence them to a greater or lesser degree, and in this sense are therefore themselves part of the system.

It is not the place of this book to discuss agroecosystems in great detail, but to point out that temperate organic agricultural systems are generally simplified over, and replace, natural ecosystems. For example in Europe, the original plant communities were complex deciduous forests, evergreen forests or steppe grassland communities with additional mountain, riverine, estuarine or coastal communities. These have been largely replaced with simplified extensive grazing systems, or more or less intensive arable and vegetable systems dominated by a few species of benefit to people. In addition, even areas traditionally unsuitable for agriculture have been extensively modified by, for example, drainage, in order to modify them and make them more suitable for agricultural production.

From an ecological perspective agricultural systems are therefore held or managed at an initial or early stage of (ecological) succession. That is, they are typified by shorter or longer periods of bare soil with simplified plant or animal communities (often one or a few crops, a few species of livestock) as compared to the climax grassland, bog/peat or forest systems that they replace. Although there are good reasons for this – primarily the reduced competition with other species and the increased proportion of production useful to people – this also has consequences for the ecological stability and sustainability of these systems, and in many ways makes them more vulnerable to pest and disease outbreaks (*see* below).

Agricultural landscapes are also modified in another important way, in that agricultural use is now the predominant landscape feature. Rather than crops being isolated elements in the landscape, it is more likely that the relatively wild areas that form reservoirs of biodiversity are themselves isolated in a 'sea of crops'. It is well known from ecological 'island' theory that such isolated areas have fewer plants and animals (the smaller they are, the fewer they have), and those that they do have are more vulnerable to extinction. Once again this has consequences for ecological stability and the sustainability of farming systems in that it removes areas that can potentially buffer agricultural landscapes against the loss of diversity and species. This is especially felt when it comes to natural background biological control (*see* Chapter 4).

A typical UK horticultural landscape with large 'blocks' of crops.

Diversity, Stability and Abundance

Diversity and stability are difficult concepts to define from an ecological point of view, and are the ongoing subjects of a certain degree of controversy as to their definition, importance and relationship. However, there is growing evidence that a diverse, complex ecosystem (with high biodiversity and lots of biotic links) provides a buffering function that prevents any one species coming to dominate or any one species becoming overexploited. This occurs because of well known and increasingly understood ecological mechanisms, such as competition between species (for space, food and mates), active antagonism, parasitism and predation as well as mutualistic interactions. Abundance, or the population of organisms, is obviously affected by these factors, and is generally considered to be regulated in some way by them, especially by factors such as parasitism that increase in intensity as a population increases (*see* Chapter 4 for a discussion of biological pest management). It should be borne in mind that it is possible that a stable population can still be high enough to cause significant damage to a crop, and that abiotic factors can have a powerful mediating effect on these biotic factors.

Complex systems are therefore generally more 'stable' because diverse links allow for compensation within the system, which tends to even out any large population changes of any one species or to accommodate any changes in resource availability. These diverse systems also seem to utilize energy more efficiently and be more productive over the long run as demonstrated by (for instance) long term and detailed studies on complex prairie grass and tropical forest communities. These complex systems tend to be more resilient – meaning that they persist (and yield) over time, at least within a certain range of conditions – and for this reason maintaining complexity is arguably a guiding principle in organic farming systems.

Pests and Diseases in Agroecosystems

Pests and diseases are by definition organisms that occur in sufficient numbers to cause loss of crop yield and/or quality. From an ecosystem perspective, and following on from the previous discussion, pests and diseases can thus be thought of as arising as a consequence of lack of complexity in agroecosystems, which allows them to reproduce and exploit resources unhindered in some way. From this perspective, a range of ecological factors predisposes agricultural systems to pest and disease outbreaks, and understanding and addressing these will help in designing effective pest and disease management programmes for organic systems.

As alluded to above, many of the factors boil down to one, that is, reduced diversity/complexity (and hence stability) in farming systems. The factors that reduce diversity and predispose agroecosystems to pest and disease outbreaks are outlined under the following headings.

Monocultures

Natural ecosystems are normally a complex, dynamic mixture of plant species that, after some time, tend to come to a recognizable 'climax' community or 'dynamic balance' for any particular ecological zone. As described above, agricultural production usually replaces this community with one dominant crop at a field level in any one season, and a few dominant crops at the farm and landscape level. These monocultures often extend over large areas, especially as agricultural production becomes more specialized, and represent a large homogenous resource or 'target' for pest and disease species adapted to exploit them.

Such monocultures create an associated impoverished flora and fauna with pests and diseases adapted to the monoculture, able to rapidly take advantage of large food sources, whilst predators and parasites, which need resources other than the crop, are more likely to die away and disappear. The modified conditions may also allow previously rare or unknown species to proliferate and become pest species as natural predators and parasites are removed from the system.

The Introduction of Crops into New Areas

Most crops are actually grown out of the areas of origin in which they evolved in tandem with their pest and disease complexes, and to which they therefore often display some immunity. This can create new pest or disease problems when new pests or diseases adapt to crops that have no genetic immunity to them – this is what occurred with the Colorado potato beetle (*Leptinotarsa decemlineata*) on potatoes (*Solanum tuberosum*) for example. There is also the tendency for pest and disease species to arrive with the crop, whereas the beneficial organisms that suppress them tend to be left behind in the areas of origin because they are higher up the food chain and often dependent on a more complex web of interactions. This allows pest and disease species to multiply rapidly once removed from their areas of origin unless abiotic (physical environmental) factors intervene.

Selective Breeding

The effect of monoculture is exacerbated by the limited number of different varieties grown or acceptable for standardized markets, which implies a limited genetic background for most crops. Crop varieties are often bred, either directly or indirectly, to produce saleable yield, and disease and insect resistance is often only incidental to this main objective. In many cases breeders use pesticides that will in any case mask resistance traits, or assume that they are available for use to protect varieties once they are released for sale. Limited resistance is often also coupled to increased attractiveness of plants bred for yield, as the increased resources these represent are also attractive targets for pests and diseases.

Reduced Rotations

Coupled to monocultural production, crop rotations also tend to become much simplified as production becomes more specialized, so that the 'target' for pests and diseases extends over both space and time, increasing the likelihood of outbreaks as pest and diseases increase in numbers and severity over the seasons. Other rotational-related factors, which tend to be altered detrimentally in specialized agriculture or monoculture, include the simplified arrangement of crops, and the intensified management, and consequent reduction of abundance and composition, of non-crop vegetation in the surrounds and wider landscape.

Crop Management Practices

Many cropping systems are essentially disrupted on a seasonal or annual basis as ground is prepared and crops sown, tended and harvested. For instance, growing annual crops usually involves extensive tillage between crops and/or seasons, which will cause a more or less large disturbance to the ecological community at a field level. It can be argued that in this case

The management of annual crops can cause large-scale disturbance to ecological communities.

those species (often pests and diseases) that are adapted to survive in the absence of their hosts, and to reproduce rapidly in the presence of their hosts, are likely to prevail. In contrast, those that survive poorly under such conditions, often predators and parasites, and reproduce slowly will not fare so well. In this case there is a balance to be developed between disrupting the life cycles of pests and diseases and disrupting other ecological factors (such as natural enemy survival), which could potentially make the agroecosystem more stable and resilient.

Changed Agricultural Practices

It should not be overlooked that many pests or diseases increase in importance when agricultural practices are changed. A major change in practice over the last thirty years has been the use of pesticides, which has had major ramifications throughout agriculture. In this case overuse and misuse of insecticides has been documented to cause resistance in pest populations over the medium to long term, and resurgence (or 'bounce-back') in the short term in many cases. It has also allowed previously minor pests to become major pests, as either competition is reduced or beneficial organisms are removed from the farm system. Herbicides and fungicides are also known to alter the nutritional balance within plants, and in some cases make them more palatable to pest and disease organisms. The same is true with the use of soluble inorganic fertilizers: these have become much more widespread, and are well known to increase levels of soluble nitrogen in crop plants, thus removing what is often a limiting factor to insect increase in natural ecosystems. Coupled to this, the removal or modification of many features such as hedgerows, woodlands, water courses and ponds removes the buffering effect of these areas from the landscape.

Chance Events

It should also be borne in mind that despite these factors predisposing agricultural landscapes to pest and disease outbreaks, sometimes chance events alone will be sufficient to cause large fluctuations in pest and disease numbers between seasons. For instance, weather events and temperature are important factors in the build-up of many pest and disease populations, and these vary in an unpredictable way between seasons and years. Nevertheless, it can also be argued that these fluctuations are more likely to be extreme in the absence of the buffering effects of diversity in the farmed landscape.

Agroecosystem Boundaries

It is also important to realize that the definition of any particular agroecosystem is subjective. Normally boundaries are drawn at the appropriate level for the elements being studied within an agroecosystem. This is another way of saying that farm systems have open boundaries, and animals and resources flow across them at whichever level they are drawn. Organic farmers and growers usually draw the line at the farm boundary for convenience, and because this is the level to which they can practically manage and have influence. There is also a real sense of the farm being seen as a living entity with its own characteristics and conditions (for example, unique rotations and enterprises). But this should not hide the fact that these boundaries are leaky, insofar as any sold produce exports nutrients from the farm system, and pests and diseases often arrive from outside the farm boundaries.

Ecological and Organic Pest and Disease Management

Organic and ecological pest and disease management aims to minimize outbreaks and/or maintain pest and disease populations below the levels at which they cause damage. It follows from our previous discussion that this is best done by treating the underlying causes of these outbreaks – that is, the reduced complexity of agricultural systems – rather than the symptoms – typically, the increased numbers of pest or pathogenic organisms. This underlines a significant feature of organic and ecological approaches to pest and disease management, which is that they give pre-eminence to 'preventative' as opposed to 'curative' approaches. The latter types of approach tend to treat the symptoms but largely leave the causes in place and are therefore, in the long run, likely to be limited in effectiveness, however effective they might be in the short term.

A whole-farm ecological and organic approach to pest and disease management therefore needs to address root causes. Indeed, organic farmers and growers are interested in building resilient systems, which means they are aiming to optimize yields and production within a stable, sustainable system rather than maximizing it in an unsustainable one. Under these circumstances we have argued that the available evidence suggests that biologically diverse systems are more likely to be resilient in that they are more likely to be stable under diverse pressures, including pest and pathogen attack. This is especially true if economic and social aspects are included in the definition of agroecosystem resilience.

Organic management practices therefore seek to manipulate the farm environment to make it less favourable to pests or diseases at the same time as making it more attractive to natural enemies or pest antagonists by building in (bio)diversity. Organic farmers and growers should integrate pest and disease management practice into their crop management routine, and the techniques for doing this are discussed in the following chapters (especially Chapters 3 and 4). Proactive techniques that build (bio)diversity in cropping practice are central to organic pest and disease management programmes. At the forefront are those that are concerned with the management of soil fertility and soil organic matter content, but other aspects of crop husbandry, and even non-crop management decisions regarding features such as field boundaries, water courses and ponds, are also important.

An ecological understanding of pest and disease outbreaks is only one side of the coin. A more complete management approach will also need to address the dynamics of individual pest and disease attack to specific crops – for example carrot fly (*Psila rosae*) attack to carrots (*Daucus carota*), or rust (*Puccinia porri*) on leeks (*Allium tuberosum*) – as the behaviour or strategies employed by pests or pathogens to attack crop plants is also important to designing management techniques.

THE DYNAMICS OF PEST AND DISEASE ATTACK

Apart from agroecosystem effects as discussed above, the dynamics of the interaction between individual pests and diseases and their crop plant hosts are also important, both for understanding the damage they do, and in designing management practices to prevent it. In effect, ecological studies define why pest or disease outbreaks occur whilst behavioural studies define how it occurs. Both sets of knowledge are important in defining what to do about it, that is, in designing effective pest and disease management programmes and practices.

Such detailed information often arises at the interface between ecology and behavioural studies, and in the case of pests, is often referred to as behavioural ecology. Knowledge about this dynamic can lead to management practices that are more effective in disrupting the invasion process, although the detail will vary with different pest and/or pathogen and crop combinations. It could be argued that much current research on pest management focuses on this dynamic, and so it should not be forgotten that the context for pests and pathogen behaviour is within the agroecosystem. Implementing a solution to a pest problem is not so effective if it simply creates another problem in another part of the system, such as, for example, declining soil fertility, increased weeds, new pests or even 'reduced sustainability'.

Leek rust has a complex life cycle and relationship with its host plant (the leek).

At an individual behavioural level, all pests and pathogens are organisms that require resources to successfully develop and reproduce. Resources include food (or nutrients), water, and the space in which to develop. Resources are always limited to some degree (otherwise we would be knee deep in aphids in a very short time!), and pest and disease organisms have to compete with and/or consume other organisms to find and get access to them. In order to avoid or minimize competition and increase their success as individuals they have evolved into specific 'niches' where they have some advantage in competing and gaining access to resources, so increasing their fitness or, in other words, their chances of survival and reproduction.

In the specific case of pests and diseases of crops, they are adapted to find, attack and consume (to a greater or lesser extent) the crop plant itself. They gain the energy and resources they need by consuming the plant directly (for example herbivores) or diverting the plant resources and physiology to their own ends (for example viruses). The details vary between the various classes of pest and pathogen, and these are discussed in summary form below. Specific plant-pest/pathogen interactions are discussed in more detail in later chapters.

Pest and Pathogen Behaviour

Plant herbivores (pests) and pathogens need to locate their hosts in a more or less complex environment. Insects and other pests are mobile and often proceed through a series of selection steps: finding the right habitat, finding the host(s) within the habitat, finding and choosing a susceptible host. In choosing a suitable host a wide number of plant characteristics can potentially play a part in their choice, including morphological features (shape, colour, size of leaves), allelochemicals (plant chemicals that serve to attract or repel pests) and nutritional factors (especially the balance of nutrients, such as soluble nitrogen in the plant tissues). Pests generally possess a more or less sophisticated sensory or nervous system to deal with this complexity, and in invertebrates at least, behaviour may be rigid or instinctive, with only a small leeway for modification through learning. In such cases a break or interruption to any part of this chain can often put a pest off and prevent them attacking a plant. In vertebrate pests the capacity for learning and modifying behaviour can be considerable, and they are capable of overcoming obstacles through behavioural modification.

In contrast, plant pathogens are often passively dispersed in the environment. They usually produce a large number of 'spores' or other propagating structures that they spread widely, relying on chance to encounter a suitable host. They are usually microscopic and therefore have at best limited powers of locomotion, and do not have a nervous system. In this case chemical cues and the nutritional state of the host plant are often critical factors in the invasion process, which is largely mediated through chemical interactions between the invading pathogen and plant tissue. The pathways and interactions involved can be every bit as complex as animal behaviour, and much is only now being uncovered through detailed scientific research. For example, some pathogens, such as viruses, have evolved to exploit vector (animal) hosts to transport them between plant hosts, thereby increasing the efficiency of transmission and infection.

PEST DYNAMICS

In behaviour terms, herbivore (pest) populations need to locate suitable plants and then colonize or eat them. Once located and eaten, plant nutrients are processed and used to provide energy to survive, locate new hosts and/or reproduce. Ecologically, pest populations will come into some dynamic balance on crop plants, depending on the sum of the various mortality and reproductive rates of individuals and the balance of immigration and emigration rates.

Pests cover a wide range of size scales, from tiny insects and nematodes all the way up to large animals such as deer. They are generally classified as herbivores in ecological terms. The details of how they locate and consume crop plants vary greatly, but they share some common characteristics in that they devour parts of, or whole, plants and are more or less active in seeking them out. Some pests, namely monophagous or oligophagous pests, are adapted to exploit a narrow range of plant species or types (usually within a specific family, sometimes one species); others, so-called polyphagous pests, are more generalist, exploiting a range of species and types. In either case they are often specifically adapted to locate, consume and digest their preferred host plants.

Pests also pursue a range of life strategies to exploit plant resources. Some pests have different life strategies at different stages in order to better exploit host plants. For instance butterflies, nectar feeders, seek out host plants on which to lay eggs for their caterpillars, which are voracious herbivores devouring leaves. Others essentially pursue a similar life strategy at all stages, for example grasshoppers, aphids and deer. However, despite following a similar life strategy, they may employ different behaviours to track plant resources, or move to locate them. Migration is a typical strategy to move on to different plants in different seasons (for example aphids or deer). Some pests may employ both strategies, using different life stages to exploit different resources and to migrate to find new hosts (for example caterpillars and moths).

The management of insect pests will depend on affecting the dynamic between pest and host plant. In particular farmers and growers will seek to reduce immigration, increase mortality and reduce reproduction of pest individuals. Less emphasis is generally placed on reducing emigration as this might lead to problems elsewhere in the crop, although it can be a viable strategy if pests can be induced to leave the crop altogether. The effectiveness of any management tactic is therefore mostly dependent on affecting the rate of immigration, mortality and reproduction of the pest, and these tactics are discussed in more detail below.

Cabbage white adults locate host plants for herbivorous caterpillars.

Reducing Pest Immigration

The immigration of pests on to crop hosts depends on the pests locating and settling on the crop plants; management programmes will generally aim to reduce immigration of pests into the crop. Pest immigration is usually broken down into two separate phases: locating host plants and settling on host plants.

Locating Host Plants

Successfully locating a host plant is a primary behaviour adaptation for all herbivorous pests. Pests of annual temperate crops often have to disperse from their overwintering sites to find more or less distant new crops. Although at the mercy of the wind and weather to some extent, many insect pests employ complex visual and olfactory senses to locate and land on suitable host plants. Mammal herbivores have sophisticated behaviour patterns that enable them to locate plants to eat. Large herbivores (when unconstrained) and birds can migrate large distances to locate specific plant hosts at appropriate times of the year. The plant-locating senses are combined in complex behaviour patterns that essentially comprise a 'decision-making chain'. Disruption of this chain is a primary goal in organic pest management, although difficult to achieve in many cases.

Most theories and explanations of host-searching behaviour consider that pests are likely to be disrupted where the crop plant is hidden in some sense, or dispersed among other vegetation, implying that crops in small, well dispersed fields, in strips or intercrops are likely to be more 'invisible' to pests: that is, they gain 'associational resistance' by being hidden within diverse stands of crop and non-crop plants, and are hard for herbivores to find. Other associational resistance effects may include the masking of host odours if they are diluted or mixed with odours from other plants (although insect senses are often very chemical-specific). Visual effects can also disorientate pests: for example, aphids colonize plants that stand out against a dark or brown background, which can be disguised by something like a carpet of weeds or clover.

Actual damage to crops in this phase of attack is unlikely to be significant, although when the animals are large in comparison with the plants, trampling damage may be evident. In this case trapping or other sampling methods may be needed to indicate that crop pests are present and seeking out crops.

Settling on Host Plants

Once an insect or other pest has located a crop plant it also needs to find the crop palatable in order to settle and begin feeding. Herbivores are often quite specifically adapted to identify palatable host plants that are likely to be suitable for their use, and employ often quite complex sensory decision-making chains for this purpose. For instance, insects often detect chemical cues on plants using taste or smell sensors that are located in their mouth, antennae or even on their tarsi (feet); these sensors can be attuned to quite specific chemical cues from the plant.

Taste is not the only cue that can be used to assess a plant's suitability, and factors such as texture or shape might also be important in specific circumstances. For instance, many herbivores are adapted to exploit specific plant parts such as seeds or flowers, and shape and/or colour can be important cues to herbivores in these cases. Plants can also use or alter such cues to repel pests (*see* Plant Defences below).

Some damage due to the pest can become evident at this stage, although damage to the crop yield is normally limited. For instance, some insects might taste the plant by probing with the mouthparts, or mammals might take a taste, both of which processes might cause a small amount of damage. Of more concern at this stage is the possibility that as herbivores move from plant to plant tasting and sampling they can transmit pathogens; for instance, potato aphids can transmit potato viruses as they move from plant to plant.

Management techniques have concentrated on repelling herbivores at this stage, and many plant secondary chemicals (*see* Secondary Compounds below) have been implicated in repelling herbivores. For instance, a wide variety of repellent effects has been reported for companion planting or intercropping systems – for example interplanting herbs such as basil (*Ocimum basilicum*), garlic (*Allium sativum*) and fennel (*Foeniculum vulgare*) in crops – but many of the systems have not been verified, and/or the reasons for their effect are not understood. Similarly ranges of plant extracts such as chilli extract are reputed to deter insect, bird or mammalian herbivores. The effect of such substances will depend on the herbivores having sensors capable of detecting the repellent substances, and for the substances to actually be repellent to any particular species.

Increasing Pest Mortality and Reducing Reproduction

Ecological theory and observation has accumulated a great deal of knowledge about herbivore (pest) mortality and reproduction and its consequent effect on herbivore fitness, however it is beyond the scope of this book to discuss it in detail. Although it has an important bearing on pest management, the detail can be seen as the flip side of the same coin. On the one side is the opinion that herbivores are limited by food resources, but especially food availability and quality, and it is through these so-called 'bottom-up forces' in the food web that herbivorous (pest) populations are regulated. On the other side is the opinion that predators and pathogens, so-called 'top-down forces' acting through the food web, limit herbivore numbers. In reality, both of these 'forces' come into play as individuals colonize and feed on plants, and elements of both are summarized in the discussion below.

Once settled on a host plant the herbivore will begin to feed on and utilize it. At this stage individual mortality of herbivores will be a result of many factors, including the nutrient content of the plant, plant resistance or defences, and pest counter adaptations to plant defences. Predation and parasitism of herbivores will also be important in individual survival. Reproduction of pest species is also likely to be affected by the same factors, namely plant nutrition, plant defences, herbivore counter adaptation, herbivore predation and parasitism. In both cases the herbivore will have to consume plant tissue or products, digest them and use them to both grow and develop or to reproduce.

Herbivore Nutrition

As animals, herbivores have broadly similar nutritional requirements to all animals, that is, a range of proteins, carbohydrates, fats, vitamins and minerals. In common with all animals, herbivores can break these chemicals down and use them to synthesize a range of other chemicals which are used to store and provide energy, or as the building blocks of tissues. Many, however, require specific complex nutrients that they cannot synthesize. Insects, for example, need certain essential amino acids, steroids, vitamins and linolenic acid, and mammals have similar requirements.

Herbivores have two essential problems: firstly, only 10 per cent of the energy available in plant tissue can be utilized by herbivores due to the inefficiencies in processing and converting complex molecules (mainly fat and carbohydrates) into energy; and secondly, some nutrients are in short supply and therefore limiting to growth and development. For example, protein (often measured by nitrogen content) is usually limiting in plant tissues where it is present in much lower concentrations than in animal tissues. In a similar way other essential nutrients, some of which have been mentioned above, can also be limiting.

Herbivores, depending on the plant tissue they feed on, may therefore need to consume large amounts of tissue in order to accumulate sufficient energy, protein or other nutrients for development. Indeed it is well known that insects consuming nutritionally poor food will compensate by eating more. Plants have also evolved to prevent herbivores accessing nutrients by, for instance, decreasing the nitrogen content of tissues, storing proteins in essentially indigestible forms, and storing proteins as toxic substances, for instance in seeds, that kill herbivores that consume them.

Aphids need to process a lot of plant sap in order to obtain sufficient nutrition.

Other nutrients, such as sodium, may be reduced in plant tissue and/or stored in different plant tissues. Nevertheless herbivores have in turn coevolved to be able to exploit specific plant nutrient stores, and this explains why many insect herbivores are specialists, essentially attacking one or a few related plant species, or attacking plants in particular ways.

Plant Defences

Of more importance in limiting herbivore numbers are plant defences. Indeed much variability in plants can be put down in one way or another to adaptations to prevent predation by herbivores. Such defences can be present at all times – so-called constitutive defences such as thorns – while others are produced in response to herbivore attack – so-called induced defences. The range of recognized plant defences includes physical barriers, secondary compounds and mutualistic defences.

Physical Barriers A plant's physical barriers can be on all scales from leaf hairs (trichomes) to thorns. In the former case such hairs may prevent small insects moving around on and gaining access to plants, while the latter may prevent large mammals from grazing on whole leaves. In a continuation of this theme, leaves or other plant parts may also be more or less tough and resistant to chewing. Grasses, for instance, often sequester silica in leaf tissues to reduce the effectiveness of grazing. In other cases leaves may be shaped to disguise themselves, or be an awkward shape for eating and processing. Leaves or other parts may even be shed (abscission) as a form of preventing herbivores attacking a plant.

Secondary Compounds These compounds act within plants as either toxins that poison herbivores or repellents and deterrents that discourage them from feeding. When they affect herbivores in this manner they are sometimes referred to as allelochemcals. There is a wide range of secondary chemicals including cyanogenic compounds (that produce hydrogen cyanide when plant tissues are damaged), tannins (that bind to plant proteins preventing digestion), coumarins (the constituents responsible for the fiery taste of peppers), terpenoids (widely distributed in all plant families and recognized as growth inhibitors) and flavanoids (respiratory inhibitors widely distributed). Some compounds may be synergetic, enhancing the effects of other chemicals.

Apart from the more or less toxic effects of secondary chemicals as mentioned above, others may alter herbivore behaviour or directly affect herbivore reproduction. For example, some plants (red clover (*Trifolium pratense*)) have chemicals that appear to mimic mammalian hormones such as oestrogens that could potentially have effects in these herbivores, and other plants have psychoactive chemicals (for example cannabinoids) that are likely to modify behaviour more or less drastically. Insect moulting-like chemicals (ecdysones) have also been found in plants, and these have the potential to disrupt insect development.

It is ironic that many crops (chillies, herbs) are grown for the properties of their secondary compounds that are used for their culinary and/or health properties. Many of these compounds are present at low levels in plants at all times (for example glucosinolates in brassicas) but many are produced, or their concentration greatly increased, by herbivore attack (for example jasmolins). Indeed, the flavour of organic crops might well derive, at least in part, to the fact that they are open to a wider range of herbivore attack, and therefore induced to produce more secondary chemicals than conventional crops where insecticide treatment reduces herbivory to a minimum.

In a counter move, herbivores have also adapted to exploit secondary chemicals which may become 'attractants' to specialized herbivores. Cabbage white caterpillars are, for instance, attracted to feed on leaves with high glucosinolate content which would deter other non-specialized herbivores from feeding on their brassica host leaves. In yet a further twist, the caterpillars themselves have evolved to sequester these chemicals in their bodies in order to deter predators such as birds from attacking and eating them. From a management point of view, these chemicals can therefore sometimes serve as cues to specialist pests at the same time as being deterrents to generalist pests.

Mutualistic Defences It has become increasingly clear that plants have also evolved to attract pest predators and parasites when under attack. Feeding damage has been shown to release volatile chemical cues that are used by insect parasites to home in on their prey. For instance *Cotesia* spp. parasitoids use such cues to locate caterpillar pests. Fungal endophytes living in and on grasses, and causing no noticeable damage, do however produce toxic alkaloids that can kill herbivores or modify their behaviour (for example ergot).

It is at this, the feeding and reproduction stage, that the symptoms due to pest attack become visible and at which most damage is likely to be done. It is therefore the stage at which most farmers and growers are also prompted to take action. However, in many cases this is akin to bolting the stable door after the horse has fled in that it is more difficult to dislodge an established population than to prevent one building up in the first place, and many organic pest management techniques aim to prevent pest populations attaining high enough levels to cause substantial damage. This is also the stage at which pesticide applications are applied in conventional agriculture, and partially explains their popularity in that they are able to correct a situation to some extent, even at this late stage, by rapidly increasing pest mortality. This option is not generally open to organic farmers and growers.

Dispersal and Emigration

Dispersal is obviously important to crop pests in order to find new host plants to consume and to increase their numbers. It is also important for pests to disperse in order to find overwintering sites or other resources that they might need to maintain themselves. Farmers and growers will be interested in limiting the spread of pests within crops, or even between crops, but the methods used will be similar to preventing immigration of pests into crops. The use of fleeces and other barrier methods in organic farming has paradoxically created the possibility that pests are prevented from emigrating from crops and are trapped on the crop, allowing the population to increase dramatically in size. Some management methods attempt to disrupt the emigration to overwintering sites, mainly by removing overwintering hosts or shelter, but such methods have generally proved impractical to implement on the large scale needed, or to be disproportionately disruptive to agroecosystems. For example, removing overwintering hosts for aphids has ramifications not only for other species dependent on these plants, but also potentially for aphid natural enemy populations as they are left without host aphids on which to prey resulting in a reduction in their number.

DISEASE DYNAMICS

Disease-causing organisms are generally microscopic, but like herbivores, depend on locating, colonizing and exploiting host crop plants in order to complete their life cycles. Unlike herbivores they are not generally able to actively move and locate host plants (except in a very limited sense in some cases) and so depend crucially on survival in the absence of host plants and passive dispersal between hosts. Once they encounter a host plant they generally invade it and either destroy the plant to mobilize nutrients, or more subtly divert nutrients to their own ends without immediately killing the plant. Ecologically the balance between pathogens and plants has been seen as a disease triangle, which encompasses the factors necessary for a pathogen to spread: that is a virulent pathogen, a susceptible host, and a suitable environment.

Pathogens are generally microscopic and have traditionally been classified into three groups: obligate parasites, nonobligate parasites and facultative parasites. A more modern classification equates these life strategies to biotrophy, necrotrophy and hemibiotrophy, which more neatly explains the manner in which these organisms exploit host plants (*see* Chapter 9). However they are classified, these different modes of feeding have consequences for the life strategies employed by specific pathogens and therefore for the management methods that are likely to be effective in controlling them.

As disease-causing pathogens are mainly dependent on passive and chance dispersal between hosts, they all produce a large number of propagules in order to increase the likelihood of encountering a susceptible host. Propagules can include spores, living organisms and/or resistant resting stages. These propagules are usually distributed in wind or water and are important for pathogen distribution within and between crops. Obligate parasites can only survive on living plant hosts and consequently have evolved appropriate strategies for survival in the absence of host plants through the use of resting spores or other resistant and/or dormant stages or, in many cases, for surviving in alternative hosts at different times of the year. In contrast, facultative parasites are able to live on decaying plant tissue and so are able to survive in crop and other plant debris in the absence of the living host. Nonobligate pathogens are able to survive on living or dead host tissue to varying degrees, and represent a range of intermediate life strategies between obligate and facultative parasites.

For all pathogens there is a series of steps by which crop plants become infected, and farmers and growers need to bear these in mind when choosing management strategies. In the first phase the pathogen makes contact with the host, known as 'inoculation'. Once inoculation has taken place the pathogen needs to penetrate the host, after which it will need to establish itself within the host and develop. At an appropriate stage the pathogen will reproduce itself and produce propagules for dissemination to develop a secondary wave of infection. Successive waves of infection will allow an epidemic of disease to develop in the crop and between crops in a region. In many cases pathogens will also at some point produce specialist structures or behaviours that will allow them to bridge the gap between crops.

Management of plant disease has focused on preventing and/or interrupting the development of epidemics. In general, organic management methods will need to concentrate on avoiding epidemics building up and preventing pathogens entering crops, as there are few curative treatments available.

Inoculation of Crop Plants

In most cases the pathogen will come into contact with a susceptible host more or less by chance. Inoculation agents include (fungal) spores, (virus) particles and/or living entities (bacteria, mycelia). Inoculation occurs as a result of dissemination of such propagules, and the source will determine the types of management strategies that might be applied. Soil-borne pathogens are traditionally the hardest to manage as the inoculum is present and protected in soil, and this is one reason that rotation is commonly practised

Charlock as an alternative host for clubroot, a brassica pathogen.

in organic farm systems, to leave as long a gap as possible between susceptible hosts in order to either starve out the pathogen or give a chance for it to be preyed upon. Other sources of inoculum include crop debris (facultative parasites), alternative hosts (obligate parasites), and introduced or volunteer crop plants. Viruses have evolved to be transmitted by alternative hosts and/or other vectors which can serve to directly inject the particles directly into the host plants. In this case it is the vector that needs to be managed rather than the pathogen. Propagules may also be transmitted by vectors such as insects in a more or less passive fashion.

Penetration of Crop Plants

The propagule will need to penetrate the host plant tissue in order to establish an infection in the host, and this is often a difficult step in the process. Environmental conditions (such as moisture, temperature) on the host often need to be favourable to penetration, and plants have often evolved resistance mechanisms to prevent it. Sometimes these factors can be manipulated by farmers and growers to manage plant infection.

Penetration of the plant may involve direct mechanical entrance of the plant cuticle with or without the production of cuticle-dissolving enzymes, and this is observed in many fungi. In contrast many pathogens (especially bacteria) enter plants through wounds or other openings such as stomata or lenticels. In some cases a single propagule is capable of penetrating a plant and establishing an infection (biotrophs), whereas in other cases a critical mass of invading pathogens is needed (necrotrophs). The management of the crop and soil may make crop plants more or less susceptible to these types of invasion.

Propagules, but especially fungal spores, may require a number of environmental conditions to be met before they germinate and attempt to penetrate crop plants. Such conditions may involve complex chemical sensing of the environment and plant surface. Many require certain environmental cues or conditions to allow successful breaching of the host plant cuticle. For instance, many fungi require the presence of moisture on the leaves or high humidity in order to penetrate plant defences, and it may be possible to manage crops to keep such periods to a minimum.

Leaves can also have quite complex biotic communities on them, and there is some evidence that the interaction between pathogens and leaf-dwelling microorganisms can have an effect on the ability of pathogens to invade plant tissue. In fact compost teas and other plant amendments are often claimed to stimulate this community to suppress pathogens on leaves and thus protect plants from invasion. This is, however, an area of complex and developing science, and the conditions under which these amendments are effective is far from being well defined. It should also be noted that under some circumstances it is conceivable that penetration of plant pathogens may even be aided.

Plants utilize a variety of mechanisms to avoid pathogen penetration. Leaves are often coated with thick waxy cuticles that resist water and make it difficult for propagules to adhere to. They also resist penetration, and may be fortified with toughening substances such as silica. Natural openings may also be positioned to be difficult to access, or defended so that the infective agents are unable to penetrate the plant. Some of these factors can be selected for in plant breeding programmes, but may be unsuitable for commercial varieties.

Establishment in the Host Plant

Once the pathogen has penetrated the host plant it will need to establish itself and exploit the plant to grow and reproduce. Facultative and non-obligate parasites (in the main necrotrophs or hemibiotrophs) use a range of digestive enzymes that (more or less) rapidly kill and destroy plant cells, causing the familiar necrotic symptoms (rots, large leaf spots and blotches) of this type of infection. The pathogen then feeds saprophytically on the resultant nutrient-rich soup. In contrast, obligate parasites (biotrophs) invade the plant tissue without killing it. They develop a range of inter- or intracellular structures to absorb nutrients directly from living tissue or cells, and divert plant nutrients to their own ends. They often produce hormonal disturbances in the host plant that can be seen as distorted tissue, often symptomatic of the pathogen, and which aid in diverting plant nutrients to the pathogen.

Plants are not, however, completely undefended and are capable of mounting a series of challenges to the establishment of pathogens. There is often a complex mixture of enzymes and proteins produced by both the plant and the pathogen during the penetration and establishment phase, and a good deal of chemical warfare can develop at this level. Similar to their defence against herbivory, plants have evolved to alter their internal conditions – for example they can alter the pH to be unfavourable to establishment, and can produce and sequester a wide range of chemicals that are toxic or in some way inimical to pathogen development (for example saponins). Indeed, many of these compounds have found a use as medicines in treating disease in animals.

Many of these systems function in a way analogous to animal immune systems in that they detect invaders and then elicit defence reactions within the plant. Defence responses vary, but commonly include hypersensitivity whereby infected tissue or cells rapidly collapse and die in order to restrict pathogen establishment. Hypersensitive reactions are caused by a chain of chemical events that end in the release of highly reactive oxidative molecules that rapidly damage the cell beyond repair. At this stage plants often also begin accumulating antimicrobial toxins in the infected tissue. Prominent among these are the phytoalexins, but a whole range of other substances such as suberin, liginin and even silicas can be synthesized and deposited around the infected area, providing a chemical and mechanical barrier to pathogen establishment. It has also become apparent that apart from reacting at specific invasion sites, plants also begin to synthesize other pathogenesis-related (PR) proteins, including enzymes, which can inhibit pathogen establishment throughout the plant. This process is called 'systemic acquired resistance'.

Pathogens in return have evolved counter measures against plant defences. They may attempt to overwhelm the plant before it can react by producing toxic chemicals or defensive structures. This is the strategy of many necrotrophs. Most pathogens have also evolved some ability to detoxify plant defence chemicals. This is taken to extremes in biotrophs, which often have a limited range of host species and have consequently become entrapped in an arms race with their hosts. In many cases, for example lettuce downy mildew, this arms race has led to the development of specific races of hosts and pathogens with particular patterns of virulence and susceptibility well recognized by plant breeders and involving

Downy mildew on lettuce; a race between host resistance and pathogen virulence.

the defence mechanisms outlined above. The consequences of this for varietal resistance are discussed in Chapter 3.

Once established within a plant the management options against pathogens are very limited in organic farming systems due to the systemic nature of the infection. In fact management methods are likely to involve containing the disease epidemic in some way, often by destroying the infected plant or plant parts.

Dispersal of Pathogens

Once established in a host, pathogens begin to use the resources gained to form reproductive structures and to reproduce themselves. Reproduction results in the production of infective propagules, which pass on to infect new hosts. Propagules are often characteristic of the pathogen, and include a wide array of types including fungal spores, virus particles and individual bacteria. Within the crop season they are normally the result of asexual reproduction and can thus be produced rapidly in large quantities. In order to contact new hosts these propagules need to disperse, and plant pathogens have evolved a large number of sophisticated life strategies to achieve dispersal with a high probability of future infection (allied to a low survival probability for each individual). It is, however, true to say that most dispersal mechanisms are more or less passive, being predominantly by means of air or water circulation in one form or another.

Intracrop dispersal of pathogens is predominantly by means of localized air currents, rain splash and/or movement in ground water, and under appropriate conditions epidemics of disease can rapidly build up in field crops of susceptible individuals from a few foci of infection. Most pathogens are also favoured by humid rather than dry conditions. In some cases, especially viruses, pathogens are transmitted on animal vectors, either passively or as actual diseases of the animal. Management options for containing epidemics are mostly concerned with preventing the environmental conditions that allow for rapid spread and build-up of pathogen numbers. In some cases it might be possible to manage animal vector transmitters of disease, and crop management practices such as irrigation and soil cultivations can also be important for managing diseases.

Intercrop dispersal of pathogens is mainly airborne, and under the right conditions infective propagules can be carried downwind for large distances. Intercrop dispersal of pathogens also involves a range of other mechanisms such as transmission in host plant seed or transmission by vectors, either passively and/or accidentally, or actively as an animal disease phase. Agricultural practice is also a significant cause of disease transmission in practice, for instance, in soil movement and plant propagation materials. Management of intercrop dispersal is more likely to require large-scale coordination, coupled, in some cases, to forecasting and quarantine measures. However dispersal occurs, many pathogens have also evolved mechanisms for surviving in the absence of their host plants, in effect a strategy of remaining *in situ* and awaiting the return of susceptible hosts.

Survival in the Absence of Host Plants

Most pathogens have also evolved strategies or mechanisms to survive in the absence of their hosts, and this needs to be taken into account when designing crop rotations or other disease management methods. Some pathogens are essentially facultative parasites and these need no crop host to survive, having the ability to survive saprophytically in the soil on plant or other debris. In some cases they also have, in effect, such a wide host range that weeds or alternate crops also enable them to bridge the span between crops. Some of the more intransigent soil-borne pathogens fall into this category.

Many parasitic pathogens (fungi, bacteria) produce some form of resting or resistant structure that enables them to overwinter in the absence of the host. These resting structures, often characteristic to a species level in fungi and bacteria, are often formed towards the end of the crop cycle in response to changes in host plant physiology or external environmental cues. They are often also the result of sexual reproduction as opposed to asexual reproduction, and require the presence of different clones or mating strains. This is obviously a time at which gene recombination can allow new strains of a pathogen to arise, although resting spores are often relatively less numerous than asexually produced ones. Often these resting spores or structures will remain dormant or quiescent for some time, necessitating some environmental or chemical cue to germinate and commence growing. They are often present in soil, where they are vulnerable to attack by other soil-dwelling organisms, but they may also be present in crop remains or on harvested produce such as seed. Management methods can be targeted at this stage of the lifecycle, and in particular it may be possible to manipulate the environment to reduce their viability, and stimulate other microorganisms to consume them and/or remove them.

Some pathogens always need a host to survive. Obvious examples include viruses, some bacteria and obligate fungal parasites. In some cases they remain in volunteer crop plants, but in many cases they have evolved to survive in alternative hosts. In fact, some fungal pathogens such as rusts have very complex life cycles alternating between winter and summer hosts, enabling them to bridge the gap between crops and seasons. Other pathogens survive on crop plant seeds or other propagating material such as tubers, and in effect use the crop plants' own survival strategy to pass between crops and seasons. Once again appropriate management methods can focus on removing this link between susceptible host crops.

BEHAVIOURAL PEST AND DISEASE MANAGEMENT

Effective pest and disease management programmes will build an understanding of the individual interactions between pests or diseases and their host crop plants into the wider and more general strategies that seek to use an ecological approach as discussed in the previous section. In this sense organic farmers and growers are seeking to tackle pests and pathogens from the bottom up and the top down.

Interactions between the individual pest or disease and its host are often mediated by complex chemical interchanges between pest or pathogenic organisms and host plants. They often proceed by a sequence of 'behavioural' responses that need to be followed in order for infection to be successful in any one situation. At this level organic farmers and growers seek practices that manipulate and interrupt these chains of 'decisions' in order to prevent successful colonization of crops. Such detailed knowledge is the flip side of the more general understanding of the wider ecological causes of pest and disease outbreaks, and when combined will produce more effective pest and disease management strategies of value to all farmers and growers.

AGROECOLOGICAL PEST AND DISEASE MANAGEMENT

Ecological interactions in farming systems occur on many levels: between organisms within a system, between elements of a system (plants, crops, fields, farms, regions), and between organisms and elements in a complex web of interactions. This chapter has implied that pests and diseases are simply a part of this complexity. In common with all organisms within the agroecosystem, pests and diseases have their defined 'niche'. They are adapted to exploit specific circumstances in specific ways, and farmers and growers need to understand these if they are to manage them.

Organic pest and disease management programmes seek to be grounded in ecological and behavioural science, and to combine them into more powerful and effective pest and disease management strategies. Ecological pest and disease management also depends on drawing lessons from both the ecology of farming systems that make them vulnerable to attack by pests and pathogens, and from the detailed dynamics of pest and pathogen interactions with crop plants. This chapter has shown that pests and diseases are strongly influenced by these types of interaction, and that the relationship between pathogens, their host crop plants and natural enemies is central to their control and management in agricultural systems (so-called top-down influences), as is the availability of food and its quality (so-called bottom-up influences).

In many cases the underlying causes and mechanics of pest and pathogen invasion have been very well described by scientific research for many important pest or disease crop combinations, even to the level of being predictable using mathematical models. Whilst the strength of this scientific knowledge is in its detail, it is also a shortcoming from an agroecosystem perspective, in that it concentrates overwhelmingly on the narrower specifics of certain pest and/or disease crop plant interactions, and has much less to say about the, arguably equally important, interactions that occur between other elements in the system. For instance, pests and diseases can be affected by plant nutrition, soil conditions and the presence of beneficial organisms, amongst other factors. Pests and diseases are also part of the field or farm community, which is a grouping of different species that work in synergy or competition to compound effects. It is very difficult to describe whole system effects from lower level components as the models quickly become difficult to interpret as more factors are added in.

In fact, experience is often the best 'model' that we have in understanding the multiple interactions and effects that are likely to occur in any field or crop. It is for this reason that organic farmers and growers have learnt to value and develop their experience as ecosystem managers. It is arguably in this arena that organic pest and disease management comes into its own, placing the farmer or grower at the centre of the decision-making process, allowing them to make management decisions based on scientific reason and past experience. The subsequent chapters of this book develop these themes so as to build up a description of current pest and disease management practice in organic farming systems.

3

Preventative Pest and Disease Management

Organic pest and disease management relies on taking an ecological and systemic approach to farming and growing. The mainstay of this management approach is rooted in proactive cultural management measures that seek to enhance the farm agroecosystem to benefit the crop and suppress pests or diseases. Proactive management requires an understanding of the farm ecosystem and knowledge of the key pests and diseases that are likely to occur (*see* Chapters 7, 8 and 9), as well as an appreciation of the dynamics of pest and disease attack (Chapter 2). When this awareness is combined with knowledge of the range of control measures available and their likely effects on pest and disease incidence, the farmer or grower is in a position to build up a sound and effective proactive management strategy. This chapter discusses the range of potential proactive and preventative methods available to the farmer and grower. The following chapter (Chapter 4) discusses the range of more adaptive cultural and biological methods that might be necessary to manage a pest or pathogen, while Chapter 5 discusses more reactive and direct pest and disease control measures that can be taken.

The principal preventative measures used by organic farmers and growers are common to all organic farming systems and are, in fact, synonymous with good husbandry techniques in all farming practice. They can be summarized as maintenance of soil fertility, practising crop rotation, using resistant varieties, promoting conservation biological control, and taking sanitary precautions. These should be considered the first line of defence against pests and diseases in any organic system, although the details and techniques used may well vary between farms and systems. The approaches themselves will normally be beneficial not just for managing pests and pathogens, but also for enhancing many other aspects of the farm system such as nutrient flows and water recycling. Although these approaches are discussed in this chapter, in order to fully implement some of them, more detailed and specialized texts and leaflets on the specific topics will also prove useful.

SOIL ECOLOGY AND FERTILITY

Soil health and fertility is a central theme in organic farming, and a great deal of information has been produced and written about this important topic. Although we recognize that maintaining soil ecology and fertility in an organic system is vital for many reasons, here we will emphasize the importance of soil conditions in relation to pest and pathogen attack, the subject of this book. Plant vigour is obviously strongly linked to soil conditions as plants are rooted in the soil and derive their nutrients and water from it. Soil provides a substrate into which plants are anchored, whilst soil conditions and the soil environment will determine the availability of nutrients and water. Soil is also the medium in which most organisms in the biosphere (mainly microscopic) are found, and which interact with plants to create strong above-ground (or 'bottom-up') effects on terrestrial food webs. Many pests and pathogens are also found, or pass at least part of their life cycle, in the soil, and many are also strongly bound into the above-ground terrestrial food webs.

Therefore, and in practice, soil conditions will have a strong effect on the susceptibility of a crop to pests and diseases. As a general principle it is normally true that a vigorously growing plant in a 'healthy' soil, to which it is adapted, is more likely to be able to defend itself against attack or compensate for any damage caused by pests or pathogens than a stressed plant growing poorly. Healthy soils are, by definition, more likely to be suppressive of pathogenic organisms that do not generally compete well in the wider environment. However, soil 'health' is difficult to define precisely. Apart from the presence of pests and pathogens, soil structure, pH, nutrient status and soil ecology are all contributing factors to soil health, and ultimately to crop vigour and fitness. Some factors, such as pH, have well defined effects whereas many others, especially nuances of soil ecology, are poorly understood. It should also be taken into account that different plant species are adapted to tolerate, or exploit, different soil types, and this is also true of crop plants. Crop species, or even varieties within species can, to some extent, be adapted to different soil types and/or conditions, and this might need to be taken into account.

It should also be appreciated that these factors and their effects are intertwined and it is difficult to separate out causal chains. To some extent this is, in any case, unnecessary as the factors work together; for instance, raising soil organic matter will increase biological activity, improve soil structure and may increase availability of nutrients to plant roots while buffering the pH. Due to this type of complexity only a superficial description of the effect of these factors on plants can be given here, where we concentrate on the effect of the soil/plant relationship on pests and pathogens.

Soil Composition, Texture and Structure

The relative proportions of sand, silt and clay particles in soil combine to give it a characteristic texture. Soils are described as sands, clays or loams depending on these proportions. The structure of the soil is the result of the way in which these soil particles aggregate, and the number of pores (air and water channels) between the aggregations or crumbs. Texture and structure are related to some extent, and good growing soils generally have a good balance of particles (that is, they are loamy) with a good mix of crumbs and pore sizes. Soil texture and structure are among the most important factors affecting the general well-being of the crop, and they have a large influence on growing conditions. For example, well drained and aerobic soils normally promote plant growth and vigour, especially as most temperate crops are more or less adapted to such soil conditions. In contrast, sandy soils dry out quickly, whilst clayey soils can be prone to waterlogging, both of which conditions can also lead to poor growing conditions and subsequent pest and disease problems.

Crops grown in compacted or waterlogged conditions will show stunted growth as their root systems are adversely affected and nutrient availability reduced. Often such crops succumb to pest attack as the reduced growth rate precludes them from recovering from even moderate levels of pest damage. Compacted, waterlogged, and often anaerobic conditions also provide favourable circumstances for some soil-borne fungi such as *Rhizcotonia solani* that cause damping-off diseases. Anaerobic and waterlogged conditions will also affect soil microbiological activity, and often has the effect of reducing the suppressive effect of these microorganisms on soil-borne pathogens and pests.

In contrast, crops grown in excessively dry conditions are prone to wilting and to problems with transpiration and nutrient uptake. In a few cases, drought favours the pathogen in attacking the crop, but more usually drought exacerbates the symptoms in the plant. Drought can also predispose crops to pest and pathogen attacks as plants become weakened, are unable to respond defensively to the attack and are unable to compensate

Soil is the key interface in organic pest and disease management.

for damage caused. Powdery mildews are well known for preferring warm, dry conditions, although they still need periods of high humidity to initiate infection. Wilts (*Fusarium* spp., *Verticillium* spp.), whilst not necessarily favoured by drought, exacerbate drought symptoms as they interfere with water transport in the crop plant.

Whilst soil texture is inherent and not susceptible to change, soil structure can be improved and preserved by appropriate management approaches. Paradoxically both heavy (clayey) and light (sandy) soils can be improved by adding adequate amounts of bulky organic matter such as green waste compost and incorporated green manures. Structure can also be preserved by avoiding compaction of soils with machinery or even by avoiding trampling. Tillage practices can also have a major bearing on soil structure. All organic farmers and growers should have a soil management plan that outlines practices for their farms and conditions, although a detailed discussion on the range of options available lies outside the scope of this book, and more specialist soil management texts should be consulted. For the purposes of pest and disease management it should be borne in mind that a well structured soil is likely to be the most beneficial in both suppressing the presence of pests and pathogens at the same time as promoting the best conditions for good and vigorous plant growth.

Soil pH

Soil pH is a measure of the acidity of a soil solution and is commonly used as a soil descriptor. It affects the solubility of soil nutrients and the chemical reactions that occur in the soil and thus exerts a strong influence on both soil and plant individuals and communities as it mediates nutrient availability. Different crops and microbes are adapted to different soil pH conditions, but vegetable and arable crops generally prefer neutral (pH = 7) or slightly acidic soils, although excess acidity can quickly reduce the amounts of suitable nutrients available to plants and the soil microbial community. Liming very acidic soils (>6.5) is a common practice that can help to promote soil biological activity, which in turn is capable of aiding in the suppression of soil-borne diseases.

It is also well known that pH can also directly affect the expression of some common pathogens such as common scab (*Streptomyces scabies*) in potatoes or clubroot of brassicas (*Plasmodiophora brassicae*). Generally a soil pH of more than 7.0 will have a suppressive effect on clubroot, whereas common scab is suppressed at a pH of less than 5.5. It is for this reason that growers are advised to apply limestone before growing brassicas but not potatoes, from which it follows that it often makes sense to grow potatoes before brassicas in the rotation so that limestone can be applied (when necessary) after the potato crop and before the brassica crop.

Adjusting soil pH over a large area for any length of time is both an arduous and questionable practice from a sustainability point of view. It is true that raising the pH of acid soils is widely practised by the addition of 'lime' products and some of these are obtainable from sustainable sources, but in the long run it may be better to work with crops suited to the prevailing pH conditions and maintain soil organic matter to buffer soil acidity. Once again a detailed discussion of these issues is beyond the scope of this book, and more detailed texts on organic soil management should be consulted. Suffice it to say that some specific pathogens such as potato scab or clubroot can be affected by adjusting the pH, at least temporarily, and that crops growing in soils with a pH to which they are adapted are less likely to be susceptible to pest and disease attack.

Soil Nutrient Status

Crop nutrition is key to growing healthy, vigorous plants, and has a large influence on susceptibility to pests and diseases. Obviously a crop plant with an adequate nutrient supply will grow strongly and will be able to both mount a defence against pathogen attack and compensate for, and potentially outgrow, damage if it occurs. In general crops should receive a balance of nutrients tailored to their requirements, both macro-nutrients and micro-nutrients, as both excesses and deficiencies can be detrimental. Some of the micro-nutrients in particular, for instance zinc (Zn), may be essential for enzymes and other molecules that protect plants against pathogen attack. As a general rule, plants that are stressed due to the shortage of a particular nutrient are more likely to succumb to pests and diseases, and absolute shortages of these elements by themselves can lead to disease-like symptoms in plants (nutrient deficiencies). Some of the principal effects of the macro- and micro-nutrients are outlined below.

Nitrogen (N): Needed in balanced amounts and at appropriate points in the season when the plant is growing vigorously, as it is vital in protein synthesis. Crops with insufficient nitrogen will be stunted, yellowish and will not grow strongly so they will not recover from attack. In contrast, crops grown with above-optimal levels of nitrogen (whether from manure, a grass clover ley or fertilizer) produce large quantities of succulent, green growth. Such growth is often vulnerable, however, as it is more palatable to insect pests and slugs as well as to attack by fungal pathogens.

Phosphorus (P): A necessary constituent of cell membranes, it improves resistance and tolerance to diseases, and is important for root development and establishment. In particular it is important in helping to resist root and foot rots, although these are also strongly linked to environmental (wet) conditions.

Potassium (K): Also vital to cell function in plants, and the macro-nutrient most often associated with reducing the severity of diseases in plants. Although the mechanisms are not well understood, the symptoms of potassium deficiency – thin cell walls, accumulation of sugar in leaves, weak stems, and accumulation of N – are all in themselves associated with plant vulnerability to pests and pathogens.

Calcium (Ca): An essential element in the cell wall structure of plants, and helps to provide strength and, indirectly, resistance to pests and diseases.

Sulphur (S): Essential for the production of proteins, enzymes and vitamins, and thus central to plant defences against pathogens and pests. In itself it has also been implicated in having a direct inhibitory effect on pathogens (*see* Chapter 5).

Magnesium (Mg): Essential in chlorophyll, and hence basic to plant health.

Micro-nutrients: Include iron (Fe), zinc (Zn), manganese (Mn), boron (B), chlorine (Cl), copper (Cu), molybdenum (Mo). Some of these nutrients although required in only small amounts are important in protecting plants against pest and disease attack through their effects in various metabolic pathways in plants, some intimately involved in plant defences.

Different crops are adapted to tolerate and exploit different nutrient conditions, and this should be borne in mind when designing rotations so that crops that require higher levels are placed nearer to fertility building periods or receive adequate nutrient inputs (*see* Crop Rotation below). Crops will also need different nutrients at different stages of their development as they germinate, grow vegetatively, flower and set seed. Once again other texts should be consulted for a more detailed discussion of the issues around nutrients and nutrient availability, while noting that crop plants with a balanced and adequate nutrition are more likely to be able to resist pests and diseases or outgrow any damage caused.

Soil Ecology

Soil has a complex ecology and is the medium in which interactions between a multitude of soil organisms (micro- or macroscopic) take place. There is a complex flow of energy, nutrients and water through the soil structure and food web and, as such, their effects on plant diseases and pests can vary widely depending on circumstances. The soil food web is important in buffering many of the key soil indicators mentioned above: texture, structure, pH and nutrient status. Microorganisms and microflora and -fauna break down organic matter providing plant nutrients at the same time as improving aeration of the soil, improving water flow and improving water-retaining capacity through numerous chemical and physical actions.

It is becoming clear that soil food web interactions have important ramifications for pest and disease management both above and below ground. A good rule of thumb is that a vigorous soil ecology with a good balance of active organisms will act to suppress plant pathogens and even pests. This suppressive effect can be due to several mechanisms: a direct ecological effect on pests and pathogens in the soil, an indirect below-ground effect on pests and pathogens acting through diverse ecological mechanisms, and an indirect above-ground effect, mainly mediated through plants.

Below ground an active soil food web will act to directly suppress pests and diseases. Direct effects arise when soil microorganisms predate and/or consume pests or pathogens. There is a large and growing list of beneficial soil organisms that have been found to act in this way representing many diverse functional groups, including viruses, bacteria, fungi, actinomycetes, protozoa, algae and nematodes. For instance species of fungi snare and consume pest nematodes, bacteria prey on pathogenic fungi, protozoans prey on bacteria, and both bacteria and fungi prey on pest insects. Even resistant structures such as spores may be targeted and eventually broken into and consumed. Some organisms, such as *Trichoderma* spp., are known to act as antagonists against pathogens such as white rot (*Sclerotium cepivorum*) of onions, and adding it to the soil has successfully contributed to combating the disease in some situations. Indeed, much research has concentrated on looking for biological control agents that can be added to soil. However, this has not resulted in commercially effective products (to date), perhaps because it is also true that organisms added to soil in conditions that are not otherwise suitable are unlikely to persist.

Indirect effects arise through organisms interacting in complex food webs and interlinked chains so that pests and pathogenic organisms are suppressed. For example, many pathogen species are weakly competitive when living saprophytically in the soil and can be outcompeted by soil organisms that are better adapted to this way of life. Some pathogens can only attack plant roots through specific sites or when present in sufficient numbers, and other soil organisms can 'displace' them, preventing access to susceptible root entry points. Many resting structures such as fungal spores will be subject to the action of detritivores in the food chain, and will be rendered unviable over time through the general action of soil organisms. It has been demonstrated in pot trials that increased earthworm density can lead to increased disease suppressiveness of soils, in this case by a 50 to 70 per cent reduction in *Fusarim* and *Verticillium* wilts in asparagus, aubergine and tomatoes. This is due to their beneficial impact on soil microbiological activity as soil is processed through their gut (allied to improvements in soil structure).

Soil ecology can also indirectly affect above-ground food webs, mainly through plant-mediated effects. An obvious effect of soil food webs is on plant nutrition and in making plant nutrients available, which will in turn affect the plant's attractiveness and susceptibility to pests and pathogens such as aphids. There is also a body of established evidence that supports the idea that plant's defences can be stimulated below ground, which makes them less vulnerable and more able to resist attack above ground. Some of the factors involved in this systemically acquired resistance have been discussed in the previous chapter (Chapter 2).

Many of the techniques previously mentioned for improving soil structure and nutrient status also have a beneficial effect on stimulating soil ecology. For example, adding sufficient quantities of organic matter such as green waste composts or green manures to soil acts to drive the food chains while improving soil structure and nutrient availability (themselves partly as a result of increased biological activity). Soil tillage methods and techniques can also have a major bearing on soil ecology, with disturbance being generally disruptive, so that minimizing tillage operations can be beneficial in the long run. However, there might be trade-offs in any particular situation; for instance, tillage tends to decrease soil organic matter and disrupt soil ecology, but can also expose pests to predators and desiccation.

The use of soil management techniques in managing pests and diseases is discussed in the following chapter (Chapter 4), which discusses adaptive cultural control techniques in more detail.

CROP ROTATION AND DIVERSITY

From an ecological point of view, diversity and complexity are key features of natural ecosystems, which have complex nutrient and energy flows through them. Many pest and disease problems arise as a consequence of the reduced diversity in farming systems, and organic farming systems aim to replace at least some of this diversity into cropping systems to enhance pest and disease management.

Diversity can be introduced into crop systems by varying them over a period of time, and by varying them spatially over the farm area. Rotation achieves both goals. Diversification can also be achieved by diversifying enterprises within a farm, and organic farming systems are also notable in that they are often 'mixed' farming systems. Crop diversity can also be enhanced by a range of more adaptive methods such as inter- or strip cropping, the timing of management operations, and crop spacing, and these are discussed in the following chapter (Chapter 4).

Rotation

Organic systems place emphasis on creating an environment that prevents the build-up of pests and diseases, and many farmers have found that designing a good, workable rotation is key to this. Once again, whilst there are numerous reasons for designing a good rotation including nitrogen fixation, nutrient management and weed management, here we focus on the advantages of rotation for pest and disease management. The general principle of designing a rotation is that vegetables and other crops from the same family should not be grown too often or sequentially between seasons, on the same piece of land. This is because many pests and pathogens specialize in attacking specific plant species or plant families (groups of species), and rotating crops around a field or farm will act to prevent the build-up of these pests and diseases.

The mechanisms by which rotation exerts its effects are numerous, and are discussed at many points in this book. A primary effect is simple 'starvation' in the absence of the host, causing an increase in mortality and reduced or no reproduction – that is, pests and pathogens are not able to complete their life cycles in the absence of the host crop plant. Rotation also serves to remove the crop from close proximity to resting stages (for example, fungal sclerotia or insect pupae) of the pest or pathogen so that when these hatch or emerge to infect new crops they have to move some distance to locate host plants.

Other effects include reduction through competition, as pathogenic organisms generally do not compete well with free living organisms, especially in soil, and reduction through predation where beneficial organisms directly consume the pests or pathogens. It should also be remembered that some weeds are also hosts for pests and pathogens, and a varied and long rotation is more likely to reduce the abundance of any one particular species of weed, thus decreasing the likelihood that they will act as a living bridge between crops.

From the mechanisms by which it functions, it can be seen that rotation is likely to be more effective against relatively sedentary soil-borne pests and diseases such as clubroot in brassicas and cyst nematodes (*Heterodera* spp. and *Globodera* spp.) in potatoes, and much less effective against mobile and/or airborne pests and diseases which arrive from outside the field, such as cabbage white butterflies (*Pieris brassicae*) or rust (various species on various crops). Some pests and pathogens such as those that cause damping-off diseases (*Rhizoctonia solani* etc.) can have a wide host range, and rotation is unlikely in itself to be a sufficient management tool in these cases. Many soil-borne pests and diseases, for example white rot of alliums, also have very resistant resting stages, which can pass for many years in the soil without the host plant being present and are stimulated to germinate once it is. In this case a biologically active soil will help to increase the rate of predation and inactivation of these resting structures, but it is unlikely to be economically possible to prolong the rotation sufficiently to avoid the disease completely (twenty or more years in some cases).

As previously discussed, rotation is most effective when as long a gap as possible is left between susceptible host plants. There are some crops and crop families that are particularly susceptible to soil-borne pests and diseases, and these require a break in the rotation before they are grown in the same field or space. Brassicas (cabbage, swedes etc.), solanaceae (potatoes, tomatoes, peppers etc.) and alliums (onions, leeks etc.) in particular should not be grown on the same piece of land in more than one year in four. This is a requirement for the Soil Association organic standards for field vegetables in the UK, although these standards do also allow double cropping of alliums or brassicas in the same season. Other crop families are far less susceptible to soil-borne pests and diseases, but a minimum two-year gap between growing them should be observed, and in most cases as long a gap as possible is the best option. Following on below is a list of the most susceptible families and common diseases for which rotation is an essential management tool.

Brassicaceae

These include many of the most popular crops such as cabbages, cauliflower, calabrese, broccoli (all *Brassica oleracea*), turnip (*Brassica rapa*), swede (*Brassica napus*), mizuna (*Brassica juncea* var. *japonica*), pak choi (*Brassica chinensis*) and mustard (various *Brassica* spp.). In an intensive

Build-up of disease on mature cabbage leaves.

vegetable system where there is high market demand for these crops there is often pressure to grow them more often than once in four years. However, it is important to stick to the four-year rule, as growing brassicas more frequently greatly increases the chances of clubroot taking hold. It is most likely to be introduced into the farm on tools or boots, although module plants brought on to the holding are also likely to pose some risk. Once this soil-borne disease infects the soil, it can persist for twenty years, and no brassicas at all can be safely grown on the infected land during this period without a management programme. Other pests such as cabbage root fly are also likely to build up in and around fields used for brassica production, and separating crops as far apart temporarily and spatially in the rotation is likely to be beneficial.

Solanaceae

These include potatoes (*Solanum tuberosum*), peppers (*Capsicum annuum*), tomatoes (*Lycopersicon esculentum*) and aubergines (*Solanum melongena*). The most troublesome soil-borne pests of potatoes are the potato cyst nematodes (*Globodera pallida* and *G. rostochiensis*). Nematode populations in the soil are far more likely to increase if potatoes are grown more than one year in four, resulting in decreased yields of potatoes, especially on sandy soils. Other solanaceous crops can be prone to infections of wilts (*Verticillium* spp. etc.), and again, crops from these families should not be grown more than one in four years. If infection occurs, then strawberries should also be avoided. Root knot nematodes (*Meloidogyne* spp.) can also infect solanaceae, but this is not so common in a temperate climate. Potatoes are susceptible to a range of tuber 'blemish' diseases and/or rots, and rotation will be necessary to keep the level of these, in some cases cosmetic diseases, at a level acceptable for marketing.

Alliaceae
Crops such as onions (*Allium cepa*), leeks (*Allium ampeloprasum*), shallots (*Allium ascalonicum*) and garlic (*Allium sativum*) are susceptible to a range of soil-borne diseases, of which white rot is perhaps the most serious. Like club root, once it has infected the soil it can persist for periods of at least twenty years, and no alliums can be 'safely' grown during this period.

Leguminaceae (Fabaceae)
These include all types of beans (*Phaseolus* spp etc.), peas (*Pisum sativum*), clovers (*Trifolium* spp etc.), vetches (*Vicia* spp etc.) and others. Although there is no formal requirement for a minimum break period under certification rules, it is recommended practice not to grow these too frequently. Bean weevils (*Sitona* spp.) will attack many species of legume causing reduction in yield, or complete crop failure if they attack at the germination stage. The potential build-up of pests and diseases in clover leys is often neglected when planning rotations. Red clover in particular is susceptible to the stem nematode (*Ditylenchus* spp.) and clover rot (*Sclerotinia trifolii*), and ideally there should be a gap of five years between growing red clover leys. Different species of nematode infect different species of clover, so growing different types can prevent a build-up of these pests.

Apiaceae
Umbellifers such as carrot (*Daucus carota*), parsnip (*Pastinaca sativa*) and celery (*Apium graveolens*) can be susceptible to various soil-borne diseases such as cavity spot (linked to infection by *Pythium* spp.) and violet root rot (*Helicobasidium purpureum*) that can be managed by rotation. A common pest, carrot root fly (*Psila rosae*) is also a weak flier and can be managed to some extent by separating crops by as much distance as possible between seasons.

Cereals
Cereals can be prone to foot rots (various species) and take-all disease (*Gaeumannomyces graminis*), which are often linked to continual cereal growing in the same fields and/or poor soil conditions. For this reason cereal rotations should be kept short on organic farms and interspersed with other crop families. Such rotations will also aid soil management in any case.

Other Methods of Increasing Crop Diversity

Apart from introducing more complex rotations, most organic farm businesses also introduce a wider range of crops or crop enterprises to their system. Modern farming systems have tended to specialize in one or a few commodity products – for example wheat (*Triticum aestivum*), barley (*Hordeum vulgare*), oil-seed rape (*Brassica napus*) – and this has severely constrained rotations. In contrast, organic farmers and growers tend to reintroduce a range of crops. This not only helps to manage pests and diseases but is also a good risk-reduction strategy to provide a more stable yield (and income) in the face of volatile commodity markets. Introducing a range of vegetable crops as break crops into a farming system will normally increase the potential length of rotation considerably, but will have implications for farm management and costs.

Apart from crop diversification, a further level of complexity can be added to a rotation by reintroducing livestock into a farm system. Livestock can provide considerable benefits in many ways over and above simple enterprise diversification as they more naturally mimic natural ecosystems in which animals are important elements and where they improve cycling of resources (for example nutrients) and energy. They also create diversity within systems by their behaviours and in the way they exploit resources.

Diversity can also be increased by introducing a range of non-crop plants or other physical features into the farmed landscape, and this is discussed more fully below as conservation biological control.

PLANT RESISTANCE

Crop plants are not completely helpless and often display some measure of resistance to pest or pathogen attack. Resistance enables plants to avoid, tolerate or recover from the effects of attack; it is heritable and characteristic of crop varieties. Although subject to genetic control, resistance is strongly influenced by environmental conditions as the plant grows (so-called phenotypical expression) and is subject to modification by the conditions under which the plant is growing. Partly for this reason, it can only be meaningfully measured relative to other genotypes (or varieties) of a crop species. In fact, it is usually measured in comparison trials using broad indicators such as plant yield at a given disease pressure and/or by qualitative comparisons of the amount of disease at various crop stages, although such comparisons will obviously be strongly influenced by season and husbandry. This is perhaps why plant breeding is sometimes referred to as an art!

Nevertheless researchers have been increasingly able to analyse resistance with more precise quantitative measurements using modern techniques. For instance it is possible to measure plant chemical response to insect or disease attack using sensitive chemical assays, or to analyse insect behaviour on or around plants using modern electronic recording and analysing techniques. More notoriously, resistance has been linked to specific genes, or collections of genes, that breeders are able to identify with increasingly powerful molecular analysis tools, and it is suggested that these can be moved between species groups to exert similar effects, perhaps forgetting the complex environment in which genes function.

Although a detailed discussion is not possible in this text, it should be remembered that in the end resistance is expressed phenotypically and is dependent on environmental conditions. A plant growing without nutrients is unlikely to be resistant to attack or to yield highly due to its immediate circumstances, whatever its genome.

Resistance Mechanisms

Resistance is normally classified by its functional effect on the pest or pathogen, but it should be recognized that it is often not possible to characterize a resistant variety completely in this way, as resistance can be due to a range of characters; these are summarized in the sections following on below.

Non Preference

This is where the plant has modifications that allow it to escape detection by a pest, and is often due to the presence or absence of chemical cues given off by the plant, or to morphological structures such as leaf trichomes (glandular hairs). Non preference is usually relative and can depend on the presence or absence of other suitable hosts for the pest, so that when there is no choice (for example in clean monocultures) the crop plant will be attacked anyway.

Morphological Resistance

Sometimes called phonetic resistance, this involves the formation of structural barriers in the plant such as hairs, hard or inaccessible tissues or parts, and exudations such as gum. For example, tough nodes and dense pith in wheat confer resistance to stem sawflies (*Cephus cinctus*) by limiting their abilities to penetrate and tunnel in the stem. Plant hairs or trichomes are also able to confer resistance by preventing access to the underlying epidermis, or secreting gums that entangle pests such as aphids. Such resistance traits can endure as they often require comparatively difficult or large adaptations on the part of pests or diseases to overcome them.

Antibiosis

This relates to the nutritional quality of the host plant. Plants with this type of resistance can prevent the pest or disease developing normally and/or completing its life cycle. Antibiosis can be mediated through the balance of nutrients in plant tissues (especially soluble N and other micronutrients), by metabolic inhibitors and/or by other inhibitory chemicals. Of particular importance in organic farming is the effect altering the balance of primary plant metabolites on pests and diseases. Aphids, for example, are well known to prefer, and thrive on, plants with high levels of soluble nitrogen and/or high sugar. In other cases the lack, or low levels, of particular amino acids is linked with high levels of resistance to specific pests or diseases. Although it might not be practically possible to manipulate the crop development to this extent, it is certainly possible to avoid soil nutrient imbalances such as nitrogen flushes that are capable of feeding through into high nitrogen content in tissues, which in turn attracts insect pests such as aphids. It should be recognized that manipulating nutrient balances may also affect taste or marketable quality in some way as a side effect.

Chemicals that affect pests and diseases are well known, and even common; they are often referred to as 'secondary metabolites', despite the fact that the exact function of many of them is not completely understood (*see* Chapter 2). Typical secondary metabolites include glycocyanides in brassicas and glycosides in potato. Chemical antibiosis can confer a strong selection pressure on pests, and some have even evolved a preference for the presence of these chemicals in plants before they will feed on them. In many cases these types of defence can also be induced in the plant by pest or pathogen attack.

Tolerance

This implies that a particular plant or variety will give a comparable yield even though the pest or disease is attacking it. The mechanisms for tolerance include compensatory growth, mechanical strength in the face of damage, and partitioning off or diverting nutrients away from attacked areas.

Resistance Breeding

Although resistance is expressed phenotypically – that is, as an interaction between the plant's genome and its environment – it is also ultimately dependent on heritable genes and so can be bred into varietal lines by breeders and selected for. It is often easier to breed and select for traits that can be measured qualitatively, such as number of leaf hairs, stem length or quantity of a given secondary metabolite. Such qualitative characteristics are often under the influence of one or a few genes, and so this type of resistance has been called 'oligogenic', 'vertical' or 'major gene' resistance. Such vertical resistance often subjects pests and pathogens to a strong selection pressure to overcome it, and creates an 'arms' race between the breeder and the pest or disease in which the breeder incorporates genes for resistance, against which the pest or disease reacts with virulence genes of its own. At present many varieties with such resistance, such as that against downy mildew (*Bremia lactucae*) in lettuce (*Lactuca sativa*), can be expected to last at the most a few seasons, before races of the pathogen arise that can overcome the pest.

In contrast 'field', 'horizontal' or 'polygenic' resistance is a result of the actions of many genes across the genome, each exerting small effects. It often results in plant characteristics that are difficult to quantify, such as vigour, or resistance that is difficult to relate to *per se*, such as compensatory ability. However, such resistance is generally considered to be more stable and durable, as pests and diseases have to overcome a series of barriers in order to defeat it. For this reason it is preferred in organic systems. Unfortunately, breeding for polygenic resistance is less straightforward and more time-consuming than major

Lettuce variety evaluation trial.

gene resistance. On the positive side, this is the type of resistance that is likely to be expressed in land races and varieties bred by farmers under specific, usually localized, conditions.

Breakdown of Resistance

Resistance is phenotypically expressed, and as such is subject to modification by environmental conditions. The effects may be as a result of their action on the pest or disease, the crop plant or both. The most important physical factors include temperature, light, humidity, and soil conditions and fertility. Unusually high or low temperatures can lead to loss of resistance provided the pest or disease is not also adversely affected. In some cases shading has been reported as leading to a loss of resistance in both sugar beet (*Beta vulgaris*) and potatoes against the peach aphid (*Myzus persicae*) and Colorado potato beetle (*Leptinotarsa decemlineata*) respectively. Soil condition and nutrient availability indirectly affect resistance in the host plant due to nutritional factors, especially in sap-feeding insects such as aphids that are sensitive to nutrient levels and ratios in crop plants.

Other factors are also likely to be important. For instance resistance will also vary with plant age as the plant passes through its life cycle and partitions energy and nutrients in different tissues and structures. Cultural measures will also affect resistance, partly by altering the previously mentioned factors, but they can also be used to reduce the selection pressure on pests and diseases to overcome resistance by presenting them with a series of varied hurdles to overcome.

Using Resistance

Use of plant resistance is a fundamental control strategy in organic farming systems, but not always easy to implement for a variety of reasons. Breeding for resistance can be a complex process and may necessitate compromises: for instance, resistance to one pest may correlate with susceptibility to another, or be correlated with other undesirable crop traits such as unpleasant taste. Some resistance characteristics might also reduce yield or increase toxicity of crop plants – obviously undesirable in crops destined to be eaten or handled. Resistance mechanisms can also hinder biological control agents: for example, plants with dense trichomes can prevent parasitoids of insects searching efficiently, and similarly brassicas with waxy cuticles which might prevent pathogen invasion can hinder natural enemies of aphids searching on the plant.

Sometimes the use of varieties may not be linked to pest or disease management at all. For instance, many larger retail markets have tight specifications that effectively demand specific crop varieties, which in many cases are not particularly resistant to pests or diseases. Even in direct sale markets customers may prefer certain varieties or types. Continuity of production may also require the use of varieties that are not particularly resistant but produce early or late in the season. In the case of organic seed, varietal choice may be limited, and exclude those varieties with good resistance characteristics because the organic seed market is small and makes their organic production uneconomic, or the companies are otherwise not interested in the organic seed market.

In any case, new varieties, many with improved resistance, are being produced continually by seed companies, and farmers and growers should always experiment with their use. A good starting point is to consult seed company catalogues and access independent variety trial information when choosing new varieties (*see* Further Information, page 408). Many seed companies run open days where varieties can be inspected while growing *in situ*. It will be prudent to trial new varieties in a limited area for one or more seasons before introducing them into the farm rotation on a large scale.

CONSERVATION BIOLOGICAL CONTROL

Conservation biological control is the conscious management of the farmed landscape to enhance the effectiveness of natural enemies, and should be basic practice on all organic farms. Management practices are geared towards providing natural enemies with the resources and refuges that they need to survive in and around the farm and crop, and protecting them from the negative consequences of farming practice where possible. Such an approach, often called 'farmscaping', is a whole-farm approach to pest management which attempts to integrate all elements of farm practice to provide the maximum amount of natural biological control within the constraints of operating a profitable farm business. Elements within a management plan could include the use of hedgerows, use of wooded areas, sowing attractant plants, and installing ponds or other water features. More adaptive crop-related techniques such as intercropping and cover crops are discussed in the following chapter (Chapter 4), while larger farm-scale management approaches are discussed in this section.

Conservation biological control has been most explored for the management of insect pests, and it would be true to say that this is the area in which most knowledge has accumulated, as insects are subject to control by a wide range of natural enemies. There has been a growing interest, partly spurred on by organic farming methods, in investigating the role of conservation biological control of soil-borne pathogens, and the implications of the ecology of microorganisms in specific environments (for example on leaf surfaces) has been important in defining potential areas where conservation biological control may be important.

Types of Natural Enemies

All organisms have their own natural enemies, and agricultural pests and disease organisms are no exception. In fact natural enemies are often drawn from the same classes and families as the pests and pathogens themselves, and at times it can be difficult to distinguish between them. However, it is worth learning and knowing something about natural enemies so that their needs can be addressed as part of farm and crop management programmes; these are identified as predators, parasites and parasitoids.

Predators

Predators generally consume many 'prey' over their lifetime. As normally understood, predators are adapted to 'hunt' down or 'ambush' their prey. From an ecological point of view, predators are specialist consumers (carnivores) feeding on lower trophic levels in food webs. Predators are found in the third or fourth trophic levels (as secondary consumers above primary consumers or herbivores in the second trophic level), and many can also be omnivores, effectively switching between trophic levels as the opportunity provides. Ground-dwelling beetles (Carabidae) are typical predators that take a wide range of insect prey, as are stoats (*Mustela erminea*) that eat rabbits (*Oryctolagus cuniculus*). Predators may consume prey at all stages of their lifecycle – for example owls (birds of the families Tytonidae and Strigidae) – or only at certain times – for example hoverfly (insects of the family Syrphidae) larvae consume aphids, but the adults feed on nectar in flowers.

Although predation is normally thought of as a relationship between active hunters and their prey, there are other specialist types of predation that are also important in pest and disease management. For instance, fungivores or mycophages specialize in consuming fungi including pathogens, whilst bacterivores, themselves normally microscopic protozoans, specialize in feeding on bacteria. Some types of fungi are also able to trap nematodes and consume them in a specialist adaptation useful to biological control.

Spiders are generalist predators.

Parasites

Parasites are organisms that are adapted to live in or on their hosts, and derive their nourishment entirely from them. Parasites generally infect their 'hosts' by passing between them in a passive way – that is, they are transported inactively, or wait for the host to encounter them in the environment. Notwithstanding this, some of the strategies for achieving transmission can be very complex, up to and including modifying animal behaviour to make transmission more likely. Animal diseases are parasites from a wide range of groups (including viruses, bacteria, fungi and nematodes) and the most known about. However, all animals and microorganisms suffer to some degree from pathogen attack, even pests and pathogens. In fact parasites themselves often have hyperparasites that prey on them and, as can be appreciated, trophic relationships within a food web can sometimes be extremely complex.

Parasitoids

Parasitoids are specialist insect parasites that are adapted to live on other insects as part of their life cycle, normally during the larval stage. The adult parasitoids (often small wasps) actively seek out their insect prey and then infect them by laying one or more eggs in or on the host. In some cases the host(s) may be collected and placed in a constructed feeding chamber. The parasitoid larva develops on or within the host body, eventually consuming it. Parasitoids consume a single, or a few, host insects during their life cycle, and the adult parasitoids often feed on alternative resources such as nectar or pollen.

The major predators, parasitoids and parasites of consequence in the biological control of insects on organic farms are presented in Table 1, along with a brief description of their life habits and prey and/or hosts.

Other natural enemy groups are presented in Table 2, and vertebrate natural enemies are presented in Table 3. Their effectiveness and use in biological control are indicated in the tables and discussed in the following sections.

Table 1: Important Invertebrate Natural Enemies in Organic Farming Systems

Organism/ Group	**Family/ Species**	**Type and Specificity**	**Life Cycle/Habits/ Conservation**
Pathogenic viruses/ insect pathogenic viruses	Viruses including nuclear polyhedrosis (NPV), cytoplasmic polyhedrosis (CPV) and granulosis viruses (GV)	Epizootics and diseases of insects and other organisms. Often adapted to and specific to a narrow host range or group	Can cause epizootics in insect populations that kill a large number of insects when populations are high and probably very important in regulating insect populations in the wild. Usually host specific and with adaptions for transmission between hosts and survival in the absence of the host. A number are being developed as microbial pesticides (*see* text), either as living organisms or pesticidal extracts
Pathogenic bacteria/ insect pathogenic bacteria	Many species including *Bacillus* and *Paenibacillus*	Diseases of insects and other organisms. Quite often specific to a range or group of hosts	Live in soil and environment. Infections often host specific but with adaptations to be transmitted and survive between hosts. Many produce insect-toxic proteins that kill insects when ingested although the infection process may take a few days. But has been developed as a microbial spray (*see* text) specific to a few insect families (mainly Lepidoptera or Diptera)

Table 1 (Continued)

Organism/ Group	**Family/ Species**	**Type and Specificity**	**Life Cycle/Habits/ Conservation**
Pathogenic fungi/ insect pathogenic fungi	Many species including entomopathogenic fungi	Diseases of insects and other organisms. Often specific to a range or groups of host species	Often require humid conditions for epizootics to develop in insect pest populations. Spores produced and spread passively to infect new hosts although they have a capacity to survive in the absence of the host. May be more or less specific to their hosts
Parasitic nematodes	Many families including entomopathogenic nematodes Steinerneatidae and Heterorhabditidae	Parasites of slugs and other animals. Many generalist but some specialist on slugs and insects	Live in the soil and most effective in moist soil. Juvenile stages seek out and infect hosts with various directed behaviours. Entomopathogenic nematodes carry insect pathogenic bacteria that infect and kill insect hosts on which the nematodes then feed. Some have been developed as biological control preparations e.g. *Heterorhabditis megadis* against vine weevil larvae and *Strinernema carpocapsae* against soil pests. Not developed for use against vertebrates

(Continued)

Table 1 (Continued)

Organism/ Group	**Family/ Species**	**Type and Specificity**	**Life Cycle/Habits/ Conservation**
Mites (Acari)	Numerous species, mainly Phytoseiidae	General predators of mites and small insects such as thrips and whiteflies. Some specialists	Simple life cycles on vegetation and soil surface. Have been used in biological control programmes especially in glasshouses for control of spider mites, fungus gnats and thrips e.g. *Hypoaspsis miles* and *Phytoseuilus persimilis*
Centipedes (Chilopoda)	Many families and species	Generalist nocturnal predators of other (small) arthropods	Confined to moist areas with more or less simple life cycles. Generally have a toxic bite to immobilize prey (not dangerous in the UK) and feed by hunting or ambush
Spiders (Araneida)	Numerous families and species	General predators of (mainly) insects	Simple life cycle on vegetation or soil surface, hunt by ambush, hunting or capture with webs; some cultural practices and many pesticides harm spider populations. Groups (or guilds) of species are thought to make a more important contribution to insect control than single species. Consume large numbers of pests such as aphids
Harvestmen (Opiliones)	Numerous species	General predators of small insects (but omnivorous)	Simple life cycle on vegetation and soil surface. They are generally mobile and they ambush and (occasionally) hunt their prey

Table 1 (Continued)

Organism/ Group	Family/ Species	Type and Specificity	Life Cycle/Habits/ Conservation
True bugs (Hemiptera)	Hemiptera in families Anthocoridae (pirate bugs), Capsidae, Miridae, Rduviidae (assassin bugs), Lygaeidae, Nabidae (damsel bugs), Pentatomidae (stink bugs)	General predators of insects with sucking mouthparts	Simple life cycles in the same habitats as their insect prey and will benefit from undisturbed rough habitat to overwinter. Depending on size these predators will attack all stages of the insect life cycle from egg to adult. Many species of true bug are plant pests and it may be difficult to distinguish between pests and natural enemies.
Beetles (Coleoptera)	Many families including Carabidae (ground beetles), Coccinellidae (ladybirds), Staphylinidae (rove beetles), Histeridae.	General predators of insects, although some such as ladybirds specialize in certain prey types	Complex life cycles which generally prey on insects at both immature and adult stages. Some are ominivorous and/or prey on a wider range of invertebrates (e.g. slugs). Will generally benefit from undisturbed habitat, especially rough grass or other shelter including crop debris or stems, to overwinter. Some ladybirds specialize in aphid predation and others on mealybugs and/or scale insects, and are available as a biological control agent for release on to crops. Other beetles including carabid (ground) and staphylinid (rove) beetles commonly feed on aphids and a wide range of other prey

(Continued)

Table 1 (Continued)

Organism/ Group	**Family/ Species**	**Type and Specificity**	**Life Cycle/Habits/ Conservation**
Lacewings (Neuroptera)	Chrysopidae (green lacewings) and Hemerobiidae (brown lacewings)	Prey on small insects including aphids, mites and thrips etc.	Complex and complete life cycles preying on insects as larvae. Lacewing larvae, especially the green lacewing (*Chrysoperla rufilabris*) are voracious feeders of aphids, each devouring thirty aphids a day during their 2–3 week developmental period. Adults require honeydew and pollen for development of eggs, although brown lacewings are predatory as adults. Adults may migrate if food sources are not available. Green lacewings have been made available as biological control agents for release on to crops (mainly *Chysoperla carnea*) and shelter can be provided to encourage overwintering
Flies (Diptera)	Syrphidae (hoverflies)	Prey on a range of insects but mainly aphids. Larvae are thought to be significant predators controlling aphid populations	Complete (complex) life cycle. The larvae mostly consume aphids (eating 235–464 in their lifetime) but may also eat caterpillars and thrips. The adults only feed on pollen and nectar, laying eggs amongst young colonies of aphids. Hoverflies can complete 1–3 generations in a year. Can be attracted using flower strips or other plants. They overwinter in the soil or leaf material as pupae and thus benefit from relatively undisturbed areas

Table 1 (Continued)

Organism/ Group	Family/ Species	Type and Specificity	Life Cycle/Habits/ Conservation
	Cecidomyiidae (gall midges)	Aphid predators in the larval stage	Complete and complex life cycle. The larvae are an important predator of aphids (eating 5–80 aphids). Have been used for biological control in glasshouses. Adults are mosquito-like in appearance with dangling legs and long antennae.They overwinter as cocoons in the soil and pupate in spring
	Tachinidae (tachinid flies)	Larval parasites of a wide range of insect families	Complex life cycles in which the larvae are parasitic on the larvae of other insects. Some are specialist on a narrow range of hosts, and some are more generalist on a wide range of insect hosts. Generally lay eggs nearby or on host insect and larvae hatches to consume non-essential tissue, but eventually killing and devouring whole host. Some lay eggs in host. Adults are common and look like large hairy houseflies

(Continued)

Table 1 (Continued)

Organism/ Group	**Family/ Species**	**Type and Specificity**	**Life Cycle/Habits/ Conservation**
Wasps and bees (Hymenoptera)	Many parasitic families including Braconiidae, Ichneumonidae, Chalcidoidea	Parasitoids of a wide range of insect hosts. Most species specialize on a restricted range of host species	Adults are small black wasps that are often easily overlooked in the field. Eggs are normally laid in or on host insects either *in situ* or in a specially provisioned nest. Larvae develop inside host and eventually kill it. Some are important parasitoids of aphids, and infected aphids become 'mummified'. After ovipositioning, the aphid will stop reproducing within 1–5 days and it is estimated that when 20 per cent of the population has been mummified, the aphid population will decline rapidly. Overwinter in hosts or nearby, emphasizing the importance of pests in maintaining parasite populations
	Vespidae (wasps), Formicidae (ants)	Generalist predators of (mainly) other insects at certain times	Adults often prey on insects and feed them to developing larvae in colonies. Predation will depend on stage of season and nest

Table 2: Principal Vertebrate Natural Enemies in Organic Farming Systems

Organism/ Group	Family/ Species	Type and Specificity	Life Cycle/Habits/ Conservation
Amphibians	Various species of frogs and toads, some quite rare now	Generalist predators of small arthropods (e.g. insects) and other invertebrates (e.g. slugs)	Aquatic stage to life cycle (eggs, tadpoles) and so require (pesticide free) water courses and ponds. Adults also need moist conditions. Adults are active nocturnal hunters that seek shelter in undisturbed places over winter (hibernacula or pond bottoms)
Reptiles	Various species including snakes, slow worms and lizards, some quite rare or endangered	Lizards prey on small arthropods and invertebrates as do snakes. In continental areas some snake species may prey on small vertebrates	Snakes are often unfairly persecuted but can consume small invertebrates in areas where they are present and small rodents depending on location and species. Lizards have similar habits but tend to be restricted to smaller arthropod prey. Active nocturnal and day time hunters with good eyesight and/or sense of taste and smell

(*Continued*)

Table 2 (Continued)

Organism/ Group	Family/ Species	Type and Specificity	Life Cycle/Habits/ Conservation
Birds	Passerines	Various species. Generalist predators of insects and small invertebrates but many omnivorous and opportunistic	Attracted to nest boxes but will nest in properly managed hedgerows and other habitat. Mainly day feeders and can consume large numbers of insects in the breeding season. Often migrate or move southwards over winter to track food and better conditions
	Owls	Various species. Predators of (mainly) small rodents and/or small invertebrates	Need nesting sites and sufficient prey in sympathetic landscape. Can be attracted to nest boxes and suitable perches. Often overwinter in habitat in UK although mortality may be high without adequate prey and shelter
	Raptors	Various species. Predators of other birds and small herbivores like rabbits and mice although many also take small invertebrates and large insects	Need nesting sites and sufficient prey in sympathetic landscape. Attracted to suitable perches. Often overwinter *in situ* or have limited migration to more suitable areas with prey and shelter

Table 2 (Continued)

Organism/ Group	Family/ Species	Type and Specificity	Life Cycle/Habits/ Conservation
Mammals	Bats	Various species. Generalist nocturnal predators of flying insects, especially moths	Need roosting sites and sufficient prey in sympathetic landscape including open water. Can be attracted to artificial roosts. Mainly night feeders. Overwinter in a torpor in sheltered sites (tree hollows, caves, roof spaces)
	Insectivores (e.g. hedgehogs, moles, shrews)	Generalist, often nocturnal predators, of small invertebrates and arthropods	Need relatively undisturbed habitat in farm landscape for breeding and foraging. Often aestivate or hibernate over winter reviving to feed in warm spells in UK. Need undisturbed shelter to overwinter successfully
	Predators/ Carnivores (e.g. foxes, mustelids (stoats etc.) and wildcats)	Generalist predators of small mammals, and larger invertebrates and arthropods	Need relatively undisturbed habitat in the farm landscape for breeding but can range widely over farmed land. Overwinter *in situ* although mortality may be high in harsh weather

The Effectiveness of Natural Enemies

Many predators are also generalists to some extent, that is, they consume a range of prey types and are capable of switching from one type of prey to another as any prey species becomes scarce. Rove beetles, ladybirds, dragonflies and birds of prey are typical predators although they may specialize to some extent in certain types of prey. For instance ladybirds prefer aphids and barn owls voles and mice. They are often adapted to detect and catch their prey on the hoof or wing and can be very mobile. Parasites and parasitoids are, in contrast, usually specialists and infect a narrow range or even just one species of host. They have evolved many specialized mechanisms to find their hosts, overcome their defences, infect and exploit them.

Specialists tend to track pest populations and are good at finding them, or being transmitted between them, even when they are scarce. From this perspective they tend to be better at keeping pests or diseases below levels at which they cause significant damage. In contrast, although generalist predators help suppress pest populations they are less likely to maintain them below this threshold as they switch prey when one type becomes scarce. However, the concerted actions of many different predators (and parasites) as found in biodiverse farm landscapes is more likely to maintain pest populations below damaging levels as they exert an effect in different places in the farm ecosystem and at different times in the farming year.

Beside the concerted effects of the behaviour of a range of natural enemies there is the remaining question of why diversity *per se* should be important for more effective control of pests and pathogens by natural enemies. It is here that ecological theory is most contentious and difficult to unravel as the effects of natural enemies must be understood in complex ecological systems. It is generally agreed that, at least for herbivores, a combination of natural enemies being favoured by diverse systems (the so called 'enemies hypothesis') coupled to the fact that herbivores are disfavoured when their host plants are spread diffusely ('resource concentration' hypothesis) leads to regulation of herbivore populations. These have already been discussed in relation to 'top down' and 'bottom up' regulation of insect herbivores in ecosystems.

Other hypotheses have been advanced for disease epidemics in diverse populations which seek to explain pathogen regulation through the presence of susceptible hosts coupled to changing physical environmental conditions (e.g. temperature and humidity). These models imply that host diversity (genetic, temporal or spatial) can help to reduce epidemic development in many circumstances. Rather less has been advanced to discuss pathogen regulation by biological factors although they are clearly important. For example, research has demonstrated that a range of microorganisms can parasitize plant pathogens and sometimes can be effectively applied to control them. Predation and (hyper)parasitism of pathogens in the soil has also been demonstrated despite the complexity of soil food webs.

Despite such theories and examples there is still no way to predict easily which systems are likely to be more effective in encouraging natural biological control with any degree of certainty. For example these hypotheses do not fully take into account the likely effects of further trophic levels on the natural enemies themselves. However, it is generally agreed that, given the complexity of agricultural ecosystems, it also follows that pests, and especially diseases, are suppressed in diverse systems.

It should also be remembered that although a diverse ecosystem is generally perceived as favouring natural enemies and as being beneficial in keeping pest or disease populations under control there has been a long evolutionary history of adaptation and counter adaptation between prey/hosts and their natural enemies. On a larger landscape scale many pest or pathogen life cycle strategies are, at least partially, driven by the need to escape their natural enemies, and thus, at least from this perspective, diversity may allow pest/pathogen populations to persist as they track resources around the landscape and escape from their natural enemies. In recognition of this, organic pest and disease management does not seek to eradicate pests or pathogens (an impossible goal) but instead to manage them at tolerable levels. Implicit in this principle is the acceptance of some level of pest or pathogen population in the farm system. However, the contention is that under such circumstances there are likely to be less frequent outbreaks of pests or pathogens because they also allow natural enemies to survive and exert their controlling interest. It would however be true to say that such ideas remain difficult to test on a landscape level and there is a wide range of ecological opinion on the matter.

Encouraging Natural Biological Control

The factors responsible for encouraging natural enemies, and therefore encouraging top-down control of pests and pathogens, in diverse systems is due to the better provision of the different resources needed for natural enemies to complete their life cycles. These are more likely to be present in a diverse environment and are likely to include some combination of the following:

Sufficient prey or hosts: natural enemies need sufficient and (sometimes) alternative prey to maintain their populations.

Supplementary foods: natural enemies often need supplementary nutrients, at least in some stage of their life cycle e.g. nectar, pollen and honeydew.

Suitable habitat: natural enemies are often adapted to specific habitat for searching for, and finding mates, and this often needs to be relatively undisturbed, at least in key stages of their life cycles.

Suitable abiotic conditions: natural enemies are likewise adapted to certain environmental conditions e.g. temperature, humidity.

Refugia: natural enemies need protection from their own natural enemies.

Hoverfly adult feeding on non-crop plant.

Hibernation or resting sites: natural enemies often need resting and/or overwintering sites to pass from one season or crop to another.

Protection from adverse farming practices: as already discussed modern farming landscapes modify or remove many resources that natural enemies need to survive (as above) and/or directly or indirectly poison them with pesticides.

Use of Conservation Biological Control

Various methods can be used to attract and maintain natural enemies in the farmed landscape including provision of shelter, alternative prey and alternative resources. This provision of resources should be underpinned with a sound organic soil management regime and an environmental management plan that links the various resources together in the farm landscape. Management practices providing these resources are described below from the perspective of the farmed landscape and more fully explored at the field and crop level in the next chapter (Chapter 4). It should be remembered that a certain amount of experimentation might be needed to find the right balance of elements needed to best support a range of populations of beneficial organisms. The costs and benefits of various tactics will need to be taken into account (*see* Chapter 6).

Shelter: zealous hygiene can remove shelter and other resources that beneficial organisms need. Undisturbed leaf litter, log piles and unmanaged (and awkward) field corners all provide areas for hibernation or tiding natural enemies (and other organisms) over lean periods. At the landscape level areas of shelter might include copses, small woods or areas of wetland. Some natural enemies have specific requirements for shelter and/or food and providing such resources can encourage natural enemies in the landscape e.g. nesting boxes for bats, owls or wasps.

Owls and kestrels eat small rodents like mice and voles and sensitively placing nesting boxes in and around fields can encourage them to nest. They can take large numbers of mice with owls capable of killing 2,000 or more per year (4–6 per night) and this is likely to help reduce rodent damage in field crops in some situations.

Alternative prey: provision of nursery or wild areas will ensure that there is at least a supply of prey for predators and parasites around the farm. Nettles for instance often support aphid populations and can nurture ladybird populations early in the season when there are none on crops.

Flower-rich habitat: augmentation of beneficial habitat can be used to provide resources for natural enemies. Providing flower-rich headlands or other landscape elements rich in flowers provides a good source of nectar and pollen that will be beneficial to many insect predators and/or parasitoids (e.g. hoverflies, parasitic wasps). Flower-rich habitat can also be used to provide a range of other resources like shelter or alternative prey to the same natural enemies.

Soil: soil management techniques should aim to enhance the health of soil in organic systems and a healthy biologically active soil is an important driver of farm ecology (*see* above). An organic soil management plan will aim to provide regular inputs of organic matter to the soil from green manures, mulches and composts that will not only stimulate biological activity but also promote better soil structure. The enhanced biological activity will not only suppress soil-borne pathogens and pests but will also affect the above-ground part of plants, by for instance inducing general plant resistance to foliar pathogens and pests, but will also help to drive above-ground food chains that include both beneficial organisms and natural enemies (*see* Chapters 4 and 9 for more details).

Environment: field elements combine at the farm or landscape level to create a more or less complex environment. A diverse landscape with many elements provides the best habitat for encouraging beneficial organisms at all levels in the food web. Landscape is made up of a number of factors including the size of fields, arrangement of hedgerows, rotational mosaic, fallow areas, meadows etc., and these elements should be arranged to give maximum diversity. At this level the landscape can be thought of as made up of a number of habitats (e.g. small copses, ponds, hedge lines) and these should be managed so that they are spread across the farm and are diverse enough to encourage a range of beneficials. They should also be arranged in a way that allows natural enemies to move around the landscape. Isolated elements increase the risk of natural enemies going locally extinct in any one part of the landscape necessitating reintroduction. Access to water is also often important within the landscape and thought should be given to managing this aspect of the farm to provide such resources for natural enemies.

Advice on the elements and their management, which can lead to successful conservation biological control, is often a target of government intervention policies in agriculture. A number of non-governmental organizations (RSPB, FWAG) are also concerned with promoting the conservation value of farmed land and can often provide good advice to farmers and growers who want to pursue their options. In some cases government grants may be available for their implementation. For instance in England (UK region) Environmental Stewardship comprises the Organic Entry Level Scheme (OELS), the Entry Level Scheme and the Higher Level Scheme. Under the entry level scheme a number of environmentally beneficial management options are available which earn points for farmers or growers who take them. Once a threshold has been achieved payment is made under the scheme. Certifying a farm as organic automatically carries a points reward under this scheme although the threshold for triggering payment is also raised for organic farmers and growers under the OELS.

CROP AND FARM SANITATION

One of the key ways of avoiding problems with pests and diseases in organic systems is to reduce the chances of introducing them in the first place. Pests and diseases can be introduced in a multitude of ways including in seed, as soil adhering to machinery, clothing, equipment, vehicle tyres, and transplants, or they may be harboured in crop residues and dumps. It is worth cataloguing the various inputs that are brought on to a farm, either intentionally or inadvertently, and then producing a plan to manage such 'risks'. Generally the more closed a farm system the lower the risk of importing pests and diseases. Even if only done on an informal level such a risk analysis can help to make staff aware of potential problem areas.

Seed Hygiene

Many pests and diseases can be carried on the seed, and healthy seed should be the starting point in any organic growing system. In conventional systems, the easiest and cheapest way to reduce pathogens and insects in seed is a prophylactic treatment with a fungicide or insecticide. There are far fewer options for managing and reducing seed-borne disease in organic systems. One option is to test the seed for presence of disease and reject infected batches. However, this is far more costly than treating the seed, and requires large samples to produce meaningful results.

Organically acceptable seed treatments for vegetable systems are usually based on hot water, heat or steam. It is often a fine balancing act to apply these treatments at a level where the pathogen is reduced without killing the seed embryo. Celery leaf spot (*Septoria apiicola*) is a particularly devastating disease borne on the seed. Hot water treatment is used by seed companies to reduce disease levels in the seed, although there have been teething troubles, with some plant raisers initially experiencing significantly reduced levels of germination. In contrast, aerated steam treatments have been highly successful (as effective as a fungicide application) in reducing *Fusarium* spp. infection on wheat seed, and have been used in commercial seed production.

Selecting disease-free potato seed is of paramount importance in organic systems, as using infected seed not only risks infecting the current crop but can introduce diseases to the soil that can persist for years. Seed used should be of at least the certified (CC) grade if growing for ware use and at least basic grade (AA, Elite or Super Elite) if growing for seed. More information about requirements of grades is available from Defra. Home saving of potato seed is not generally recommended in organic systems as this greatly increases the chances of spreading infection. The two most serious diseases to look out for on potato tubers are black scurf (*Rhizoctonia solanii*) and skin spot (*Polyscytalum pustulans*).

Transplant Hygiene

Many organic vegetable crops are grown from transplants, and particular care should be taken that these are not a way of introducing diseases into the system if they have come from outside. Any plants imported into a holding should be accompanied by a phytosanitary certificate to

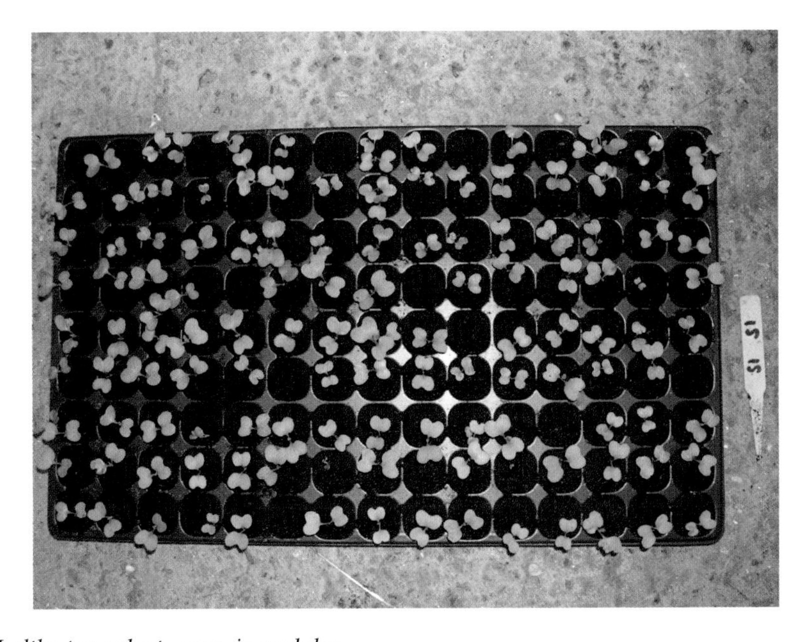

Healthy transplants sown in modules.

confirm that they are disease free. Diseases such as clubroot can easily be transmitted through soil, and organic standards insist that transplants exported to another holding must be grown in a soil-less medium. Good hygiene practices are also important for plants raised within the holding. Module trays and pots should be cleaned and disinfected every year, with an acceptable disinfectant, to avoid the build-up of fungal pathogens such as *Rhizoctonia* spp. and *Pythium* spp. that cause damping off.

Field Hygiene

Good sanitation practices within the field are equally important to avoid forming a 'green bridge' for pests and diseases. Trash or residue from crops that have been harvested should be rotavated in before they act as a source of pests and diseases for other crops. This can be particularly important to prevent the build-up of aphids and downy mildew in successionally grown crops such as lettuce. Volunteers can also act as a source of disease, particularly potatoes that become infected with blight (*Phytophthora infestans*). It is important to remove as many tubers as possible from the ground when harvesting to prevent subsequent volunteers, as in milder climates a frost severe enough to kill off the remaining tubers cannot be guaranteed.

When working in fields and successional crops it can be beneficial to work from the youngest crops to the oldest as these are more likely to harbour pests and diseases and therefore more likely to be a source of contamination. Similarly when equipment such as weeders or harvesters run through fields they should pass from the earliest to latest plantings and should in any case be clean and free from contamination before being introduced into the field.

Some weeds in the field can also harbour pests and diseases. Charlock (*Sinapis arvensis*) in particular is a brassica and can act as a reservoir for brassica crop diseases such as leaf spot (*Alternaria brassiceae*) and, within reasonable limits, it can be beneficial to manage such weeds.

It can also be beneficial to train workers to identify the most common pests and diseases and to in any case report suspected incidences. In this way infestations can be rapidly assessed and dealt with if necessary.

Farm Hygiene

Tools, machinery and boots are also ways of transmitting pests and pathogens between sites. In reality, a grower is unlikely to disinfect these when moving between fields in the same holding, but it should be considered essential if one field has a known problem, to avoid spreading the disease around the whole farm. Disinfecting equipment should also be considered necessary when moving between farms, especially if a contractor has carried out operations on many farms in the area.

Cull dumps, such as piles of potato tubers left in farmyards, are a major source of infection both within and between farms. Growers have some duty of responsibility to at least cover these dumps if not destroy them.

Potato volunteer plants should be removed from cull and compost piles.

Logistically as inputs to the farm system are a major source of contamination it can be beneficial to organize one access point to the farm and connect this directly to the main road so that all deliveries are made to one place. Signs should request visitors to contact the farm office and observe minimum hygiene standards (e.g. clean boots and overalls). Such hygiene should also extend to workers or other visitors such as advisers.

4

Adaptive Pest and Disease Management

Apart from preventative measures which they build into their farming systems and which have been discussed in the previous chapter, organic farmers and growers also have recourse to a wide range of adaptive cultural methods that they can employ to manage pests and diseases on their holdings. Adaptive cultural measures take a more interventionist approach to pest and disease management and are intended to further bias conditions against pests towards the crop. They are normally aimed at manipulating the environment immediately around or within the crop, are likely to be more temporary in nature, and are often used as a reaction to a developing pest or disease problem. Once again, the best adaptive methods to use in any situation will depend on knowing which pests or diseases they are aimed at, and having a good understanding of their potential effect on the crop, including an estimation of the damage they are likely to cause. This basic knowledge then needs to be combined with an awareness of the range of adaptive management techniques available, and their likely effectiveness, and these are discussed in this chapter.

Adaptive and cultural methods such as planting depth, plant spacing, intercropping and use of barriers are likely to differ between crops and seasons, and even change over time, as the farm system evolves and develops. They also often involve more immediate costs to the farmer or grower, and this implies that their use should be costed against their effectiveness and any alternatives (*see* Chapter 6). In fact, cost is likely to be an important factor in choosing which to use, and the best combinations of techniques to be used will very much depend on the specific farm situation and will often hinge on the experience and learning of the farmer or grower. Indeed, organic farming is recognized as being knowledge intensive, and the farmer or grower should expect to be constantly learning and adapting their cultural practices to suit changing conditions, both on and off the farm.

CULTURAL MANAGEMENT

Diversity is a desirable feature of organic farming systems, and from a farm business perspective is considered to increase the resilience and sustainability of the farm system. From an ecological point of view it also helps to moderate, and therefore manage, pest and disease outbreaks in crops and on the farm. For these reasons organic farmers and growers are therefore aiming to enhance the diversity of their cropping practices by taking advantage of, and using, the full range of cultural management options available to them. These practices should present any pests and pathogens with as variable an abiotic and biotic environment as possible. Using diverse cropping or cultural practices allows a range of balancing factors to come into play within the crop, which act to suppress pests and pathogens, it is to be hoped at levels below those at which they cause economic injury, although this is not always the case. Such diversity may be created at a number of levels, either within a field or crop, in the boundaries between neighbouring fields or crops, and/or over time within the season. These cultural management practices are described in more detail in this section.

However, before describing these, it should be recognized that building diversity into a crop system can be demanding on the grower or farmer, and may also demand a complex balancing act to keep farm costs in check. In a market garden, a wide array of crops planted in succession so as to satisfy market demand naturally lends itself to a diverse system, with a number of crops with different planting times and densities, as well as diverse husbandry and harvest techniques. In contrast, farmers might struggle to create a diverse cropping environment in some arable systems where planting dates and crop densities are often quite uniform over large areas. However, even in this situation, mixing varieties or growing crops in alternate strips can introduce diversity, although it should be recognized that these types of practice are not always popular in commercial systems as they can be more difficult to manage (and therefore more costly).

It is also worth remembering that (bio)diversity also offers a range of other services, apart from pest and disease suppression, that do not often appear on a costing sheet. These include increased efficiency of nutrient recycling, better microclimate regulation, enhanced water regulation, and erosion control. These are all factors that, although admittedly often not cost effective in the short term, especially during any conversion period, will, given time, come to play a major positive role in ecological and economic sustainability of any organic farm system.

Tillage and Soil Cultivation

The foundation of pest and disease management on an organic holding is soil management. This aims to create a biologically active soil which works either directly or indirectly to suppress pests and pathogens. This has been described extensively in the previous chapter (Chapter 3) and will not be discussed in more detail here, apart from where tillage practices are used to control pests and pathogens more directly. It is important to understand that many tillage practices which invert the soil can have both beneficial and negative effects, and for this reason should be used with understanding. In general the best principle, from the point of view of pest and disease management, is to disturb the soil as little as possible compatible with good husbandry, and to work towards minimum tillage systems if at all possible.

Ploughing and tillage are used to work the soil and prepare it to receive the crop. In organic farming tillage is also widely practised to manage weeds, the effect of such tillage operations being to turn or disturb the soil to a greater or lesser degree. Ploughing will generally invert the soil completely to some depth, whilst harrowing will disturb, but not necessarily turn, the soil depending on how it is carried out. Rotovation techniques are commonly used in horticultural crops, and will also thoroughly mix the soil and any debris or crop remains to a set depth. Obviously inversion, mixing and any major movement of soil will disrupt the soil flora and fauna to a greater or lesser degree. This is likely to interrupt and break many linkages in the soil food web (as previously described) and thus potentially remove some of the suppressive effects of this web. On the other hand, such movements can dislocate and remove pests or pathogens to depths at which they cannot easily infect the crop, or will expose them to desiccation and/or predation on the surface. The flocks of birds that traditionally follow the plough can potentially devour many pests and ploughing buries the resting stages of many crop pathogens at too great a depth for them to successfully reinfect the crop.

Tillage is an important method of direct control for some pests, such as leatherjackets (*Tipula* spp.) and slugs. The larvae of leatherjackets overwinter in the soil, and cultivating the land can help expose large numbers of these larvae, when they are either desiccated or eaten by predators such as ground beetles or birds. In such cases the timing of tillage can be

Leatherjacket damage to lettuce.

important, too. Grassland should be broken up in early spring to increase the time of exposure before the crop is planted. If the ground is severely infested, planting should be delayed until after midsummer.

The effects of soil cultivations on slug control are complex. It is generally agreed that inversion techniques such as ploughing or discing can cause large reductions in slug populations through their physical action in destroying slugs. This can be particularly effective if the timing of operations corresponds to times when the slugs are breeding. However, inversion techniques can also result in burying some of the population at deeper depths from where a few protected individuals can make their way to the surface over a period of weeks, only to attack any vulnerable plants at a later date. Plants are most vulnerable to slug attack during the period of emergence, and the structure of the seed bed can influence this. Consolidated seed beds are better than a loose cloddy structure for reducing slug attack. The way in which machinery is used can also affect the severity of attack. For instance, when a direct drill has not been covered properly, slugs can crawl along the line of the drill and down the open seed slots, destroying seedlings as they emerge.

Bed structure can also have implications for disease control by providing a barrier to pathogen infection. For instance potato tuber blight can be reduced by ridging up the potato crop so that the extra layer of soil acts as a barrier to spore infection, especially in the period after defoliation. Indeed, for this reason (and for reasons of weed control) organic potatoes are normally grown in single row ridges rather than three row beds, as is commonly practised in conventional systems.

Apart from the direct tillage effects on pests or pathogens seedbed preparation can also have strong indirect effects throughout the cropping period. Tillage is traditionally used to prepare a good seed bed which allows the crop to establish well and allows roots to grow strongly and make firm contact with the soil. Many other factors (light, nutrition, etc.) can be altered by tillage operations and may also come into play. For instance, plant nutrition may be enhanced as the rate of mineralization is increased (at least temporally). In some cases crop residues may produce compounds (during decomposition) that may also have an effect. The general principle is to allow crops to develop from a good base and therefore be more likely to avoid or resist infection with pests and/or diseases. A good seed bed can also allow the crop to grow vigorously in appropriate soil conditions which will also lead to naturally resistant crops. In contrast, tillage carried out inappropriately may destroy soil structure, allow compacted soils to develop and/or create pans that may increase the crop susceptibility to diseases, particularly foot and root rots.

Minimum Tillage

Minimum or conservation tillage has the effect of allowing a build-up of organic matter in the soil, and this can have repercussions for pest and disease management. On the positive side it usually creates an environment that is more beneficial to soil-dwelling natural enemies such as rove beetles and spiders than tilled systems. It is also true that the soil food web and ecology can remain in place, and this allows the build-up of a beneficial flora and fauna including earthworms. On the negative side the build-up of crop or other residues can act as a source of inocula for new crops, provide a refuge for pathogens, and/or a food source for facultative pathogens. There is a growing range of information on minimum tillage under organic agriculture, and this should be consulted for a more detailed discussion of the pros and cons of adopting such a system. It is, however, true to say that, taking a broad view, and in the long run, it will probably be found to be essential when designing ecologically sustainable farming systems, as the advantages will come to outweigh the disadvantages.

The Timing of Planting

Planting time can be manipulated to avoid pests and pathogens. Sowing at the best time to minimize the chance of pest or disease infestation is a very common practice in organic growing systems, and in some cases it might be possible to grow crops in conditions in which the pest or pathogen is absent. For instance, it may be possible to avoid peak egg-laying periods of insect pests, or avoid conditions such as periods of high humidity in which pathogens can invade crop plant tissue. More sophisticated planting regimes might be chosen to better synchronize crop growth with natural enemy life cycles (which in turn will impact on pest life cycles), to grow crops at times when alternative and preferred hosts for pests or diseases are present, and/or to trap pests in part of the crop (*see* Trap Cropping, p.103) but may require more complex planting regimes. The general strategies for manipulating crop timing with respect to pest and disease pressure are:

Delaying sowing until the risk of pest infestation has passed: This process is commonly carried out by growers of maincrop carrots. The first generation of carrot fly has subsided by the end of May, and for this reason organic carrot growers often delay sowing until after then, when the risk of pest infestation is much reduced.

Planting early to maximize crop growth before the risk of infection: The risk of infection from potato blight (*Phytophthora infestans*) increases throughout the growing season from June onwards. For that reason, organic potato growers try to grow the crop as early as possible, so that if the crop does become infected it has completed most of its growth and yield production. Other methods may also allow crop development to be brought forwards. For example in the case of potatoes chitting, the seed can help to advance maturity, and in other vegetables sowing and transplanting modules can also give the crop a head start.

Avoid certain periods completely: Depending on site and climate, some growers may find that it is not economic to grow certain crops at certain times of year, as they consistently suffer large economic loss. This is often the case with late planted lettuce crops, which frequently succumb to infection with downy mildew (*Bremia lactucae*) in late September or

October. Leafy salad brassicas also suffer from inundations of flea beetle (*Phyllotreta* spp.) in high summer, and many growers avoid planting at this time.

Tactical planting: At some sites there may be certain periods of the year when crops can be grown with very few pest and disease problems. At the end of July, aphid populations crash and remain low through August before beginning to rise again in September, reducing the chances of aphid infection on leafy crops such as lettuce or brassicas. However, growing out of season may need to be balanced with likely market demands.

Notwithstanding this, crops also have optimum requirements for germination, establishment and development. For example, lettuce germinates best in cool conditions, and squash develops better at higher temperatures. In these cases choosing planting times at which plants will germinate and grow quickly will allow plants to grow quickly through susceptible periods and/or allow them to develop to better resist pests and pathogens. In fact this might be a better option than growing crops out of season, which may bring other problems: for example, it will increase the resources needed to successfully bring them to harvest (protected cropping in winter can suffer from a lack of light and heating, which may have to be artificially supplied).

Apart from the likely higher costs of raising crops out of season, changing the timing of planting may conflict with the demands of the market, and this needs to be balanced up when considering when a crop should be grown. On the one hand producing out of season may lead to poor sales due to low demand or increased storage costs; on the other, growing crops early or late may create a new (and premium) market for produce, which can cover the monetary costs of production. For example, organic carrot growers may wish to grow early in the season when there is high carrot fly pressure in order to capture the early market, in which case other direct methods – such as covering the crop with fleece – may have to be employed.

Plant Spacing

Plant spacing alters the micro-environment within a crop, the growth pattern and duration of growth, and the degree of competition between crop plants. All these can have a potential influence on the incidence and severity of pests and diseases. Organic growers traditionally sow at slightly lower densities than conventional farmers and growers, as this reduces the competition between crop plants and allows the development of stronger, more resistant plants which are also likely to be able to compensate to some extent for damage. However, in some circumstances, for example where damage is anticipated at germination, it is sometimes expedient to sow at higher densities to compensate for any plant removal or failure to establish. In this case crops may need to be thinned if establishment is better than expected.

Planting at larger distances will usually produce a more open canopy, promote air circulation and lower air humidity. This can help reduce the sporulation (and spread) of pathogens and reduce successful plant infection thus lowering disease incidence, which more than compensates for yield reduction due to lower density. Avoiding excess applications of nitrogen-rich fertilizer or manure will also help reduce excess canopy growth (and reduce plant attractiveness to pests and pathogens). In contrast, more open crops can be more susceptible to attack by some pests. Pigeons (*Columba* spp.) will preferentially attack a brassica crop with a more open canopy, as they have space to land in between the plants and so feel safer. Aphids and other insect pests also prefer to land on crops surrounded by, or at least contrasted against, bare earth and can be attracted to crops with a more open canopy.

In choosing plant spacing, the most important factor that would affect the success of the crop should be considered, and this should also be balanced against the economic and other implications of changing plant spacing (especially on yield). In many cases other factors might outweigh any pest and disease management benefits, as other crop management practices will be more important, such as being able to weed crops mechanically or to harvest them efficiently.

Crop Nutrition

Organic rotations aim to feed the soil rather than the crop. They rely on fertility building periods, and different crop types to exploit different phases in the fertility cycle. Green manures such as grazing rye (*Secale cereale*) are used to hold nutrients and prevent them from leaching, whilst leguminous green manures such as vetch and various clovers are used to fix nitrogen. Grass clover leys are often used as two-year breaks in organic arable and/or vegetable rotations, and may be down for much longer periods in livestock systems. Once ploughed in, the nutrients held in green manures and crop residues become available to soil microorganisms and plants as they are broken down. It is a feature of organic systems that these nutrients become available much more slowly and over a longer time period as compared with the addition of soluble mineral fertilizers.

Whilst green manures, coupled to the crops in rotation, can do much to maintain soil health and nutrient status, some nutrients, especially phosphate and potassium, are likely to diminish over time as they are effectively removed in harvested crops. Micro-nutrients may also become limiting in some systems. Nutritionally stressed plants are likely to be more susceptible to pests and pathogens as well as giving lower yields. In this case organic farmers and growers need to resort to additional nutrient inputs. The preferred inputs are organic composts or animal manures, which help to replace the removed nutrients and also support soil biological activity. Occasionally recourse may need to be made to processed, but organically approved, fertilizer inputs (such as chicken pellets), either to provide a boost to the crop because of unforeseen circumstances (such as excessive leaching), or to correct nutrient imbalances

Crimson (left) and Persian (right) clover are green manures that also support beneficials.

and/or micro-nutrient deficiencies. In both cases the routine use of such supplements will require justification and derogation from a control body.

Composts are made from recycled farm waste or green waste imported on to the farm for the purpose. In order to prevent the spread of plant pathogens, all material should be properly composted, preferably in a windrow, to ensure adequate temperatures are achieved to destroy either living pathogens or their resting stages. In the UK, bought-in composts should be composted to PAS 100 standards, and this, and similar standards that exist in other countries, should ensure that the required composting processes have been achieved. Similarly animal manures should be thoroughly composted to destroy any pathogens that may have passed through the animals or bedding. If it is not possible to compost animal manures they should be stacked for a least six months for the same reason (and this is required under Soil Association Organic Standards).

Cropping operations should be timed so that nutrients from green manures and other supplements become available at the right stage of crop development, and are not leached out of the system before the crop can use them, or are only available after the crop can use them effectively. Compost and animal manures should be applied at the appropriate stage in the rotation, and either ploughed in or applied as compost mulch. It should be remembered that plants with nutrient imbalances, but especially excess of nitrogen, and lack of potassium, are vulnerable to pest or pathogen infection. Much research has also demonstrated that composts have a suppressive effect on soil-borne pests and diseases, and that on occasion this extends to protection of the above-ground parts of the plant. Whilst it is difficult to use this effect consistently as it is very dependent on field conditions, it is a major advantage of using composted materials.

Polycropping

In temperate organic systems crops have traditionally been grown in monocultures. In recent years both growers and researchers have increasingly investigated the benefits of growing simple mixtures, as this practice potentially brings benefits to pest and disease management on both a field and a regional scale. The evidence that plant mixtures have lower numbers of pests and/or pathogens than do pure stands has been discussed in Chapter 2, where it was suggested that higher natural enemy populations persist in diverse mixtures due to more continuous food sources (nectar, pollen and prey) and favourable habitat or microclimates. In tandem with this, mixed crops present a more dilute resource for the pests and/or pathogens (which have to work harder to find or encounter them), and consequently pest or pathogen numbers tend to be reduced in mixed stands.

Techniques include intercropping, undersowing, strip cropping and companion planting, and although different terminologies are used, these practices often overlap in practice. With this caveat in mind the various practices are briefly discussed below.

Intercropping

Intercropping can provide a diverse crop environment, and the term is really a catch-all phrase indicating the practice of growing more than one crop in the same place or field at the same time. Intercrops can include row intercropping, mixed intercropping, relay intercropping, strip intercropping, and even completely random plant mixes. Classic mixed intercropping involves the cultivation of different crops in the same field, often in the same or alternate rows, and often on beds or ridges. Some crop types have even coevolved to be cultivated together, the best example being maize-beans-squash, which benefit or tolerate the intercrop environment, especially in the tropics where light is not generally such a limiting factor.

There has been a large number of studies on the value of intercropping for preventing or reducing pest and/or disease incidence in crops. The evidence is at best mixed, with some crop combinations seeming to offer protection while others do not. In some instances combinations actually seem to be worse, as for instance when aphid virus vectors land and 'taste' a variety of plants whilst looking for a suitable host, causing more widespread infection. On balance it seems that crop combinations should be carefully chosen to be compatible, so there is not excessive competition making one or more of the intercrops more susceptible to pests or diseases. It is probable that a great deal of experimentation remains to be done with crop combinations in temperate organic systems, but work in the tropics has shown what can be done where combinations of maize, cover and forage crops and weeds have been used to design a so-called 'push-pull' system that considerably reduces the incidence of maize stem borers.

In temperate vegetable systems evidence is sparse, and observed effects are often quite situation-specific. For instance, mixed intercropping of carrots with onions has been demonstrated to reduce attacks by carrot fly on carrots and onion thrips (*Thrips tabaci*) on onions, compared to those on

Intercropping carrots and leeks.

carrots and onions in monoculture, but this depends on the ratio of carrots to onions in the mixture. In one set of studies on this intercrop, greater numbers of ground predators (carabid and staphylinid beetles) were observed in the intercrop, and the intercrop was more effective before the onions bulbed. It appeared that fewer carrot flies entered intercropped plots, indicating that both higher predation rates and the masking effect of volatile chemicals given off by the onions were probably instrumental in reducing pest attack. In this case intercropping with French marigold (*Tagetes patula*) was ineffective, although this plant has been shown to give off a volatile chemical that deters nematodes from attacking other crops.

Intercropping also aids pest control efforts by reducing the ability of the pest insects to recognize their host plants. For example, thrips and white flies are attracted to green plants with a brown (soil) background, ignoring areas where vegetation cover is complete, and some intercrops have spatial arrangements that produce a complete vegetation cover that is unfavourable to these pests. In a similar vein, random plant mixes and even mixes of varieties have been shown to be better at delaying and reducing disease incidence in crops, especially cereals. This is because the passively distributed spores have less probability of encountering a susceptible host in more complex mixtures. Such mixtures are regarded as effective in protecting less resistant varieties, but it is often more difficult to harvest and market mixtures, which explains their slow uptake by farmers and growers.

Undersowing

Undersowing is commonly practised in temperate organic arable crops as a means of establishing leys after cereals, but is becoming increasingly common in vegetable and other systems. It is really a form of relay intercropping, with one crop (the ley crop) being sown after the cereal and allowed to develop shaded by the cereal where it does not compete too much. In some circumstances this has been shown to reduce pest or disease incidence. For instance, red or white clover has been shown to reduce cabbage root fly (*Delia radicum*) damage when undersown in brassicas (for example cauliflower, Brussels sprouts). Unfortunately without careful placement or management, yield can also be substantially damaged, and although a great deal of research has been aimed at designing systems and/or machinery to achieve this, it is probably fair to say that none has achieved widespread acceptance for use in commercial systems. Another promising approach in these circumstances is the use of less vigorous clover types, such as subterranean clover (*Trifolium subterraneum*) and bird's foot trefoil (*Lotus corniculatus*), which could be potentially more widely used in undersowing regimes.

Strip Cropping

Strip cropping is common in vegetable production systems, where, for instance, alternating beds can be sown. These often fit in with small-scale complex horticultural production systems which have more complex and flexible marketing arrangements. Such strips will increase crop diversity and are expected to reduce damage by pest and disease to the crops. However, little research has been done on the effectiveness of strips, but they are probably not as effective as intimate intercrops. It has been shown in some cases that planting green manure strips alternately with vegetable crops can help to maintain higher levels of natural enemies in and around the vegetable crop than would otherwise be the case, and that when cut, the natural enemies can be encouraged to move on to the crops.

Companion Planting

Companion planting is also really a form of intercropping. The phrase normally refers to intercropping in small scale or home production systems. 'Companion plants' are often not crop plants in their own right and may be flowers or herbs; they are usually planted at much lower densities than the crop. It has been demonstrated that some combinations of crops grown with a background of 'companion' plants have lower levels of pest infestation than crops grown in bare soil – but once again, much evidence is anecdotal rather than rigorously tested. As with intercropping, it is also probably true that their effectiveness is situation specific, perhaps working in one location and/or season but not another. Furthermore it can be quite difficult to avoid excessive competition, and the consequent yield reduction, with crop plants.

Popular companion plants have included marigolds (especially African marigolds *Tagetes minuta*) and garlic. These are reputed to have allelopathic, mainly repellent effects on pests such as aphids, root flies and nematodes when planted between, for instance, tomatoes, carrots or brassicas. Some of the mechanisms for these actions are well known, for instance marigolds produce a root secretion containing thiopene that deters nematodes.

Birdsfoot trefoil sown as a companion plant to cauliflower.

One method of companion planting currently being developed is sowing 'green manure' companion plants in modules with brassica crop plants. The idea behind this is to target the companion plant where it has most effect, without building up a high enough population of companion plants that they cause competition with the crop. It is also cheap to carry out operationally, if seed sowing is carried out mechanically. Bird's foot trefoil (*Lotus corniculatus*) is one species that has been tried commercially, but others are currently being developed.

Weed Management

Weed management will also strongly influence intercrop diversity. When left between plants or between rows, weeds have been shown to reduce pest damage or attack. However, weeds will also undoubtedly reduce yields if left to compete with the crop during critical periods, and farmers and growers should take care with this method and not let weeds reach the point where they are producing seeds or other multiplicative structures. Some growers allow non-competitive weeds such as chickweed (*Stellaria media*) to grow after the critical weeding period has passed in carrots as a means to reducing carrot fly attack. Weeds may also have other beneficial effects as alternative resources for natural enemies and soil-dwelling microorganisms, and these are discussed elsewhere.

Trap Cropping

The aim of trap crops is to provide an alternative attractive site for pests or pathogens away from the main commercial crop. They can be used either to provide an alternative attraction for a pest which is seeking the crop, or to stimulate a pest or pathogen to germinate or otherwise infect an immune host on which it cannot complete its life cycle. In either case the trap crop is usually destroyed to ensure the destruction of the pest or pathogen.

Classically trap crops have been used to attract pests that are actively seeking the crop, either to consume it and/or to lay eggs in it. They should therefore be more attractive to the pest than the actual crop in some respect. The trap crop will often be in the same family as the main crop, but a species that the pest finds more attractive, although it can also be a sacrificial planting of the actual crop sited in some way to be more attractive. Examples include the planting of turnips as a trap crop in order to attract cabbage root fly away from a commercial calabrese crop, which has been demonstrated to reduce damage to the calabrese, and the planting of sacrificial (early) beds of carrots around a main crop to attract carrot root flies, which are weak fliers and preferentially settle on outer beds on the upwind side of the field. In the latter case, these beds can also be harvested early before damage becomes manifest, as a means of recuperating some yield.

In this form trap crops have to be managed carefully both to provide an attractive target for the pest and in order that they don't allow the pest to actually multiply and exacerbate the infestation. For example, many pests preferentially settle on the edges of fields (where they first encounter or are blown against crop plants), and trap crops are often more effectively placed around the crop to be protected. The width of the trap crop will need to be balanced between the mobility of the pest and the amount of yield that can sensibly be foregone. If the trap crop is to be destroyed, then this needs to be done without affecting the actual crop. An example of a trap crop that can be difficult to manage is nasturtiums planted with cabbages: the nasturtiums are often preferred to cabbages as egg-laying sites

Nasturtiums can be used as a trap crop for cabbage white butterflies. (By courtesy Maggie Haynes)

by cabbage white butterflies, but unfortunately the creeping habit of nasturtium combined with the mobility of the butterflies and caterpillars make it a difficult combination to manage effectively on anything but a small scale. Similarly a row of mustard (*Sinapis alba* or *S.nigra*) planted between brassicas can serve as an attractant for flea beetles, but its effectiveness is likely to be limited by the mobility of the pest.

Sometimes a trap crop can be used to clean up a pest from a site before the commercial crop is planted. The objective is to stimulate a pest or pathogen to come out of its resting phase and then break its life cycle, so that it dies without reproducing, either because it cannot complete its life cycle or because the trap crop is destroyed. This requires detailed knowledge of the life cycle of the pest in order to ensure that pest numbers are reduced and not exacerbated by planting the trap crop. One such example is *Solanum sisymbriffolium,* a trap crop which is sown before potato crops to clean up potato cyst nematode. This plant produces a root exudate that stimulates the eggs of the nematode to hatch, but the pest is unable to feed off the roots of the trap crop, so dies before completing its life cycle. One drawback with this crop is that it requires at least 15°C for germination, so the growing season in the UK is limited to midsummer.

In practice, the financial benefits through reduced pest infestation need to be balanced against the considerable costs of growing a sacrificial trap crop. In some cases the cost of growing the trap crop could be offset by putting it to other uses – for instance, the turnips could be grazed by sheep, which are less likely than a packer to reject a crop of turnips if their appearance is spoiled by cabbage root fly damage. Many green manures (legumes) are also being investigated as potential trap crops, and in this case will obviously have the potential to act as fertility builders as well.

Timing of Harvesting

Like the time of sowing, the timing of harvest can also affect pest and natural enemy populations. Timing can be arranged to avoid periods of damage, to reduce carryover populations or even to suppress or encourage natural enemies. The host crop is often also the site where pests overwinter or pass through a resting stage, for example in stems, on fallen leaves or other crop debris, therefore destroying the crop remains will also remove the pest. However, if natural enemies are also present they will also be destroyed, and in some cases it might be expedient to leave overwintering sites for them.

Strip-harvesting practices can also affect pests and natural enemies. On the one hand, whilst harvesting in strips might give natural enemies a chance to pass from one part of the crop to colonize another, it also gives pests the same opportunity. Harvesting practices can also promote the spread of disease spores, and strip cropping might therefore aid airborne disease dispersal.

Once harvested, managing residues can also play an important part in promoting on-farm biodiversity. The removal or destruction of crop residues also destroys many natural enemies which overwinter in or on or nearby pest victims. For example, it can sometimes be beneficial to leave brassica residues, as parasitic wasps are present on the dried stems. Sometimes plants also revive in spring and flower, presenting a source of nectar for parasitic wasps. Sweetcorn (*Zea mays*) left *in situ* has also been observed to provide overwintering sites for ladybirds, and no doubt many other beneficial insects.

Although post-harvest pest management is not discussed in detail in this book, there is a clear potential for pests and pathogens to pass from one season to another on stored planting material, or produce stored for consumption. All seeds and/or planting material, such as tubers, should be harvested from healthy plants and cleaned prior to storage. Seed can often be tested for pathogen loading, and this needs to be done where there is any doubt. Seeds can often be at least partially cleaned by heat-treatment methods that remove pathogens borne on the seed coat, although viruses are not so easily removed. Planting material such as tubers or bulbs are difficult to clean because pathogens, and especially viruses, are often resident within the plant tissue, although pests might also be present. In this case it is often expedient to buy clean planting material at frequent intervals, if not every season.

OTHER CULTURAL CONTROL METHODS

Apart from manipulating the crop and enhancing diversity within the cropping environment, farmers and growers have increasing access to more direct cultural control methods. Many of these methods represent temporary or adaptive measures which can be taken against a specific pest or disease threat. They are best described as physical or mechanical controls, but normally stop short of the chemical methods described in the following chapter (Chapter 5). They can involve considerable cost both to the farm business, often due to the labour necessary to implement them, and/or the wider environment, so should generally be used with caution. The most popular methods include barriers such as crop covers, mechanical control methods such as picking and squashing, and scaring and traps, especially against rodents and other vertebrates.

Crop Covers

Crop covers can be a very useful tool for organic growers for a number of reasons: to advance crop maturity, extend the season, avoid hail damage, and protect against frost in addition to the exclusion of pests. Choice of the correct cover for the intended purpose and careful management is essential for effective use. An ever-growing variety of fleeces, meshes and nets is available to growers.

Fleece

Fleece was originally developed to advance the maturity of vegetable crops, but can also be very effective at excluding pests as well as protecting against frost. Lightweight fleeces of around 17g/m^2 are the norm for general use, providing 2 or 3°C of frost protection with heavier weights of 30g/m^2 protecting down to –5 or –6°C. Fleece is the cheapest crop cover option for growers, although quality can be an issue with some brands tearing easily. Some fleeces use recycled material but often have a shorter life span, which makes it difficult to balance the environmental costs. Reinforced edges help to prolong life, as can increased hair strength – provided this does not also reduce light transmission. On the down side fleece is flimsy and easily damaged by large vertebrates such as deer and raptors trying to access small animals trapped or sheltering underneath. Any damage will increase the likelihood of pests gaining entry to the crop, and damaged sheets should only be reused where crop advancement is of more importance than pest exclusion. Fleece should be dry when removed for storage, and stored out of reach of small rodents which love to make nests out of it.

The best use of fleece is early in the season to make use of the temperature lifts gained underneath. This can be too much for summer crops, however, and leaves that come into contact with the fleece can become scorched. If possible crops should be uncovered in dull, though not bright, conditions. There are other disadvantages to fleece: it can promote soft growth, which will be more vulnerable to autumn frosts. In cauliflower the use of fleece can delay maturity in winter crops by up to seven days, as curd initiation is triggered by the accumulation of units of relative cold. A major disadvantage from a grower perspective is the difficulty of seeing what is happening underneath fleece, and of inspecting crops. Weeds can also advance at a pace under the fleece, so if the farmer or grower is not careful it is all too easy to miss crucial weeding windows.

Enviromesh on brassicas.

Mesh

Mesh is more expensive than fleece but the cost can be spread over several seasons, with some manufacturers justifiably claiming a life of up to ten years. Consequently the total amount of resources used is much reduced. There is considerably less crop advancement or frost protection under mesh, but as there is more airflow and less humidity as compared to fleece, this should mean fewer disease problems. The tighter the mesh size the less air flow, however. Care is needed to secure the edges because as the crop grows, the edges can easily ride up, providing entry points for pests. The mesh size should be chosen according to what pests need to be excluded: thus 1.3mm mesh will exclude cabbage root fly, carrot fly, aphids and most caterpillars, with the exception of diamond back moth. This size reduces flea beetle activity but 0.8mm mesh will exclude them altogether, 0.17 × 0.37mm mesh will exclude thrips. For some crops it may be necessary to use hoops to prevent damage to the crop, particularly with the heavier meshes on delicate plants such as baby-leaf spinach.

Nets

Nets have a much larger mesh size than fleece and mesh, and are only normally effective as a barrier in excluding much larger pests. They are not normally laid directly on crops as the crop would grow through them, and are usually supported above the crop in some way. For example, bird netting generally needs to be held above the crop not only to prevent damage due to its weight on plants, but also to prevent the crop growing through the larger mesh and becoming entangled in it. Birds also learn to land on netting and peck through it, and so the net needs to be suspended some distance above the crop. The cost of caging large areas with netting is likely to be prohibitive and uneconomic unless the crop is of very high value. Fleece or mesh may make an adequate temporary substitute in some cases.

Using Crop Covers for Pest Control

The most important principle of using crop covers to exclude pests is that the cover must be in place before the pest infestation (including egg laying) takes place. It should be borne in mind that some pests have been observed to lay eggs through mesh on to leaves that are touching the cover; covers should not be used on crops where the pest is already in the soil. Note that it is easy to trap pests such as flea beetles under the mesh if this is put on late. It is also important to check that transplants are pest free prior to planting and covering; if possible use pest forecasting to predict risk, and judge accordingly the best time to apply covers.

The cover must be, and must remain, intact until the target pests are no longer a threat, as any tears can let them in. The edges should be well sealed, preferably by burying the edge, or at least the edge exposed to the prevailing wind; this can be by done by ploughing in the edges, weighing them down with bags of soil or using pegs, which are available commercially. Care is also needed when removing covers for physical weed control as this can expose the crop to pests.

Crop covers are commonly used on carrots to exclude carrot fly, and on brassicas, not only to keep off cabbage root fly, flea beetle, mealy aphids and caterpillars but also larger pests such as birds and rabbits. Lettuces can be covered to prevent root aphid damage, for example, or leeks for thrips. The balance is between the potential economic damage of the pest, and the cost and effort involved in covering the crop. There are also considerable costs associated with crop covers, not just in the material but also in the handling. Costs of handling covers, mainly to allow weed control, can be high and have been reported as varying from £45 to £300/ha for calabrese and even up to £700/ha for carrots in the UK.

There can be disadvantages to using crop covers as a routine pest management tool. In this book we strongly emphasize the system approach and the role of biodiversity in crop protection, but when using crop covers, many pest natural enemies are excluded along with the pest. If pests are able to gain entry without their natural controls, then the problems can potentially be worse than if the crop had been left uncovered. Slugs can also be more of a problem under covers, as the humid microclimate is more favourable to them. There is also a risk of pathogens surviving on crop covers, and diseases being spread between crops when covers are transferred. It is therefore best not to reuse covers immediately on the same crop type, and if possible to allow them to dry out in the sunshine prior to storage. So, while for many situations crop covers are necessary and useful, alternative strategies should also be investigated and used where possible.

From a broader perspective crop covers also represent a challenge to the sustainability of organic farming systems. With the mantra 'reduce, reuse and recycle', we should first question whether the cover is necessary, or if good biodiversity and habitat management measures can deliver instead. If covers are unavoidable then it is better to use a product that will last longer, while ensuring it stays intact long enough to do its job. There is also work under way to develop biodegradable crop covers, some of which are already commercially available. This is driven by the increase in costs of disposal, as it is now illegal to burn or bury waste plastic covers on farm. Organic growers do not want a product that degrades too quickly, however, as it could put the crop at risk.

Mulching

Mulches have been shown to affect plant growth as well as pest and disease incidence. Both sheet and living or organic mulches can have various effects; these are not discussed in great detail in this book, but are summarized below for their effect on pests and diseases.

Sheet (Plastic, Cardboard or Paper) Mulches

Sheet mulches are used to cover the ground with an impenetrable layer and are normally used for weed control in clearing ground and in slow-growing crops; they may also alter soil conditions and affect crop development. Plastic mulches can, for example, slightly heat soil and improve crop development, although soil may be drier underneath the plastic. Rolled paper mulches generally degrade rapidly in wet conditions and may actually cool the soil. Such effects are likely to alter the periods in which crops are most susceptible to pests and diseases, and can help reduce the length of these periods as well as accelerate maturity, so delaying the period of risk.

(Clear) plastic mulches can also be used to heat the soil in the absence of the crop, and elevated temperatures can have the effect of sterilizing the soil to some extent, thereby removing or reducing pests and pathogens. The use of plastic mulches for this purpose should be treated with caution, however, as it is also likely to affect the soil ecology to a greater or lesser extent, and may in fact result in the 'bounce back' of pathogenic organisms.

In the case of plastic mulches different colours have also been shown to affect not only plant growth but also pest and disease incidence. Red mulches tend to stimulate root growth and blue mulches leaf growth, due to the properties of the reflected light. This is turn can indirectly affect infection by pests and diseases, but can also have a direct effect. Reflective (silver foil) mulches, for example, tend to deter pests such as aphids from alighting on the crop, and yellow mulches have been shown to have similar effects in some cases but not in others. In principle, longer wavelengths of reflected light might be expected to attract insects into the crop and shorter ones to deflect them. Some pathogens are also known to show reduced sporulation when some wavelengths are predominant or others excluded, however results in polytunnels covered with such plastics have shown mixed results.

Organic or Compost Mulches

These mulches are used to cover the ground with a loose layer of organic material. A wide range of materials can be used including straw, green waste compost, manure, and others. Such materials in themselves are used widely in organic farming systems to promote soil health, and have many beneficial effects including the stimulation of soil biological activity, which can in turn suppress many soil-borne pathogen and pest species. As with plastic mulches they can also form a barrier between the plant and soil, so preventing pests or pathogens completing their life cycles, either because

Experimental coloured plastic mulches.

they cannot escape from the soil and infect the foliage, or because they cannot drop from the foliage to the soil to complete part of their life cycle.

Living Mulches

Living mulches are effectively a form of intercropping, undersowing or companion planting, procedures that have been described in a previous section. Living mulches may be cut and left on the soil surface, in which case they will have a similar effect to organic mulches, as described above.

Mechanical Controls

Direct mechanical control of pests and pathogens has largely fallen out of favour in larger scale farming systems, although it can have a place on smaller holdings. Generally these methods can be summed up as physically removing the pest or pathogen from the crop, or destroying them *in situ*. Labour costs are likely to be high as the work is slow and workers likely to regard the task as more menial than most. In some cases there have been attempts to scale up mechanical pest control measures, but on the whole they have been regarded as inefficient at best and have not been widely adopted, or developed, in contrast to mechanical weed control which is widely used on organic farms. The various methods are summarized below.

Hand Picking

A time-honoured method of removing pests is to pick them off plants or the ground and destroy them. Slugs and caterpillars are frequently removed from small-scale plantings in this way. In order to be effective the pests should be removed and/or destroyed. Squashing *in situ* is another variant of this method, and aphids are often dealt with in this way on a small scale as whole colonies can be squashed at once. Plant diseases may also be removed by defoliating or pruning plants, and such methods can slow epidemics from developing as a source of in-crop inocula are removed, although it is likely to be a delaying tactic at best.

Hand picking is very labour intensive, and only as efficient as the motivation of the picker. Many pests are likely to be overlooked as they will be hiding under leaves or are too small to deal with, and the process is therefore likely to need repeating many times. It is only likely to be justified on a small scale and in valuable crops at some vulnerable stage of development or when a crop needs 'rescuing'.

Dislodging and Shaking

Dislodging pests is really a variant of hand picking, and can be quicker in the sense that the vegetation or whole plant can be dealt with in one go and the process can potentially be mechanized. This is likely to be inefficient with mobile pests, but will be aided if there are many ground predators likely to consume them as they try to find their host plants from the ground. In some cases mechanical devices have been developed that can be run over the crop to dislodge pests and capture them. Flea beetles, for instance, jump readily when the plant they are feeding on is disturbed, and running across the top of the crop with a flexible rubber tube or with cloth strips mounted on booms or trollies can induce this behaviour. Mounting vertical or horizontal plates covered with grease can be used to trap the leaping beetles, but the plates need to be cleaned regularly and are not very efficient. Colorado (potato) beetle can also be induced to drop by disturbing plants, and devices have been developed that literally comb them out of potato crops by running comb-like collectors through the crop to disturb the plant, stimulating them to drop into collection areas from which they can be removed and destroyed.

In small-scale systems it may also be feasible to dislodge pests using a high pressure jet of water, which under the right circumstances or in a polytunnel might also aid irrigation and watering. Aphids, for instance, can be dislodged with such a jet of water, but these methods are not generally very effective, for various reasons. Air-blowing devices have also been developed, which literally blast insects off the crop and some even heat the air to kill the pests, so overcoming the problem of the insects returning to the plant – although this is at some risk to heat damage to the crop. Recent developments have also seen claims that 'hot air' blowers can help not only to dislodge pests, but also to stimulate plant defence systems against disease-causing organisms as well.

Heating, Steaming and Flaming

Some pests or pathogens can be physically burnt off plants, but this is a highly risky procedure unless practised in a trap crop (*see* above) as the plants are likely to be as susceptible to the treatment, if not more so, than the pests. In a few cases flaming might be justified, and organic growers often remove potato haulms once blight has reached 10 per cent of crop foliage in order to prevent the disease getting into the tubers. Other methods, such as mechanical flailing, might be better from an environmental point of view because steaming or flaming uses large amounts of gas fuel. Stubble burning was widely practised in the past in the UK, and was probably effective in reducing populations of overwintering crop pests in the surface or trash layers of fields, although the impact on natural enemies was also likely to have been high and negative. Such methods are in any case now restricted by law due to widespread public nuisance.

Steaming, wet heat and/or dry heat has been shown to be effective in reducing the external pathogen loading on seed, and is one area in which organic farming can benefit from these technologies, as concentrated batches can be treated quite efficiently.

Other Methods

A wide range of other methods has been tried for use against pests and pathogens in farming systems. Most have remained at the experimental stage, for example microwave, infrared and ultraviolet radiation, and many would in any case probably be inadmissible for use in organic farming systems.

A widely used system in domestic situations is electrocution, whereby insects are attracted to a device which electrocutes them. Another device has been developed for use in glasshouses, which shines ultraviolet light into the crop as it travels down the row. It is said to be effective against powdery mildew and potato blight when they are directly visible to the uv light. Such methods are at present unlikely to be viable or effective on a field scale.

Water and Irrigation

Water is essential for normal plant development, and lack of water can lead to stressed plants that are more susceptible to certain pests and/or diseases. Irrigation is vital for polytunnel production, and can be important in outdoor field crops under dry conditions, especially in field vegetable crops. The key to adequate water for the crop in the field, apart from sufficient rain, is moisture conservation in the soil. Many of the methods previously described for increasing soil organic matter will also preserve soil humidity. Other methods include mulching (*see* above) and even contour farming. Many farms, especially vegetable growers, will also provide irrigation at crop establishment and in drier periods.

Once again it is not the place of this book to describe irrigation systems or techniques in detail, but to note that irrigation can have both an indirect and a more direct influence on pests and diseases. Indirectly, a well managed irrigation plan will supply the crop with adequate moisture to develop and consequently encourage strong vigorous plants that are more resistant to pest and pathogen attack and infection. Such plants are also more likely to compensate for pest or disease damage, although the resultant dense canopies may encourage higher humidity and disease to some extent. There is some evidence that water-stressed plants are more susceptible to pest attack, and certain diseases are also favoured in these conditions, which should therefore be avoided.

More directly, irrigation can increase or decrease disease risk in crops. Overhead irrigation is likely to increase humidity in and around the crop, and create the conditions that many pathogens need to infect crop plants. Pathogen spores or propagules can also be spread by water drops splashing through the canopy. Consequently limiting overhead irrigation to periods where the leaves and canopy have time to dry out can be beneficial. This can be achieved by, for instance, watering in the morning and allowing the canopy to dry out in the afternoon before the onset of dark, at which point high humidity is likely to be prolonged through the night. Conversely drip irrigation systems are much more economical with water and do not create such humid conditions (although they may be more costly to set up). In some specific cases irrigation can reduce the risk of soil-borne disease as dry soil is less biologically active. For instance, potato scab can be mitigated by adequate soil moisture at the onset of tuber initiation, and this can be supplied by irrigation in dry conditions.

Pests are likely to be less affected by irrigation in temperate climates, although in drier areas green fields could potentially act to attract pests in otherwise 'brown' areas. Pests can be affected either physically by being struck by water, or as a consequence of the change in environmental conditions. The physical impact of irrigation water is unlikely to be very significant in removing pests from crops, although high pressure sprinklers have been shown to dislodge pests such as aphids. One area in which water can be effective in reducing pest infestation is in polytunnels and protected crops. Mites in particular, but also aphids and whitefly to some extent, are affected by either wet leaves or jets of water. Regularly damping down polytunnels can help to keep mite populations in check as they prefer dry, dusty conditions, whilst whiteflies and aphids are disturbed by strong streams of water.

The inundation of fields with water as an irrigation or pest management technique is not regularly practised in temperate organic systems, but innundation is likely to impact and reduce the numbers of ground-dwelling predators, although it may also drown soil-dwelling pests. Inundation techniques may also assist in spreading pests and pathogens into water courses. Flowing water courses may need to be sensitively managed to reduce the risk of spread of pathogens if irrigation water flows back into ditches, as it can carry infected plant material or the pathogens themselves, which may then be reintroduced into uninfected crops as more irrigation water is extracted.

Fencing

Fencing and electric netting are generally used against larger-sized pests as a barrier to access and free movement. In some cases netting can be electrified, which can increase its effectiveness, although a system for delivering a high voltage, low current shock will need to be put in place. Netting and fencing can be time-consuming to set up properly, and can increase the time taken for other crop management operations if they need to be removed to carry them out (for example, weeding).

Fencing is used to keep large herbivores out of fields or parts of fields. Rabbits are the principal herbivore pest in the UK, although deer and badgers are also common. Permanent fencing is costly to install and needs to be strong to resist rabbit teeth and badger claws, although semi-permanent rabbit fencing made out of chicken wire bent over and dug into the ground on the outside edge is quite effective. It is vital to remove animals from within the protected area for the permanent fencing to be effective. In the long run such semi-permanent fencing can be more cost-effective than movable (electric) fences, which need to be constantly repositioned within the field, although this flexibility may also be required on occasion.

Movable internal field fencing is often also electrified. In this case it needs to be carefully positioned, and the ground under and around it cleared to prevent vegetation shorting out the current, which will reduce its effectiveness. It also needs to be pegged down to prevent animals squeezing under it. Although a deterrent, electric fencing can be surprisingly ineffective on well insulated mammals such as rabbits or determined

Movable and electrified rabbit netting (with rabbit burrows).

ones such as badgers. Electric fencing can be run off a car battery charged by a solar panel, which may reduce environmental impact but is unlikely to be economically viable (in the short term) if an electricity supply is located close to the field.

Other barriers can be erected in fields to prevent the movement of pests or to control the movement of pathogens. Hedge lines and other features will control the flow of air in and around the field and may cause pests or pathogens to be deposited in certain areas or on the lee side. Such traps will have to be worked around on the farm but might be modifiable to change airflow in a beneficial way. In some cases pests can be discouraged by vertical barriers strung across a field. Carrot flies, for example, generally fly close to the ground and are typically weak flyers, so a mesh or fleece barrier 1–2m high with an overhanging lip on the windward side can trap or filter these pests out of the wind or lift them over the crop (if crop strips are not too wide).

Scaring

Scaring is a time-honoured method of deterring pests capable of associating mechanical or visual stimuli with 'danger' or 'peril'. However, many of the techniques are of questionable value against these animals, as they are also generally capable of learning. Some are also unacceptably cruel for modern usage. The traditional scarecrow has been replaced by more sophisticated methods such as flashing lights, gas guns, balloons and kites. When used in conjunction with netting or traps they can be more or less effective against birds, rabbits and other vertebrate pests. However, their utility is often reduced over time, as animals learn that they pose no threat and become habituated. The most effective methods are likely to evoke instinctive survival or flight behaviours, which keep the animals moving or skittish. For

Movement-sensitive device that aims a jet of water at vertebrate pests.

instance helium-filled kites bobbing in the wind and mimicking birds of prey are generally quite effective against birds as compared to gas guns, to which the birds (but not the neighbours) quickly become habituated.

Farmers and growers occasionally back up deterrents with shooting, but once again animals will quickly learn when scaring is likely to be lethal or not, and shooting in itself is rarely effective, taking at best a few individuals for a high cost in time. Scaring, like trapping, is subject to animal cruelty legislation, and some sensitivity is likely to be necessary if farmers and growers regularly host customer open days or are close to urban areas.

Traps

Traps can be used in organic systems as a method of pest control. They can be used on a wide range of animal species, although their effectiveness depends on the proportion of the population that can be captured and removed. In general this is not very high, as in practice the number of traps is limited and pest populations (especially invertebrates) can be very large. There is also a certain amount of labour involved in maintaining and emptying traps. A range of animal welfare laws and other countryside legislation needs to be taken into account when trapping animals.

Traps have been traditionally used against vertebrate pests where they have been useful in supplementing diet (or if you can't beat them, eat them!). A range of traps is available for trapping rabbits, rodents and birds which may be made more effective by baiting. Normally such traps can only trap a small proportion of the population in question, and are probably not particularly effective as many of the pests have high reproductive rates. In organic systems it may be permissible to bait rodent traps with proprietary poisons in and around buildings, although clarification should be sought from any control body, and this makes such traps much more efficient. Such measures may anyway be necessary (or even occasionally ordered) under public health regulations.

Invertebrates can be trapped. Such methods are usually more useful for signalling periods when pests are active and indicating periods when other control methods might need to be employed. The beer trap for slugs is a very well known method of control. The smell of alcohol signals fermenting vegetative matter to the slug and lures it into a death of drowning in alcohol. Its effectiveness in actually controlling slug populations remains open to question, and there may be more than an element of vengeance in using this method. Anecdotal evidence also suggests that not even slugs will touch the economy brands of lager.

Other traps commonly used are sticky traps against flying insects. These are normally bright yellow and hung above the crop in glasshouses or polytunnels, although carrot fly traps are also available for use in carrot crops. They are useful in monitoring the presence of crops, but especially in field crops, are unlikely to have a significant effect on the pest population. It is also important that the traps do not contain any insecticides not approved by the control body.

Some insect traps have been made a lot more efficient by including pheromone lures in them. For instance, click beetle traps can be used to lure adult click beetles in order to monitor their presence in a field. Pheromones can only be used in traps that are being used for monitoring purposes and not as a means of controlling pest populations. It is not clear, however,

Proprietary slug trap.

where the line between 'very intensive monitoring' and population control lies, but once again in many cases even a large number of traps is unlikely to have much effect on pest populations. The same is true of traps using other methods to lure insects, such as light traps popularly run at night to capture flying moths.

Hunting

Hunting has also been a traditional method of pest control with a dubious history. Traditionally such methods have been very effective where a bounty has been placed on the pest species and its rate of reproduction is fairly low, and indeed some species have been hunted to extinction by such methods. Currently hunting is subject to a range of contradictory and complicated legal enforcements in most countries where temperate organic agriculture is practised. In any case it should not be unnecessarily cruel if practised on organic farms, and it should be expected that effectiveness will be limited, not least because the hunters have a vested interest in maintaining at least some prey alive to continue the population from one season to the next.

BIOLOGICAL CONTROL

Biological control is often understood as the intentional use of beneficial organisms or natural enemies to control pests and pathogens. However, in its widest sense, biological control will be occurring naturally in all farming systems, including organic ones, as a consequence of the relationships within existing food webs and between organisms in the agroecosystem. In fact, this contribution has largely been unrecognized and unappreciated, and it is one of the strengths of organic farming that it has been placed at the centre of management concerns.

Three broad approaches to utilizing biological control within farming systems have arisen over the past few decades as ecological science has begun to underline the importance of natural ecological control mechanisms in controlling pests and pathogens. They are conservation biological control, inundative biological control and imported (or classical) biological control. The primary goal of organic farmers and growers should be to promote the former. That is, they should seek to enhance the environment for natural background biological control, and to practise conservation biological control. A whole-farm approach to doing this has already been described in Chapter 3; here we will concentrate on discussing implementation at the level of the crop or field.

Inundative biological control involves the release of natural enemies into the crop as a temporary measure to suppress pests and diseases, and this is also briefly discussed, albeit as more useful in protected crops. Inundative control is more likely to be disruptive of the crop system than conservation control, and more likely to be applied repeatedly, although control should be expected to persist for some time under the right conditions.

Classical biological control is mainly the preserve of professional scientists and institutions because of the complex statutory requirements for such programmes. Generally predatory or parasitic organisms are sought in areas from which pests, pathogens and/or crops originated, tested under controlled conditions, and then released in the new area. It is probably true to say that such programmes have had a limited impact in Europe where many crops have been present for some time, but have been more successful in other areas where temperate organic systems have been developed with imported crops (often from Europe).

Conservation Biological Control

At a field and crop level it is important to be proactive in taking measures that ensure that natural biological control acts to suppress pests and pathogens. Biological control is always present as part of natural food webs in natural populations, and the aim of farmers and growers is to utilize and enhance this 'background' control to their benefit. Many approaches and management techniques have been developed to achieve this practically at field level and within crops, but whichever methods are used, the aim is to compliment the whole-farm approaches to conservation biological control previously discussed. In this section we discuss both crop-specific measures which are likely to rotate with the crop and therefore be more temporary, as well as field approaches which are likely to remain in place between seasons whatever the crop.

Crop Approaches

A range of cultural management methods that create diversity at the scale of the crop plant has been discussed in the previous section, and included measures such as intercropping, undersowing, minimum tillage and using organic manures. All these, and many other cultural measures, can promote survival of natural enemies by either avoiding disrupting their life cycles or providing resources that promote their survival, although on occasion they may also adversely impact them. It should be stressed that cultural operations in crops should aim for the minimum feasible amount of disruption to natural enemy populations in order to put conservation biological control on as sound a footing as possible.

Notwithstanding this, there is a range of practices available that can help to promote natural enemy activity in and around crops, and these are described below.

Choice of Variety and Crop Type Variability between varieties and the consequent effects on natural enemies have been mentioned previously. Natural enemies, and especially insect parasitoids, are sensitive to the size and shape of plants and prefer hunting in specific 'niches'. Different varieties can potentially help or hinder natural enemies both in their structure and form. For instance hirsute (hairy) varieties hinder not only pests but can also hinder parasitoids or predators searching for their hosts. Greater leaf areas also reduce the searching efficiency of predators and parasites as they have more area to cover. Some plants may even secrete allelochemicals that attract natural enemies into the crop when they are attacked by herbivores. It may therefore be possible to choose varieties that favour natural enemies with the same caveats as choosing resistant varieties.

In a similar vein it has also become apparent that epiphytic organisms that live on plant leaves and other surfaces can help to suppress pathogen infection of plants, and there is some evidence that such populations are mutualistic – that is, they have evolved with the plant as part of its defence system. Certain varieties may thus be less susceptible to pathogen invasion due the presence of these natural anatagonists. Whilst it is premature to be able to pick varieties based on this trait, it should at least be borne in mind when carrying out other cultural management practices, such as applying soft soap, which are likely to disrupt leaf ecology.

Sensitive Weed Management Many natural enemies (and other desirable species) use weeds either as shelter or as alternative sources of food. Weed control should be targeted only at the most competitive weeds, and should be aimed at critical periods when weeds compete with the crops, and at preventing weeds from seeding or increasing in numbers. A list of weed species and their desirability in terms of their biodiversity value is presented in Table 3, although it should be appreciated that there is the potential to experiment with a wide range of weed species over and above the limited range presented in the table. Of course the desirability of having weeds present in any situation will also need to be judged for their likely impact on yield and as part of the whole-farm management plan. Environmental schemes may value their presence on farms, especially rarer arable weeds, and they may thus also contribute to stewardship points (in the UK) and/or environmental subsidies.

Weeds can attract beneficials into crops.

Table 3: Some Weed and Field Margin Species Beneficial to Natural Enemies and Pest Management

Plant Species	Description	Other Notes
Pollen and nectar sources provide a food source for many adult predators (e.g. hoverflies, lacewings) and also parasitic wasps, and should be present throughout the season		
Yarrow (*Achillea millefolum*)	Herbaceous perennial, flowering June–August. It has a well developed fibrous root system and is a natural component of chalk grassland	May also act as an alternative host of carrot fly so needs careful monitoring if growing near carrot crops
Hogweed (*Heracleum sphondylium*)	Herbaceous biennial. Grows on fertile, heavier soils. Flowers June–September	As above
Cow parsley (*Anthriscus sylvstris*)	Herbaceous biennial or short-lived perennial. It is native in grassy places, hedgerows and wood margins. Flowers April–June	As above
Wild carrot/Queen Anne's lace (*Daucus carotta* L.)	Herbaceous biennial. Flowers June–August. Prefers well drained soils	As above
Corn marigold (*Chrysanthemum segetum*)	Less common annual weed, occurring on acid or calcium-deficient arable soils. Flowers June–September	
Alternative hosts provide additional food resources for pests and help to sustain predator populations at a sufficient level to provide control in the target crop		
Common nettle (*Urtica dioica*)	Herbaceous perennial. It is native on riverbanks and in hedgerows, grassy places. It thrives in nutrient-rich, open-textured soils in the pH range 5–8	As with other alternative hosts, need to be careful that pests do not move into crop once weeds are cut down
Redshank and relatives (*Persicaria* spp. *Polygonum persicaria*)	Summer annual. It occurs on most soils but especially on moist, rich, acid peaty loams	Can be a particularly vigorous weed, which would outweigh any advantages in pest control

(Continued)

Table 3 (Continued)

Plant Species	Description	Other Notes
Corn spurrey (*Spergula arvensis*)	Annual weed occurring on light sandy soils deficient in lime. It is found mainly on soils with a pH between 5.0 and 6.0	Only competitive against crop at high populations
Wormwood (*Artimesia absynthium*)	A herbaceous perennial, growing to 1m tall. Thrives on arid, well drained soils, rich in nitrogen	Reported as a trap crop for flea beetles
Fat hen (*Chenopodium album*)	A summer annual weed. A vigorous competitor growing more than 1m tall and commonly found in potato or root crops. It grows best on fertile soils and is common on sandy loams and frequent on clay, but less numerous on calcareous soils and gravel	Particularly attractive to black bean aphid (*Aphis fabae*)
Wild radish (*Raphanus raphanistrum*)	Annual or biennial weed that occurs on cultivated and rough ground. It prefers a nutrient-rich, lime-free, sandy or loamy soil	Other brassica weeds (e.g. charlock) may have similar effects. Need to take care that it does not act as green bridge for other problems such as soil-borne diseases (particularly clubroot)
Refuge habitats for predators encourage generalist ground-dwelling predators (e.g. carabid beetles, spiders) and overwintering sites for others (e.g. ladybirds)		
Yorkshire fog (*Holcus lanatus*)	A tufted, perennial grass, native on rough grassland. It can tolerate a wide range of conditions including waterlogged soils but also moderate drought. It grows best at a pH range of 5–8	

Table 3 (Continued)

Plant Species	Description	Other Notes
Cocksfoot grass (*Dactylis glomerata*)	A tufted perennial grass. Thrives on well drained soils and is tolerant of low nutrient levels	
Fescues (*Festuca* spp.)	The fescues encompass a wide range of perennial tufted grasses suitable as a refuge for predators. Red fescue (*Festuca rubra*) is one example that can tolerate many climates, although it thrives best in full sunlight	
Non-competitive background weeds can act to confuse pests. Low-growing, prostrate weeds that are not too vigorous are likely to be more suitable		
White clover (*Trifolium repens*)	Perennial, native in grassy and rough ground. It is most frequent on soils of pH 5.0 to 8.0 and is common on clay soils but is rarely found on peat. As it is nitrogen fixing, it has a competitive advantage against other weeds on low fertility soils. It spreads through an underground stolon structure	As with other weeds, can compete against the crop if not controlled
Chickweed (*Stellaria media*)	This annual weed is one of the most common in the UK. It occurs on a range of soil types but is most abundant on lighter soils. It is indicative of high soil nitrogen levels	

Green Manures and Flower Strips Green manures and flower strips can be grown on margins or on strips running across fields, and provide both food and shelter for parasitoids and predators. Flowering plants in particular can be very important and they should be planted to bloom at times the natural enemies can make most use of them. Insect parasitoids (mainly small parasitic wasps) in particular feed on nectar as adults and need this resource to develop their eggs and provide energy for hunting their prey. Hoverflies and lacewings likewise need pollen and nectar sources to develop. Simple flowers or small ones with pollen and nectar within easy reach are best, and umbellifer flowers and composites are particularly good for beneficials.

Although many attractive flowering species grow naturally in crops or field margins (Table 3), there is also a wide range of potential mixtures and plants that can be sown, and these are presented in Table 4. Pollen and nectar mixes might include clovers, vetches, bird's foot trefoil and sainfoin (*Onobrychis viciifolia*). In many cases these can also be mixed with grasses and other attractant plants, which all play a role in bringing natural enemies into crops. Species should be chosen that extend the flowering season as long as possible, and especially at times when wild flowers are likely to be scarce such as early spring or late autumn. Green manures can provide structure to field landscapes and provide refuges for natural enemies in the absence of suitable crop habitat. Some of the more commonly used ones are presented in Table 4.

Attractants can be intercropped to bring pests and parasites directly into the crop, although this might necessitate more complex management arrangements for crops. In some cases it is easier to leave permanent beds within fields rather than resow each year, especially in field margins, and to guarantee early flowers. Many species such as crimson clover, phacelia, coriander and corn marigolds can self-seed; others such as fennel are perennials that should grow for at least a few years. In some cases it is also possible to allow crops to flower. Umbellifer crops such as angelica, carrot, fennel, parsnip, chervil and parsley are particularly good for this.

Table 4: Beneficial Plants for Organic Farming Systems

Plant Species	Cultivation Details	Other Notes
Attractant plants providing a pollen and nectar source for beneficial insects and which may also serve as green manures		
Corn marigold (*Chrysanthemum segetum*)	Drill seeds into bed in April–May at a rate of 1–2kg/ha (0.1–0.2g for each 1m bed). Drilling rows can be varied to fit in with the mechanical weeding set-up for the surrounding crop. Strips can be planted within beds or sown at field margins	Corn marigold will flower between July and September attracting hoverflies, other predators and parasitic wasps

Table 4 (Continued)

Plant Species	**Cultivation Details**	**Other Notes**
Coriander (*Coriandrum sativum*)	Drill in May to June at a rate of 10–15kg/ha (1–1.5g for each 1m bed). A fine seed bed is required, germination is slow, and the soil should be kept moist. Drilling rows can be varied to fit in with the mechanical weeding set up for the surrounding crop. Will grow best on light to medium soils with good drainage. Strips can be planted within beds or sown at field margins	Coriander is normally grown for the leaves or seeds but it has been shown to be very effective in attracting hoverflies when it flowers. Most varieties will begin to flower within eight weeks, and continue to flower in succession for up to eight weeks
Fennel (perennial) (*Foeniculum vulgare*)	Sow in May/June at a rate of 10kg/ha. As with coriander it grows best on lighter, free-draining soils	The seeding type of fennel (not to be confused with florence fennel grown for bulbs) attracts large numbers of hoverflies when it flowers. Fennel will begin to flower approx. 3 weeks later than coriander, but the duration of flowering is longer, making it particularly suitable for controlling pests later in the season such as cabbage aphid
Phacelia (*Phacelia tanacetifolia*)	Sow from May to July by drilling seeds into a bed at a rate of 10kg/ha (approx 1g for each 1m of bed). Phacelia will grow on most soil types but thrives best in moist soil and will flower from July to September. Strips of Phacelia can be planted around the field margins or as beds within the crop to augment predator numbers	Very attractive to bees, hoverflies and other pollen- and nectar-feeding insects. Plants grow vigorously and are good for smothering weeds. Will generally self-seed, so permanent beds can be left *in situ*

(Continued)

Table 4 (Continued)

Plant Species	Cultivation Details	Other Notes
Green manures that provide habitat and other resources for beneficials		
Bird's foot trefoil (*Lotus corniculatus*)	Small seed sown at 10–15kg/ha; requires fine seed bed and good moisture for germination and establishment. Best sown March to August	Nectar source for many insects and larval food plant for others. Has been grown in modules with brassicas to reduce cabbage root fly damage
Buckwheat (*Fagopyrum esculentum*)	Sow in May with drill rate of 80kg/ha (8g for each 1m of bed) at a depth of 1cm. It will regrow if cut at the onset of flowering	Flowers rapidly (six weeks after sowing) but not frost hardy. Useful as a green manure for preventing nutrient leaching (doesn't fix nitrogen)
Crimson clover (*Trifolium icarnatum*)	Broadcast in March–April or August–September at a rate of 15kg/ha. Should not be sown deeper than a few mm	Will flower 6–10 weeks after sowing. Tends to die rapidly then set seed after flowering. Useful as a green manure
Mustard (*Sinapsis alba*) or other brassicas	Sow in March–September at a rate of 20–25kg/ha	Flowers rapidly (six weeks after sowing). Not frost hardy but also useful as a green manure cover crop. Should be avoided in vegetable rotations with brassicas to avoid pest and disease problems
Persian clover (*Trifolium resupinatum*)	Broadcast in March–April or August–September at a rate of 10kg/ha. Should not be sown deeper than a few mm	Will flower 6–10 weeks after sowing. Longer flowering period and slower to set seed than crimson clover. Useful as a green manure
Ryegrasses (*Lolium perenne* and others)	Usually sown in mixtures with vetch or clovers at around 30kg/ha in spring or autumn	Perennial ryegrass is often used with clovers for longer term leys although varieties exist for use as shorter term green manures

Table 4 (Continued)

Plant Species	**Cultivation Details**	**Other Notes**
Rye/grazing rye (*Secale cereale*)	Sown at 150–180kg/ha in late summer or autumn	Good overwinter crop and green manure
Sainfoin (*Onobrychis viciifolia*)	Best sown in a chalk and dry soil in late spring. Drilled to depth of 2–3mm at 70kg/ha. Slow to establish and may take two seasons to reach full production	Excellent fodder crop
Sweet clover (*Melilotus officinalis*)	Sown at 15kg/ha in summer months. Requires fine seed bed and moisture for good establishment	May be challenging to grow
Vetch (*Vicia satvia*)	Broadcast in autumn to overwinter or as summer green manure at 70–125kg/ha	Forms thick dense stand which is attractive to predators and prey. Good green manure and N fixer
Yellow trefoil (*Medicago lupulina*)	Small seed sown at 10–15kg/ha; requires fine seed bed and good moisture for germination and establishment. Best sown March to August	Often undersown in cereals as non-competitive
Plants that provide habitats for overwintering predators and also for beetle banks		
Grasses including cocksfoot (*Dactylis glomerata*), Yorkshire fog (*Holcus lanata*), red fescue (*Festuca rubra*), fescues (*Festuca* spp.), timothy (*Phleum pratense*)	Beetle banks are best established in September. Plant strips 2m wide and consisting of at least 30% tussock-forming grasses such as cocksfoot. Broadcast seed at a rate of 25kg/ha immediately after cultivation	Recolonization of predators will be more rapid in smaller fields. Dispersal can be improved by breaking up larger fields with beetle banks

Once again it may be possible to make use of environmental schemes such as the Environmental Stewardship options in the UK Organic Entry Level Scheme, although the management stipulations may be quite onerous; for example, nectar flower mixes will earn points depending on areas sown with the stated intention of boosting the availability of essential food sources for a range of nectar-feeding insects, including butterflies and bumble bees. In this case the mixture should contain at least four nectar-rich plants (for example red clover, alsike clover (*Trifolium hybridum*), bird's foot trefoil, sainfoin, musk mallow (*Malva moschata*), common knapweed (*Centaurea nigra*)), with no one species making up more than 50 per cent of the mixture by weight. The mixture should be sown in blocks or strips of at least 6m wide at the edges of fields and in areas not exceeding 1ha and no more than 3ha per 100ha. Half the area should also be cut to 20cm between mid June and early July (to encourage late flowering), and the whole area to 10cm in late September to October. It should not be grazed in the summer or used for access, turning or storage.

Minimal Use of Organic Sprays and Products Sprays and other products are not harmless and should only be used sparingly when grave economic loss is expected (*see* Chapter 5). For example, insecticidal soft soap will harm hoverfly larvae in aphid colonies, and sulphur used as a fungicide is known to affect parasitic wasps negatively. Acceptable organic pesticides should be formulated to biodegrade quickly in the environment. For example, natural pyrethrum breaks down in sunlight within twelve hours after being sprayed, although it is a non-selective insecticide that will affect beneficial as well as pest insects during the period in which it is active. In addition to the active ingredients, the formulations used in sprays are also likely to affect the microbial communities on plant leaves, which may subsequently affect their resistance to pathogen invasion. If used, measures should be taken to restrict the side effects of organic pesticide applications. Methods include spot spraying the areas where the pests are prevalent to avoid blanket use of products, and choosing the least toxic products in any situation.

Skylark and Fallow Plots Square, uncultivated plots can be left to develop in crops by leaving unsown areas. They are mainly used in cereal production systems by turning the drill off to leave unsown 10m strips one or two times per hectare sown. Such plots are beneficial to many farmland bird species, including skylarks, but they are also important refuges for natural enemies which can forage in the surrounding crops. Any weeds that develop in the plots can also be a source of alternative resources for natural enemies. Skylark plots can earn points under the UK Organic Entry Level Scheme, in which case they need to be at least 4 × 4m and located in fields of more than 5ha. Mechanical weeding must not be carried out in the plots.

Overwintered Stubbles Overwintered stubbles can be an important source of food for farmland birds, and also an important refuge for natural enemies. Once again, cereal stubbles are often left to overwinter, but there is some merit in leaving other crops down as residues. For instance brassicas can flower early in the spring and provide resources for natural enemies as they emerge from the previous year's crop, and before they disperse to find new crops. Care obviously needs to be taken to manage such regimes so that they do not provide a bridge for pests and pathogens to pass to the new crop. Once again they often attract payments under environmental schemes, and in the UK are eligible under the Organic Entry Level Scheme.

Field Approaches

Field elements can provide resources for natural enemies, and management techniques for hedgerows, field margins and other field areas have been developed which support in-field biodiversity and also encourage natural enemy survival. Such approaches should link into a wider farm management plan for conservation biological control as previously discussed (*see* Chapter 3) and to any approaches taken to make the crop more attractive to natural enemies (*see* above). Many of these approaches may imply higher costs in on-going field management operations, although this should be balanced against the pest management function and wider environmental gains. Some of the measures are also likely to be eligible for support money under various agricultural schemes designed to enhance the environment and/or promote organic farming methods.

Some of the more common approaches are described in the sections here below.

Beetle Banks Beetle banks have become widely used in the UK and other EU countries. Annual cropping in particular is disruptive to field ecology as compared to more stable perennial systems, and conditions in a field can be dramatically altered year on year; permanent strips or 'beetle banks' can help to stabilize the ecology of a field. Such perennial islands can supply food and other resources for beneficial organisms as well as overwintering sites for predators and parasites. They also provide habitat for ground-nesting birds.

A beetle bank is a simple earthen bank, often created by simply ploughing two furrows together or by using bed-forming equipment; these banks are sown with perennial grasses and allowed to grow. In this respect tussock-forming grasses such as cocksfoot (*Dactylis glomerata*) and Yorkshire fog (*Holcus lanata*) are ideal for providing sheltered and warmer microclimates for predators. They need to be minimally managed, normally by mowing infrequently, as flail and rotary mowers can be lethal to a range of beneficial organisms. They provide overwintering and alternative habitat for ground beetles, rove beetles, spiders and other natural enemies, and such sites are critical in encouraging beneficial predators to disperse earlier during the spring months when pest populations are just beginning to increase. Placing beetle banks at 50–200m intervals across large fields can also help reduce the distances over which natural enemies have to travel

Beetle banks can be used to break up fields, and protect and encourage natural enemies.

to reinvade new crop habitats, and allow them to colonize a crop or newly tilled spaces more quickly once they are replanted.

Beetle banks are best established in September by planting strips 2m wide with at least 30 per cent tussock-forming grasses. Mixtures are normally broadcast at a rate of 25kg/ha immediately after cultivation to form the beetle bank. They are often a management option under environmental schemes, and in England are eligible under the Organic Entry Level Scheme. As might be expected, this stipulates that they must be between 2 and 4m wide and about 0.4m high. After the first year, when several cuts may be necessary, they may only need to be cut to control scrub encroachment.

Margins and Buffer Zones Uncultivated or conservation margins, especially if rich in flowers, provide shelter and food for beneficials. Grasses, especially tussocky grasses, support a wide range of insects and can be important in their own right, but also provide overwintering sites for beneficials; on occasion up to 3,000 beetles and spiders per square metre have been counted. These margins also provide refuges from field cultivation operations, and also potentially provide alternative feeding sites.

Other types of margin strip, including strips for game birds and wild bird seed mixes, may also be cultivated, offering a diverse mixture of species that can be utilized by beneficial species, both insects and birds. These margin strips are all encouraged under environmental schemes, and in the England Organic Entry Level Scheme will attract points, depending on their areas.

Rough Habitat Zealous hygiene can remove shelter and other resources that beneficial organisms need. Undisturbed leaf litter, log piles, and

Margins should be managed to provide food and shelter for beneficials.

unmanaged and awkward field corners all provide areas for hibernation or for tiding natural enemies (and other organisms) over lean periods. Such areas effectively provide nursery or wild areas that ensure a supply of prey for predators and parasites around the farm. Nettles, for instance, often support aphid populations, and can nurture ladybird populations early in the season when there are none on crops.

Hedgerows Hedgerows are a common sight in and around fields, and provide a year-round resource for beneficials. They are usually habitat for numerous plant and animal species besides natural enemies. Generally the longer a hedgerow has been established, the larger the number of associated species, provided it is also protected from pesticides. Hedgerows are generally managed by cutting, and this has been shown to reduce the number of resident beneficial species so should be carried out sensitively, at the appropriate time of year, and in rotation around the farm on a two- or three-year cycle. Managing hedges with different structures and species will support a greater range of natural enemies, and when filling gaps in a hedge there is an opportunity to select shrub or tree species that support larger numbers of invertebrate species.

Hedgerows also support a large number of flowering shrubs and herbaceous plants along their bottoms, and these can be a rich source of food (pollen, nectar, leaves) for both predators and prey (Table 3). Certain species associated with hedgerows are particularly good at attracting predators, especially umbelliferous weeds such as yarrow (*Achillea millefolium*), hogweed (*Heracleum mantegazzianum*) or cow parsley (*Anthriscus sylvestris*). Tufted grasses such as cocksfoot (*Dactylis glomerata*) or Yorkshire fog

(*Holcus lanata*) also form good refuges for generalist predators such as carabid and staphylinid beetles. They also provide a good habitat for ladybirds to overwinter. If the hedge understorey is degraded it can be resown with flower-rich grass mixes. Tussocky grasses, for instance, are particularly important for overwintering insect and spider predators. For this reason it may be advantageous to leave areas around the field margins unmown.

Ditch and Pond Management Many natural enemies need access to water and ditches, which should therefore be managed sensitively with this in mind. Ditches with natural vegetation and slow flow are more likely to be attractive to beneficials than sanitized and/or lined channels with fast-flowing water. Many amphibians like access to stagnant water for breeding, and there is benefit in renewing natural dew or field ponds that might in any case be areas of the field that are wet or otherwise difficult to manage.

Augmentation Beneficial Habitat Some natural enemies have specific requirements for shelter and food, and providing such resources as nesting boxes can be of great benefit in other ways – for example, nesting boxes may be used by bats or wasps. Access to water is often important in these cases, too (*see* above). Owls and kestrels eat small rodents such as mice and voles, and sensitively placed nesting boxes in and around fields can encourage them to nest. They can take large numbers of mice, with owls capable of killing 2,000 or more per year (four to six per night), and this is likely to help reduce rodent damage in field crops in some situations.

Inundative Biological Control

Inundative biological control involves enhancing the background level of any biological control that might be occurring within the crop. There are two basic approaches to inundative control: one is to release a specific biological control agent directly into the crop, and the second is to directly modify the environment in some way that attracts and concentrates the agent or natural enemy in the area where control is needed. Inundative methods of biological control would normally be expected to increase temporarily the number of parasites or predators in and around the crop, and so increase the mortality of the pest or pathogen, consequently improving crop yield or quality in some way. Once the biological control agents have exerted an effect their population will often die away or fall back to previous levels, unless conditions have changed or the pest or pathogen population is sufficient to maintain the natural enemy population at some higher level.

Releasing Biological Control Agents
The release of natural enemies in and around the crop in an effort to enhance the background biological control that might be occurring is particularly important for protected crops, which normally lack such agents unless they are deliberately introduced. It is much less important for field crops, where they should already exist in the background as a consequence of conservation biological control methods. Such introduced

biological control agents are in any case likely to disperse when released into open field systems, and are unlikely to be as effective as they are in protected crops where they are more confined.

The range of biological control agents that can be bought as proprietary products has increased dramatically in the past decade. Natural enemies, which can be predators, parasites or parasitoids, are artificially multiplied and then sold in a manner suitable for release. Some of these products are used more like biological sprays or pesticides, and these are discussed in more detail in the next chapter (Chapter 5). Others are released and potentially are able not only to attack the pest or disease, but also to multiply in the field and provide continuing protection, at least for a limited time, before gradually dispersing.

In order to be effective, biological control agents bought for use should be applied strictly according to the manufacturer's instructions. They are likely to be less effective, or even completely non-effective, if they are used outside the stated environmental conditions, and if the pest is not present when they are introduced. They are more likely to be effective if the various approaches to conservation biological control, as previously discussed, have also been undertaken. Some of the most commonly used biological controls are described below, but the list is now too long to describe them all, and in any case will be subject to change as new products are made available. For this reason, manufacturers' websites should be consulted for the latest information.

Insect Predators A range of insect predators is now commercially available and potentially of use in small-scale systems, although their cost is likely to be too great on a larger field scale. Examples of predators include ladybirds and lacewings, available as eggs, larvae or adults, which can be released on to crops to control aphids. These types of predators are probably more usefully encouraged by the various conservation approaches discussed above. Insect predators such as gall midges, ladybirds and even predatory bugs may also be used in protected cropping (provided they do not escape) to control aphids, whiteflies and mealybugs, among other pests.

Mite Predators Mite predators are available and regularly used in protected and/or glasshouse cropping to protect crops against spider or other mite pests, and normally function effectively once established and as long as the prey are present. Some will also control thrips that are occasional pests of greenhouse crops.

One of the most commonly applied biological controls in protected cropping is *Phytoseiulus persimilis* against red spider mite (*Tetranychus urticae*). Predatory mites are received in a vermiculite mixture, which should be stored at 5–10°C in the dark, and used within eighteen hours of receipt. This mixture is then sprinkled directly on to the plants, at the recommended rate, applying larger amounts where there are pest 'hot spots'. It is most effective in humid glasshouses where the temperatures are above 21°C, as this allows the predator to breed more quickly. In

Ladybird larvae, a biological control agent, devouring an aphid.

hotter, drier glasshouses, alternative species should be used, such as *Amblyseius californicus*.

The predatory mite *Hypoaspis aculeifer* is used to control sciarid fly (*Bradsysia*, *Lycoriella* and *Sciara* spp.). Mites are dispatched in a mixture of vermiculite, which should be stored for not more than two days before application. They should be applied to the soil surface (never to the plant), which should be open structure, high in organic matter and moist (but not soaking wet) in order for the mite to be effective.

Insect Parasitoids These parasitoids are on the whole small parasitic wasps, supplied as pupae or adults for release on to crops. *Trichogramma* spp. wasps have been used in the field to parasitize the eggs of pest insects (usually Lepidoptera), and some parasitoids of caterpillars (for example *Cotesia* spp.) are also available. The effectiveness of these parasitoids in the field is likely to be very dependent on environmental conditions and the pest status, and applying them is probably not worthwhile unless extensive conservation biological control measures have also been taken.

It is in protected cropping that the most sophisticated delivery systems have been developed, for releasing and maintaining parasitoids over a cropping season. Parasitoids are available for the control of aphids, whiteflies and leaf miners, and usually work effectively if applied under the right conditions. For example, *Encarsia formosa* is commonly used against glasshouse whiteflies (for example *Trialeurodes vaporariorum*); it is purchased as pupae stuck to card strips, which are then hung on the plants, releasing the adults. The twenty-four-hour average temperature in the greenhouse should be at least 17°C to allow the predator to thrive. As with other biological control measures, it must be stored away from light at temperatures of 8–10°C, and used within one to two days of receipt.

Mite predators in protected cucumbers.

Entomopathogenic Fungi These fungal diseases of insects have been effectively employed against aphids and whitefly, mainly in protected cropping, but also increasingly in field crops. They are more effective in a humid and warm environment, which is also favourable to plant disease and hence may not always be suitable for use. They are discussed in Chapter 5 as they are more often applied as microbial sprays.

Parasitic Nematodes Parasitic nematodes are also increasingly used as sprays against pest insects, and are consequently discussed in Chapter 5. They are effective against a number of soil-dwelling pests including root flies, and for this reason have great potential. Some nematodes are also effective parasites of slugs under a range of conditions. As of yet some of these treatments are unlikely to be affordable on a field scale.

Antagonistic Microorganisms Enhancing disease management through the use of natural antagonists is the least understood of the potential biological control systems available. Whilst it is understood that microbial communities on leaves and in the soil can act in ways that suppress pathogenic organisms and prevent them infecting crop plants, the specific details are often hard to unravel and in any case are complex. Some biological sprays applied by organic farmers and growers, and in particular compost teas and other microbial-containing solutions, are reputed to work to bolster and/or provide nutrition to both epiphytic and soil microbial communities and hence protect crop plants. In practice it has been very difficult to establish the conditions under which these products can be consistently made and effectively applied, although there is a great deal of on-going research on this subject.

The Concentration of Natural Enemies

Natural enemies may be concentrated in and around crops using many of the conservation biological control methods discussed above. However, there is also increasing interest in artificially increasing natural enemy numbers in various ways, some of which are described here below.

Artificial Foods Artificial foods have been tried to attract predators and/or parasitoids. Normally applied as sprays, these foods provide an abundance of some limiting resource – usually carbohydrate as sugar solutions. Although this has been tried with predators such as green lacewings, it is probably not as cost effective as conservation biological control in raising background levels of predation.

Pheromone and Attractants Some research work has focused on the possibilities of attracting predators or parasites into crops by spraying them with pheromones or other attractant chemicals. Such attractants can also be used to manipulate the natural enemy population by, for instance, moving them on to field boundaries to overwinter rather than remaining in the crop to be harvested (and probably killed in subsequent cultivation). At the moment such techniques remain speculative, but a possibility – although it is difficult to imagine that they will add much to a well managed conservation biological control farm plan.

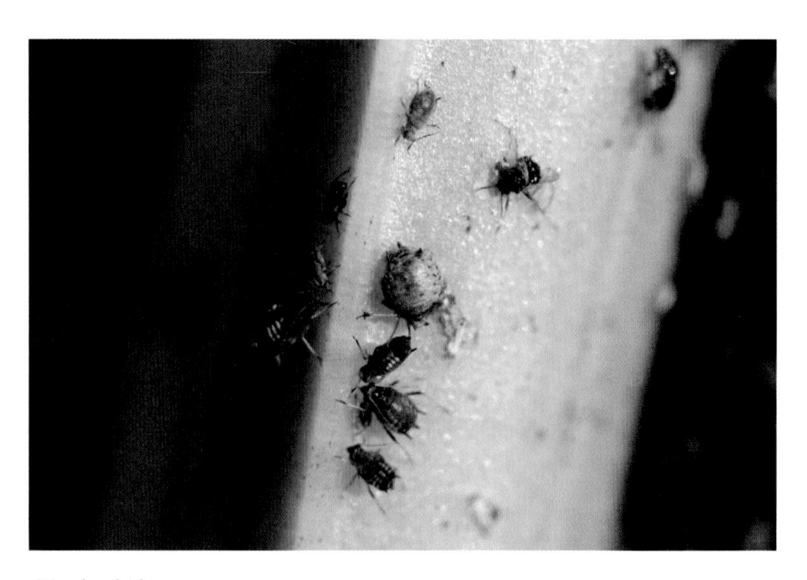

Parasitized aphid.

Releasing Hosts Paradoxically many studies show that, and many biological control programmes in protected cropping rely on the fact that, the pest is present. Encouraging a small number of pests and maintaining their population can act to improve biological control in the medium to long term, as a continuous supply of hosts allows the natural enemy populations to build up.

Classical Biological Control

The original research on biological control was spurred on by the idea of looking for natural enemies of pests in their area of origin either from which crops had arrived, or from which the pests had escaped (often with the crop). Despite the fact that there have been some spectacular successes with finding and releasing natural enemies in new areas, there have also been some notable failures (for example cane toads) that have led to worse problems; this activity is now tightly regulated, and only practical with a large research effort and back-up. There is limited scope for farmers and growers to intervene at this level, and once released, classical biological control agents are expected to spread and exert their influence without particular help from them. However, even in these circumstances, conservation biological control would be expected to help.

5
Direct Pest and Disease Control

Occasionally organic farmers and growers may need recourse to more direct control measures against pests and diseases. Most of these measures are based around the use of proprietary commercial products which are manufactured, and which are brought on to the holding for the purpose of pest and/or disease management. They include biological control agents, microbial and microbially derived pesticides, botanical pesticides, oils, soaps, dusts, pest growth and behaviour modifying substances, inorganic pesticides, organic synthetic pesticides and a range of other amendments that defy simple classification. Their use is only justifiable in organic farming situations where economic damage is inevitable and will put the farmer's or grower's livelihood in jeopardy. This is because many are biocidal, many are not particularly selective, and most are likely to have various negative ramifications throughout the farm system, possibly exacerbating longer-term problems. This is especially true where they impact negatively on soil ecology and farm biodiversity.

It is in this arena that the most conflicts arise between practical farm necessities and organic principles. Experience has shown that the period of conversion to organic farming methods is one point at which many farmers or growers look to substituting conventional inputs for 'organic' inputs, and that it takes some time to adopt and settle into an organic and agroecological approach to pest and disease management (as described in the previous chapters). However, once this period has passed, many farmers and growers find that their need for such inputs diminishes. In a similar vein, even many experienced farmers and growers need to retain a sense of 'control' over a developing pest or disease problem, and recourse to inputs often satisfies a physiological need to 'do something' even though many interventions are often ill timed or ineffective. It is only from experience that farmers and growers will be able to judge which are those situations where they really do have to take corrective measures, and which methods are likely to be the most effective if they do.

Given this, most organic standards define which methods and materials can be used, and many proscribe those that can't. Farmers and growers should therefore always err on the side of caution and consult their organic control body where there is any doubt about using one of these direct control techniques. On the whole, a pragmatic approach to the use of such products is best, and if farmers or growers are finding that they are relying on such products, they should examine ways in which their use might be avoided or curtailed within the farm system by looking at the more preventative and adaptive approaches discussed in the previous chapters.

PRACTICAL CONSIDERATIONS

Although many of the commercial products described in this chapter act directly on the pest or pathogen organism in order to kill it, others act more indirectly to achieve their effect, either achieving a kill through an intermediary organism or by modifying the pest or pathogen behaviour in some way. In the former case these products are normally described as 'pesticides', and are usually subject to national laws in their production, marketing and use. In the latter case, these products are often referred to as 'biological control agents' or '(organic) amendments', and are not necessarily subject to the same rigorous testing and marketing standards as pesticides. In this chapter we have joined the two together because their use and application share many features in common: that is, they are normally bought in, they work on a one-off basis, and they need to be spread on or around the crop in some way, often using specialist equipment. They are, however, often subject to different legislation in their use, and specialist advice may need to be sought on this.

As a primary consideration, any pesticide that is used, 'organic' or not, should also be allowable under any relevant pesticide application laws of the country concerned. For instance in the UK, all such products need to have been tested and approved for pest control use by the government Chemical Regulation Directorate (CRD) before they can be sold and/or used as pesticides. Suffice to say that they should also be used in accordance with their manufacturer's instructions. Non compliance with instructions or misapplication may be a legal offence in some countries for which the farmer or grower is likely to be liable, depending on the circumstances and the product in question. In many countries, including the UK, home-made pesticide products are illegal as they do not comply with these requirements. As a final hurdle, these products will also need to be allowable under the organic standards of the certifying or control body to which the farmer or grower belongs, which in turn will have to comply with any relevant government legislation on organic production.

Amendments and biological control agents are usually subject to different legislation, depending on the product and its mode of action. The most reputable will be made by companies that rigorously test them and supply adequate application instructions. In this case, normal marketing and trading standards legislation applies. However, it should be recognized that many of these products will not have undergone the same rigorous testing for efficacy as regular pesticide products, and their modes of action are not

Appropriate and well calibrated equipment should be used for application.

always fully understood because their pest and disease control effects are, at best, indirect. In any case, all such products should be used in accordance with the manufacturer's instructions. Use outside of the conditions specified by manufacturers is likely to lead to poor or even no effect.

As these products are bought in they will also need to be treated as economic farm inputs and their use properly costed in order to gauge their effectiveness. This aspect is treated in more detail in the next chapter (Chapter 6), but cost effectiveness is more likely if the pest or disease is prevented from attaining levels where it causes economic damage to the crop. In practice, this means monitoring the crop in order to treat it in a timely manner and under the right environmental conditions. If the product requires specialist application equipment, for instance a sprayer, this should be used in order that the product is delivered to the relevant target. Such equipment should also be calibrated to deliver the right dose in the right way.

BIOLOGICAL CONTROL AGENTS

There are many biological control agents now available for use, not only in organic systems, but in conventional and IPM systems, too. These may include viruses, bacteria, fungi, insects or arachnids. The use of these has been previously discussed (Chapter 4) regarding the situation in which they are used to augment natural background levels of natural enemies. However, and increasingly, biological control agents are being developed which are used as one-off spray products that directly impact on the pest population in a more or less rapid way and then die back. The agents being developed in this way include nematodes, fungi, bacteria and viruses. Normally they would be expected to disappear once the pest population has crashed, rather than persisting in order to give long-term natural biological control. In this section we discuss the use of nematodes and fungi, whilst microbial (bacteria and virus) products are considered in the following section.

When bought in, biological control agents should always be used according to the manufacturer's instructions and under the conditions specified. In some cases it may be necessary to consult specialists; alternatively the larger suppliers often offer good advice. It is always essential to know the identity of the pest to be controlled, as many of these agents are host specific. The agents need to be released at a time in the pest life cycle when it is susceptible to attack, and at an adequate rate to have an effect, which depends on the area to be covered and the severity of the pest infestation. It is generally recommended to order the natural enemies and prepare to release them more or less immediately. If storage is necessary it should be under adequate conditions so as to avoid mortality (manufacturers will specify the conditions) and for as short a period as possible. In protected cropping a combination of natural enemies is often released, and these should be chosen so that they work together rather than antagonistically. Once again specialist advice may need to be taken where the multiple release of natural enemies is made.

Nematodes as Biological Control Agents

A growing range of nematode products is becoming available to control pests, mainly insects and slugs. They are normally applied as spray solutions of nematodes, and may require specific temperatures and soil moisture to be fully effective within a reasonable time period. They may also act best on specific life stages of the pest insect or slugs and therefore need to be applied at the appropriate time. Different species are more effective against different pest species, so non-target organisms are not unduly affected. The cost of these products implies that they are more likely to be used in smaller scale high value crops rather than large scale field crops, although costs may reduce as more research is undertaken and new species are isolated.

Pest insects for which nematode control agents exist include vine weevils, chafer grubs, leatherjackets, sawflies and caterpillars. In the case of caterpillars and sawflies the nematodes are *Steinernema carpocapsae*, which enter the spiracles of the pest and once inside release bacteria that stop the larvae from feeding. This results in the death of the caterpillar. The solution should be sprayed on to leaves as soon as caterpillars are noticed and should contact the caterpillars to be effective. Repeat sprays are likely to be necessary as the nematode cannot transmit efficiently between hosts above ground.

Other *Steinernema* spp. or *Heterorhabditis* spp. are present in other preparations that control soil-borne insect pests but act in similar ways to target the specific pest. Generally the soil should be moist and warm in order for the treatment to be effective. In some cases, and under favourable conditions, the nematodes may go on to produce subsequent generations that also suppress pest populations, although it is more likely that repeat applications will be necessary. Research work is currently looking at the application of similar nematodes in the control of a wider range of soil-borne pests, including carrot fly.

Nematode preparations containing *Phasmarhabditis hermaphrodita* which kill slugs are also sold to be watered into the soil. Although a naturally occurring nematode, the application will augment the population to levels effective for slug control. It will kill the majority of slugs of all common species below the soil surface where slugs hatch and are small; it will not kill those above the soil surface, and may be ineffective against large slugs. It is best when soil temperatures are above 5°C. Applications are effective for about six weeks before further doses are required, but in wet weather, when soil is kept moist, further generations of nematodes may provide some level of control without repeat applications.

Fungi as Biological Control Agents

Entomopathogenic fungi are widespread natural enemies of insects; fungal pathogens of other organisms such as nematodes are also common. In the case of insects, under the right environmental conditions the fungus penetrates the insect cuticle and grows inside it, killing it and digesting it. Fungal infection often modifies insect behaviour so that the insect climbs to the top of plants to die; here it is glued into place by fungal hyphae, and from this position the fungus produces spores which are dispersed on the wind to encounter new hosts.

A few products are commercially available for insect control (depending on country). They can all be applied as sprays, although other delivery methods have been developed, especially for indoor crops. Although they have been mainly developed for indoor crops, they are suitable for field application; however, they are likely to be much less effective where temperature and humidity cannot be controlled as they require high humidity and moist conditions to germinate and penetrate their insect hosts. Unfortunately these are also good conditions for fungal plant pathogens to spread and infect crop plants, and organic farmers and growers would generally try to manage their crops to avoid these conditions. Some products are formulated with oils and other substances that protect the fungal spores in low humidity and from sunlight (which reduces their viability), but this does not make them markedly more efficacious. Many of the fungi preferentially infect hosts with a soft body such as aphids or whiteflies, but they can be fairly non specific, attacking a wide range of non-pest hosts with these characteristics, which might make them unsuitable for general use in organic systems.

A fly naturally infected by entomopathogenic fungus.

Commercial preparations have been made of *Beauvaeria bassiana, Peacilomyces fumosoroseus* and *Verticillium* lecanii, all of which have a wide insect host range, but in the main they have been commercialized to control pests such as aphids, whiteflies and thrips. Their use has been best developed for protected cropping. The fungus *Metarhizium aniopilae* is apparently more host specific and has been commercialized for use against insects with harder cuticles such as grasshoppers, beetles, cockroaches and termites.

Fungi have also been investigated for their potential as biological control agents against nematodes. Some species of soil-dwelling fungi have evolved to snare their hosts in their hyphae, from where they can penetrate and digest them; however, they have so far not proved effective for field use. Research has also investigated the potential of using fungi as biological control agents for use against other fungal pathogens, but although commercial formulations are available in some countries, these products are generally regarded as having limited success. In both cases these fungi are naturally present in soils, and as such, promoting general soil biological activity might be as effective as artificially adding these organisms to control soil-borne diseases.

The best known biological control agents in this respect are fungi of the genus *Trichoderma*, which have been demonstrated to have biological control action against soil-borne fungal pathogens, and even above-ground pathogens on leaves and/or fruit. The fungal hyphae grow towards the hyphae of other fungi, entwine them and then degrade their cell walls to digest their contents. *Trichoderma harzianum* has been produced as a commercial biological control for use against soil-borne pathogens in particular, and other *Trichoderma* spp. (most notably *T. viride*) have been produced for use against other diseases, sometimes as mixtures of different species. Other potential biological control agents include *Gliocladium virens* and *Coniothyrium minitans*, which have been extensively investigated for controlling *Scelerotinia* diseases as they are specialized mycoparasites of sclerotia. Commercialization of these products has been haphazard, however, and they are unlikely to be available in all countries. Their effectiveness will also depend on ensuring that other farm practices create a favourable environment for them to survive.

MICROBIAL AND MICROBIALLY DERIVED PESTICIDES

Microbes and microbial products offer a huge future potential in the control of pest and pathogen organisms. The main source of such organisms is the soil, where the complex food webs and ecological interactions occurring provide a rich vein of evolutionary possibilities that have led to the development of a (probably) large number of potential pest control products, most of which undoubtedly remain to be discovered. In their natural setting such metabolic pathways will have served some adaptive function or have evolved to exploit some resource, and for this reason we should always be cautious of extracting particular processes or products out of context for use in pest management. Furthermore it should not be assumed that all products will be suitable for use in organic systems. This is especially true of those that display broad spectrum biocidal properties, as their potential to negatively affect the farm agroecosystem will be similar to conventional pesticides. Notwithstanding this, a number of microbial or microbially derived products have gained widespread use in organic systems.

Bacillus thuringiensis (Bt)

Bt was discovered at the turn of the twentieth century in Japan, and is perhaps one of the most widely used biological control agents, both in organic and conventional farming. Several strains of Bt have now been developed to control different pests, but the most common use in organic agriculture is the control of caterpillar pests. The strain of Bt used for this purpose is *Bacillus thuringiensis* spp *kurstaki*. Other strains and/or species are marketed for control of mosquito larvae, Colorado potato beetle and beetle larvae.

Bt for caterpillar control is available in a variety of formulations but is generally used as a spray made up from a spray concentrate or granules. These contain the bacteria, but more often bacterial spores and/or toxic parasporal bodies that must be ingested by the caterpillar to have an effect. For this to occur the spray mixture must be applied at an adequate dose to the surfaces on which the caterpillars are actively feeding. Younger caterpillars are generally more susceptible, or at least require a smaller dose for an effect, than larger caterpillars. Once it enters the caterpillar gut the high pH ruptures the spores causing the release of a toxin that in turn ruptures the lining of the insect's gut. Once this occurs larvae become inactive, stop feeding, and may regurgitate liquid. Inside the caterpillar the spores pass through the infected gut into the insect's body cavity and blood, where they germinate and attack the internal body tissues and organs. New spores and parasporal bodies are formed which are released back into the environment, either as liquid discharges from the mouth and anus, or when the corpse eventually rots and deteriorates.

Unfortunately it appears that natural transmission of the disease in the wild is not very effective and so repeat sprays are likely to be necessary. This is because the spores rapidly lose their viability in the bright sunlight and the dry conditions found on leaves (the bacteria is originally a soil-dwelling organism where this is not such a great problem).

Bt is considered safe to apply up to the point of harvest and has minimal side effects on non-target species. In the UK, Bt is applied widely to brassica crops against caterpillars both in conventional and organic systems. No derogation is required for applying Bt in organic systems in the UK. As it is classed as a pesticide it should be used according to the directions on the label, and not used for purposes not authorized on the label. The correct strain should be chosen to control the pest identified, and it should be applied when pests are still small. For this reason crops need to be monitored, and multiple applications may need to be made for a continuing pest problem.

Bt is effective against cabbage caterpillars.

Manufacturers recommend repeating the treatment every seven to ten days during the period the caterpillars are present and hatching at levels at which the crop is under threat. If heavy rain falls before larvae have had an opportunity to ingest the spray it may be necessary to repeat the application sooner. They also usually recommend avoiding periods of high light intensity (sunny weather!), and state that the spray works best in the temperature range 10–20°C. It should not be sprayed on to wet leaves or if rain is expected within six hours of treatment.

There is increasing concern over pests developing resistance to Bt as it has become widely used in all forms of agriculture. For instance it appears that resistant strains of diamondback moth, an important pest of brassicas, have developed. This is likely to be exacerbated as GM crops engineered to produce the Bt toxin become more widely adopted, further increasing selection pressure on pests to develop resistance.

Spinosad

Spinosyns, of which the commercial product spinosad is composed, are fermentation products produced by actinomycetes (soil bacteria such as microorganisms) of the genus *Saccharopolyspora spinosaone*. Once ingested, spinosad causes rapid and continuous excitation of the insect nervous system, which leads to death. It is effective against a wide range of common pests such as fruit flies, caterpillars, leaf miners and thrips. On the down side it has some, albeit less, effect against beneficials such as ladybirds, lacewings, spiders and predatory mites.

For the purposes of organic application it should be considered to have a broad spectrum of activity, at least against certain groups of insects, but this includes many which contain beneficals, and the product should therefore be used with care in organic systems. On the positive side, spinosad quickly degrades in sunlight and the conditions normally existing in crop foliage. On the negative, there have been concerns raised about its toxicity to honey bees, and it is also moderately toxic to fish and aquatic invertebrates as it degrades more slowly in water; it is even slightly toxic to birds. In the UK, permission to use this substance in organic systems will only be granted as a last resort, under derogation, and this should include a strategy outlining how damage to beneficial insects might be minimized: for instance, applying it late in the evening may be one method of limiting damage to beneficials such as bees. As a commercial pesticide it should always be used in accordance with the label directions.

Viruses

Although virus diseases have potential for use against pests, or even pathogens, no commercial applications have been developed for use in organic field crops. They break down rapidly in UV light, and even formulations with UV sun screens have not been able to overcome this limitation in a satisfactory way to date. Some granulosis and nuclear polyhedrosis viruses have been developed, and are successfully used in tree crops, both commercial orchards and forestry plantations. Organic farmers and growers are most likely to encounter a granulosis virus used against the codling moth (*Cydia pomonella*) in apple orchards. Research may lead to future virus biological control agents becoming available as they are generally very host specific and are expected to have few non-intended side effects.

BOTANICAL PESTICIDES

There are many natural plant extracts that have some sort of pesticidal action on herbivores or pathogens, and the ecological basis for this has been explored in previous chapters (especially Chapter 2). Based on this knowledge, many botanical pesticides that have an action against pests and diseases have been isolated, produced and marketed. They include chemical compounds, essential oils and other mixed ingredients which are normally extracted from plants and purified in some manner to obtain the natural active ingredient. This is then usually formulated in a way in which it can be delivered to the crop, often as a dust or spray, with the aim of killing or modifying the behaviour of the pest or pathogen.

Indeed, many of the so-called 'natural' synthetic pesticides – for example pyrethroids – are based on plant-derived molecules. It should be borne in mind that many such products are the end result of a long chain of research aimed at making them more efficient but therefore usually more toxic, more broad spectrum and more persistent, and that this makes them ineligible for use in organic systems. At the same time it also needs to be appreciated that the original plant-defence molecules function in a specific context and work through complex mechanisms, often in combination with other chemical or physical defences, and therefore it should not be expected that they will have any dramatic effect when taken out of this context and placed in a different environment at different concentrations.

The same caveats apply to botanical pesticides as to other pesticidal products: that is, in most countries, for a substance to be marketed as a pesticide, it must have been approved by a competent authority. This process will approve a product not only as to its safety, but it will also test its efficacy as a pesticide under farming conditions. This approval is an expensive process, however, and for many natural products with a small potential market this cost cannot be justified, meaning that many are not in fact marketed as pest- or disease-control products but as amendments or fertilizers with some other mode of action (*see* below). The regulations do, however, differ between countries, and therefore the range of botanical pesticide products available to organic farmers and growers will vary between countries. Specialist advice will need to be sought if there is doubt as to the use of some of these products. All the products mentioned below have been used in organic systems, but for many of them their use will be restricted by control bodies which should be consulted as a final arbiter.

Pyrethrum

Pyrethrum is a natural insecticide extracted from species of *Chrysanthemum*, notably *Tanacetum (Chrysanthemum) cinerariifolium* and *T. coccineum*. The active ingredient is concentrated in the flowers and seeds, which are pulverized to extract the pyrethrins, a mixture of six or so insecticidal compounds. The active ingredient is formulated in various ways for sale, often as either a dust or a spray solution. Pyrethrins are toxic to all insects, causing a rapid knock-down as they excite and paralyse the nervous system. At lower doses insects can recover from this, but higher doses will kill them. In some cases synergists are added to the formulation, which increases the toxicity and prevents recovery from even low doses. Unfortunately some of these synergists, including piperonyl butoxide, one of the more common, are not suitable for use in organic farming systems in some countries. Where this is the case, care should be taken not to apply products with these compounds, but specifically to choose one that is approved for use in organic systems.

Natural pyrethrum is a non-persistent insecticide that rapidly biodegrades, especially in sunlight. It is, however, a fairly broad spectrum insecticide, so does also have some action against beneficial insects; it should therefore not be used on a routine basis. It has a very low mammalian toxicity but is toxic to fish, so care must be taken not to let it run off into waterways. This pesticide has gained formal approval in many countries, including the UK, and can be used in organic systems under restricted use to control aphids, blackfly, greenfly, caterpillars and whitefly, among other pests, on organic vegetables. A derogation must be sought from the control body before applying it, and it can only be used in accordance with the label instructions by law. In the case of organic farming there should also be a management plan to outline how damage to pollinators can be minimized by, for instance, spraying at times when these are not active, or by avoiding spraying when crop or flower strips are in flower.

Apart from 'natural' pyrethrins, 'synthetic' pyrethrins have subsequently been produced and developed by larger commercial concerns. These are more toxic and persistent in the environment, and are not suitable for use in organic farming systems.

Rotenone

Rotenone is the active ingredient extracted from the roots of several tropical and subtropical species including *Derris* spp., *Lonchocarpus* spp. and *Tephrosia* spp. It has long been used as a 'natural' pesticide and as a fish poison. The active ingredient, when available, is formulated in different ways, but normally as a dust or spray concentrate. Rotenone is a contact and stomach poison that inhibits cellular respiration in insects, rapidly causing them to stop feeding and eventually die. It has a broad spectrum of activity and has been used to control aphids, sawflies, caterpillars and a range of other insect pests. It is moderately toxic to mammals and very toxic to fish. It is non-persistent in the environment, rapidly degrading in sunlight and soil.

The use of this insecticide in organic systems is currently problematic and subject to review because of concerns regarding its effect on mammal brain tissues and fish, and it has been removed from the approved list of products in many countries. In the EU, for example, it has recently (2008) been withdrawn for use in all plant protection products and is no longer available for this purpose.

Neem

Neem products originate from the neem tree (*Azadirachta indica*), a tropical tree native to south-east Asia. It is most commonly used in the tropical countries where it grows, for health remedies and for pest control: for instance, dried leaves are added to storage products, or the seeds are crushed into a powder and sprinkled on to the plants. Although a large number of chemical products derived from this tree claim insecticidal or disease-control properties, the active ingredients of the most commonly cited group are azadiractins. These have been shown to be repellent to, and a feeding deterrent to insects. Other more direct effects include the suppression of egg hatching and the disruption of ecdysis (moulting between larval stages). These chemicals have also been shown to have antimicrobial properties and under some conditions to be capable of controlling plant diseases, with fungicidal action against black spot, powdery mildew, anthracnose and rust being claimed at various times. Trials have shown that nematodes are also controlled by neem applied to the soil under some conditions.

Neem can therefore be a highly effective insecticide that acts against a wide range of insects including biting and sucking insects such as aphids and caterpillars, and can also be potentially used as a fungicide or nematicide. It breaks down readily within a few days especially in bright sunlight, but is capable of being absorbed into the plant in some cases, and may concentrate in growing tissue. As it acts as both an antifeedant and an insect repellent, it should cause less harm to beneficial insects. Some suppliers import neem products into European countries either as oil extracted from the seeds or as neem cake, the remaining pulp after the oil has been extracted. The oil can be sprayed on to plants in a diluted form (although it can be difficult to dissolve), and the cake can be added to the soil against nematodes.

It should be recognized that in many cases neem is not approved as a plant protection agent, partially because it has proved difficult to extract a unique active ingredient, but also because patent applications have been seen as problematic in what is an indigenous plant in many countries. For instance, neem is not currently approved by CRD in the UK, so cannot be sold as a pesticide in the UK. In the EU's review of plant protection active substances, products containing azadiractin were also not approved. It does, however, have plant stimulatory effects (for obvious reasons), and is sold as a proprietary plant stimulant or even organic fertilizer in some cases.

Growers should be clear about why they are buying and using such products, and should cost them according to their effectiveness in increasing yield in the same way as they would any other organic amendment. Control bodies are likely to need a justification for applying such organic amendments.

Other Botanical Pesticides

Historically a range of other botanical pesticides has been used in crop protection, especially before the widespread use of synthetic pesticides. Some of these are still available for use in some countries and under some conditions, although many have been withdrawn from use, either because they are not commercially viable or because of health and safety concerns; these include the insecticides described below.

Quassia

Quassia is extracted from the roots and bark of the South American tree, *Quassia amara*, and latterly from *Picrasma quassioides* in cooler climates. It is a water-soluble contact and stomach poison, effective against aphids and caterpillars but relatively safe towards some beneficials such as ladybirds and their larvae; however, it can have a detrimental effect on the larvae of hoverflies. The extract is one of the most bitter substances available and so should not be applied too shortly before harvesting. It is not clear if commercial extracts of quassia are available, although bark chips can be obtained in many countries and the active ingredient extracted by steeping in water and boiling. Any legal requirements for the use of quassia as a pesticide should be observed; it has not been approved under the EU review of plant protection substances, so is not available for use in this region.

Nicotine

Nicotine extracts from tobacco plants contain a range of active ingredients of which the principal one is nicotine. It has a contact action that will rapidly paralyse an insect's nervous system, and can be used as a fumigant, dust or spray-based extract. It has a broad-spectrum activity against insects, including many beneficials such as bees, which, coupled with a very high mammalian toxicity, make it unsuitable for use in organic systems. In fact it has largely been withdrawn from insecticidal use except in traditional agricultural systems, and is not commercially available in most countries.

Ryanodine and Sabadilla

Ryanodine and sabadilla are both botanical insecticides that were commonly used but are no longer widely available as commercial formulations. Ryania is extracted from *Ryania speciosa*, a woody tropical plant, and is a slow-acting stomach poison which causes insects to stop feeding soon after ingestion. It has been most used against caterpillars and other chewing insects. Although it is not especially harmful to beneficials and/or natural enemies, it has moderate toxicity to mammals, and it can be more persistent as compared to the other botanical pesticides.

Sabadilla (an alkaloid) is produced from the seeds of the tropical lily species *Schoenocaulon officinale*. It is a broad-spectrum contact insecticide that also works when ingested. It has been produced in various formulations for use against thrips, caterpillars and leaf hoppers, among other insect pests. Like other botanical insecticides, it causes loss of nerve function, paralysis and death. Sabadilla breaks down rapidly in sunlight and air but is highly toxic to bees so should be used with care in organic systems, in the unlikely event that it is available for use.

ESSENTIAL PLANT OILS AND PLANT EXTRACTS

In common with botanical pesticides, essential plant oils and plant extracts are obtained from plants. Unlike botanical pesticides they are not normally applied to crops with the express purpose of causing mortality of the pest or pathogen, although this might be a side effect of their use, either directly or indirectly. Rather, they are applied to crops for their stimulatory growth effects, or for their pest and/or disease repellent effects – and by this count neem could also be listed in this category (*see* above). They have many and varied modes of action, and many have actually been the focus of research as to their pesticidal properties, but have generally not been marketed as pesticides at the end of the day. There is potentially a long list of amendments and plant oils, and here we concentrate on those that might be available to commercial growers. As previously stated, farmers and growers need to be clear about why they are using such products, and weigh any benefits against costs. In the case of essential oils their high costs are likely to preclude their use as pest management tools.

Garlic

Garlic has a long folk tradition as an insect and disease repellent, and research has at one time or another also demonstrated antibacterial, anti-fungal, amoebicidal and/or insecticidal qualities. The active ingredients are sulphurous compounds including diallyl disulfide and trisulfide with insecticidal properties although the principal mode of action is probably one of repellancy, and this can be effective against a range of pests including carrot fly and cabbage root fly. Another active ingredient is allicin, which has antibacterial and anti-fungal properties. Garlic has also been shown to be an effective nematicide and has been demonstrated to repel larger pests successfully, including rabbits, moles and deer under some conditions. Although in the past preparations suffered from inconsistent action between batches, manufacturers now claim to have developed a more consistent supply of garlic extract and this should not be such a problem.

Garlic granules have been used effectively against cabbage root fly.

Garlic preparations are most commonly sold as granules or as dilutable sprays. Timing of application relative to the stage of pest infestation can be critical, and this may explain some of the variability in efficacy observed in trials. It is generally believed that it is important to apply it before pest build-up, as its mode of action is to deter pests from seeking out the crop, but it will not cause them to leave it once they are there already. Although garlic formulations kill pest insects, nematodes and some pathogens, they could potentially also kill beneficial insects and microbes and thus are not recommended as an all-purpose spray for outdoor use. Granules are to be preferred in this respect as they can be accurately placed close to the area in which they are to act, normally the soil around the base of plants.

Product availability and approval for use is likely to depend on regional pesticide and/or plant protection legislation as well as on control body interpretation of organic principles and laws. In the UK one product is currently available as an approved pesticide and others are available in other European countries (Norway, Denmark) for control of nematodes and/or cabbage root fly. The active ingredients are included in the Annex 1 of the EU regulation 91/44 and are thus approved for use in organic farming systems.

Essential Plant Oils

These are normally understood to be volatile oils that are produced and stored in plants, often in specialized glands, and which, when released, spread and evaporate to give the characteristic smell associated with many plants. Their function is not always known but many are presumed to have some defensive purpose against both pests and diseases. Although many plants have these essential oils, they are characteristic of herbs, which are well known for their culinary qualities, health effects and/or fragrance. These alternative markets are likely to make their use as pest repellents or plant disease curative agents too costly for practical use. Some of those materials that have been demonstrated as effective in this respect are briefly summarized here below.

Citronella

An extract from citronella grasses (different species of *Cymbopogon* and also lemon grass), this oil has been demonstrated to have repellent effects against insects and mites. It also has mild antiseptic properties and is used in soaps for this purpose. Most uses of citronella are domestic, and they are unlikely to be cost effective in commercial systems.

Citrus Oils

Derived from bi-products of the citrus industry, citrus oils have been developed for use as plant amendments that enhance and fortify the plant cuticle and have anti-evaporative properties, which may help to protect crop plants from pathogen infection in some conditions. Some oils (limonene and linalool) are also insecticidal. Some commercial formulations are available.

Herb Extracts

These include the oils of plants such as thyme, rosemary, lavender, cloves and peppermint, to name a few. At various times and in different combinations they have been shown to have insect repellent properties and/or protective properties against pathogen attack. Most have too high a value to be of use as cost-effective pest-management tools, and are in any case unlikely to work effectively on a practical field scale, although some, for example mint oil, have been approved for use in the EU.

Pine oil: A large number of pine oils exist, many with mild antiseptic and/or insect repellent properties. They are often used in soaps or other cleaning products. Although displaying some potential, they are not currently allowed under the EU review of plant protection products, although they might be available in other regions.

Pepper extracts: Plants of the pepper family contain a range of active chemicals but the most well known is capsaicin and closely related waxy molecules. They can be formulated to be absorbed by soft-bodied insects (such as aphids, thrips, mites, white flies) where they stimulate the nervous system and repel them from plants, although high doses can kill the insects. They also have antifungal properties. In common with garlic, pepper extracts could also potentially deter mammalian herbivores from feeding on crops.

Other Extracts

A range of other extracts that give some protective effects to plants has been commercialized or produced. For instance, bioflavinoids act to boost the immune system of plants and encourage them to produce phytoalexins, which help to protect against pest and disease infection. Various other plant extracts have similarly been marketed from time to time, but all should be used with an understanding of what is to be achieved, and in accordance with specialist advice.

ORGANIC AMENDMENTS

Apart from those from plants, extracts from other organic sources have also been used to promote crop growth by farmers and growers. Principal among those used in organic farming systems are both compost and seaweed extracts. These have been used as plant stimulatory products with an indirect, and beneficial, effect on pest and disease incidence in crops. Obviously their use as amendments compliments, and indeed may be synonymous with, their use in soil management practices as described in Chapter 3. Other organic amendments include the use of green and animal manures, and the value of these has also been described in previous chapters as part of proactive soil management programmes which also act to manage pests and diseases.

Seaweed Extracts

Seaweed extracts have long been used in organic farming for their plant stimulatory effects. Their general mode of action is to boost plant growth and stimulate the plants' own defence systems. Research has shown that seaweeds contain phytoalexins and other plant growth hormones that directly boost plant growth, and has indicated, but not definitively confirmed, that seaweed and other chemicals act to protect against pathogens and/or interrupt the reproductive cycles of some insects and repel others. Some of the effects observed may also be due to the seaweed providing additional micronutrients that boost soil microbial activity and/or crop plant growth.

Many commercial formulations of seaweed extracts exist, most as spray concentrates that need to be diluted for application. Some dust formulations also exist which might be more efficient for application to soil in limited areas. Because of the potential increase in the demand for such products from all farmers and growers there have recently been moves to restrict their use in organic systems. In the UK only sustainably harvested seaweed can be used, amid growing evidence of the value of seaweed as a habitat, both underwater and on the shoreline. Dried seaweed or seaweed that is washed up or collected can be used, but calcified seaweed that is collected by dredging is not permitted.

Compost Teas

Much work has been done on the use of compost teas as agents for disease control. The principle behind their use is that they add microorganisms to the crop environment, and these act as antagonists against pathogens. There are four methods by which this may occur. The first is competition for resources, such as seed, root or leaf exudates which may act as a food source for the pathogen and from which the competitive organisms exclude them. The second is the production of antibiotics effective against pathogens; *Pseudomonas* spp. are most effective in this respect. Thirdly, some organisms may actually parasitize the pathogen: *Trichoderma* is one such example of a parasite that has been trialled successfully against a number of diseases (*see* previous section). The fourth is the induction of the plant's own systemic defence system by the microorganisms in a way similar to the effect of seaweed.

Compost teas are made by two main methods: either an extraction method which is fairly rapid, or a brewing method which takes at least twenty-four hours. Whichever method is used, the quality of the tea depends very much on the quality of the compost being used to make it. The compost should have reached at least 55°C to ensure that the major pathogens are inactivated and/or broken down, although some diseases such as club root and tobacco mosaic virus might survive at much higher temperatures. Using a PAS100 certified source will minimize the chances of disease transfer.

Whilst compost teas can undoubtedly have a beneficial effect against plant diseases in some circumstances, these effects can be difficult to replicate with any consistency, especially in the field. Much of this variability is due to the variation in the feedstocks used to make the compost tea, but with increasingly sophisticated controls over the composting and extraction process, more consistent effects may be able to be generated in the future.

Other Amendments

Various organic amendments derived from other organic materials have at one time or another been used to stimulate plant growth, bolster plant defences, and/or promote soil biological activity. Some of these amendments function as they break down in the soil releasing products that directly affect pest or disease organisms. For instance when mustards are ploughed in as a green manure, they break down to release isothiocyanate which effectively fumigates the soil, reducing or suppressing various soil-borne pests and diseases. Other brassica crops used as amendments can release similar chemicals with similar effects.

Organic amendments can also act indirectly to bolster plant defences by stimulating soil biological activity which, acting through soil microorganisms and roots, can make plants more resistant to attack. The mechanisms by which this is achieved have been discussed in previous chapters.

PHYSICAL AGENTS

Some materials can act physically on pests or pathogens and in some way prevent them functioning normally. Many of these agents are said to have a physical mode of action, and many are only selective in the way they are used or delivered, so care should be taken to avoid harming populations of natural enemies.

Soft (Fatty Acid) Soaps

Soft soaps have long been used as pest control agents in organic farming systems. The active ingredient is potassium fatty acid soap, in the best cases derived from natural and sustainable plant oils. It works by disrupting the waxy cuticle of insects and can also affect cell membranes; however some research has indicated there might be additional toxic effects. It has most effect on soft-bodied insects such as aphids, thrips, whitefly and spider mites, and much less effect on many beneficial insects that have hard bodies such as ladybirds, ground beetles and bees, especially if applied at times when they are not active.

Soft soap is most often used against aphids.

It is important to spray at the correct concentration (2 per cent), and it must actually make contact with the insects for it to be effective. Like other organic insecticides, it breaks down very quickly especially in sunlight, and is not persistent in the environment. However, because of its mode of action, over-application may result in phytotoxic effects such as tip burn and leaf scorch in some crops, and a few plants should be tested before treating the whole crop if in doubt. Its use may be restricted in some countries like the UK where a derogation must be sought before application, and where its use is currently being monitored to ensure that it is being used correctly. It also has ambiguous status under the EU review of plant protection products, where it was listed as not approved although this may be subject to change.

Oils

Rape seed, fine mineral and/or other vegetable oils have been marketed for their insect repellent and physical mortality effects. Depending on how they are used, they can have various modes of action, principally against insect pests. The main mode of action is to disrupt the outer layer of the insect's cuticle (itself a thin waxy layer), which affects the water-regulating and/or respiratory function of the cuticle. In this respect oils also prevent the exchange of gas through egg membranes and so are quite effective against insect eggs, and may also suppress fungal spore germination. They also block insect spiracles, again preventing the insect breathing and interfering with water regulation. Oils can also affect the feeding behaviour of insects, creating an unfavourable environment which inhibits the efficient functioning of mouthparts. As plant cuticles also contain similar waxy layers, care should be taken to prevent damage to plants by excessive application.

Oils are generally applied either directly to the pest on the plant during the growing season, or to the overwintering stage of the pest. They generally work best on small, relatively immobile and/or soft-bodied insects such as aphids, scale insects and whiteflies, or life-style stages, especially eggs, and may temporarily induce fungistatis (or interrupt development) in fungal spores. They are relatively non-persistent, and for this reason can have minimal impact on beneficials if used wisely and not applied when natural enemies are active.

Dusts and Fine Abrasive Powders

Kaolin clay, diatomaceous earths, boric acid crystals and silica gel have all been used to control insect pests, although the two latter are not acceptable for use in organic systems. They function either as abrasives, disrupting the outer waxy layer of the cuticle, or to clog spiracles and exoskeleton joints, hindering respiration and movement. They may also cause excessive grooming, as insects try to dislodge particles. They may also act as dehumidifiers, drying out insects or other small organisms, and are in any case more likely to be effective against small, soft-bodied insects or mites. Although of low environmental toxicity they have not been popular as control agents in field crops as the fine dust is not easy to distribute efficiently, and is anyway quickly dispersed in the environment, although some products have been formulated to be sprayed on. Some products have more use in protected cropping.

More recently a range of products that acts in a similar way but can also be applied as sprays has been developed. They are generally based on glucose polymers (starch) and vegetable oils, and clog or block insect spiracles as they dry out, causing respiratory failure. Once again they are only likely to be selective when applied to avoid natural enemy activity, but are recommended for use against a variety of pests including whitefly, thrips, aphids and spider mites. Such products may need a derogation in some countries or regions (such as the UK).

INORGANIC PESTICIDES

The use of inorganic (and normally historic) pesticides is gradually being phased out of organic farming practice, despite some sectors (principally those involved in fruit and potato production) arguing that they would not be able to operate without them. Although their use has been justifiable in the past, it is clearly becoming less so as their various environmental effects are becoming better understood. Many are in any case unlikely to be registered for use in the future as the market is small and the patents long expired. Like other materials in this section they should only be used as a last resort when other cultural or biological methods have failed and the crop is at risk of severe economic damage.

Copper

The fungicidal properties of Bordeaux mixture, a mixture of copper sulphate and hydrated lime, were discovered in the nineteenth century. There are a number of similar copper formulations (for example Burgundy mixture) generally applied as preventative fungicide sprays against diseases such as potato blight or downy mildew. The solution dries on the leaves of the crop plants and gradually releases soluble copper which prevents the fungal spores from germinating. Although copper is a naturally occurring element in the soil, and may be deficient in some soils, there are concerns that accumulation of copper in high concentrations may have an adverse effect on earthworms and other soil microflora. Copper can also be toxic to fish and other aquatic organisms. This is less likely to be a problem in potato crops which are grown in a rotation, but it has been shown to accumulate to levels affecting soil biota in perennial crops such as orchards or vines. It is likely to be disruptive of natural antagonistic microbial communities on leaves, depriving plants of this natural biological control.

In the UK, copper compounds can currently be applied in organic systems as a fungicide under derogation as copper sulphate, copper ammonium carbonate or copper oxychloride. The majority of the copper used

The prevention of potato blight accounts for most copper use in organic field crops.

(around 80 per cent) is applied to potato crops to control blight. The remainder is used against apple scab in orchards and some onion crops against downy mildew. The total amount that can be applied in a single season is restricted to 6kg/ha at the time of writing, and this limit may be lowered with a view to eventually phasing it out within a practical time scale, although some ongoing work suggests that much lower doses than those currently used can still be effective. In some other countries such as the Netherlands, its use is already prohibited.

Sulphur

Sulphur is most commonly applied in top fruit orchards against fungal diseases such as apple scab, but has been applied to field crops to limit fungal diseases, and in some combinations can also have insecticide properties as well. Normally applied as a spray, it has also been supplied as a dust and as sulphur candles for fumigating protected crops. It is also a plant nutrient, and since the Clean Air Act, with less dependency on burning coal, sulphur levels in the air have decreased, and some soils may actually be deficient in this element and this can have an adverse effect on many crops including cereals, oilseed rape, brassicas, peas and grass.

It has been shown to harm some types of natural enemies such as parasitic wasps and mite predators depending on how it is used. It is, however, a natural substance occurring in the soil and is thought to pose little risk to mammals, soil life or the environment. As a pesticide it is subject to the use specified on the label, and may also be subject to restriction by certification bodies. For instance in the UK the Soil Association has recently changed its standards to require a derogation for applying sulphur. This also applies to soil deficiencies, and to apply sulphur to this end, a derogation must be approved, and a soil analysis must be provided to show that levels in the soil are deficient.

Iron Phosphate

Relatively recently, iron phosphate has been approved for use in organic systems as a molluscicide. The active ingredient is contained in pellet form mixed with a bait, and this is scattered evenly around the crop in a similar manner to conventional slug pellets. It acts both as an antifeedant and a toxin to slugs and remains active for a few weeks even in rainfall. Slugs which eat the bait become immobile and die.

As a registered molluscide it should be applied as instructed on the label at the recommended application rate. Organic control bodies may require that a maximum dose is not exceeded in a single season, and that there is a plan for avoiding repeated use in the longer term, and to that effect in the UK a derogation should be sought for its use. Although it breaks down into a harmless substance in the soil (iron and phosphate) there is some question as to the effect of the carrier chemicals and bait, with some research suggesting that they are harmful to soil fauna, including earthworms. In any case, and as with other chemical controls, emphasis should be placed on systems-based cultural controls such as cultivations, traps and barriers before resorting to this chemical measure.

Potassium Bicarbonate

Potassium bicarbonate has only relatively recently been approved by CRD for use as a fungicide in the UK, but it has been in use in other countries for a longer period. It is effective against a wide range of fungal pathogens, including powdery mildew and blackspot. Its mode of action is to inhibit the growth of fungal spores, and it may have further effects, such as collapsing hyphal walls and altering pH to prevent germination of spores.

As it is essentially similar to domestic baking soda, it has very low toxicity, and is a food grade material. It should be applied with a wetting agent to be most effective, allowing it to act as a contact fungicide by damaging the cell wall membrane of fungal spores. Although regional rules as to its use may vary, it can only be applied under derogation in organic systems in the UK, and it must be shown to be a replacement for copper or sulphur, not used in addition to it.

ORGANIC SYNTHETIC PESTICIDES

Organic synthetic pesticides are not normally allowed in organic farming systems. They include a wide range of pesticides (insecticides, acaricides) and fungicides such as organophosphates, carbamates, synthetic pyrethroids and neonicitinoids. They generally disrupt pest nervous systems or pathogen metabolic processes, and often have a broad range of activity against these organisms. They are not therefore particularly selective unless at least some effort is made to target them. This is one of the principal reasons they are prohibited in organic farming systems, as their capacity to cause long-term unintended negative side effects outweighs any benefits they bring in the short term.

6

The Monitoring and Economics of Pest and Disease Management

(with Ulrich Schmutz)

In the previous chapters we have discussed the reasons for pest and disease attacks to crops, and the measures, both preventative and adaptive, that organic farmers and growers can use to reduce their severity and/or avoid them. However, there is a crucial step that we have not discussed, and that is in assessing and understanding the risks and probability of pests and diseases causing significant damage in any particular crop or farm situation. This is important because although there will always be some level of pests and/or pathogens present in the crop, the farmer or grower will need to evaluate the level of damage at which it is worth taking preventative action. In most, if not all, cases this is likely to be a value judgement depending on the circumstances prevailing in any particular farm system. However, that judgement can be taken based on a sound understanding of the risks and probability of pest and disease damage in any particular crop or season. In this chapter we discuss the range of tools, such as monitoring, forecasting and sampling, available to farmers and growers to support them in this decision-making process.

As part of the actual decision to manage or control a particular pest, farmers and growers will also need to factor in the likely costs of any control measures. On the one side is the cost associated with any damage that is likely to occur if no action is taken (the loss of income), and on the other the costs associated with taking action (the costs of the control measure). Normally the cost of control should not exceed the cost of damage, although there are some situations in which high short-term costs might be incurred to reduce long-term risks (and hence costs) across the rotational cycle. In this chapter we therefore also discuss the techniques available to evaluate the financial costs associated with pest and disease control measures.

DAMAGE AND ECONOMIC THRESHOLDS

The status of organic crops with regard to pest and disease in any situation depends on a number of factors and is to some extent subjective. It will depend on the abundance of the pest or disease-causing organism (how many are there? how many plants have been attacked?) as well as the amount of injury and the lasting damage they are causing. Injury is defined as the actual effect that pests or pathogens have on the crop plant. Damage is a measurable loss due to the injury and is normally measured as loss of yield or quality to the harvestable crop. It is important to realize that even visible injury to a crop may not in fact cause any, or very little, measurable damage (for example, some caterpillars on swede leaves). On the other hand, it can be difficult to spot injury – as with a virus or soil-feeding insects – although significant damage is in fact occurring. Monitoring and sampling crops is important in order to spot and estimate damage, and this is discussed in more detail below.

Damage

Damage is, ultimately, the important measure for farmers and growers. Damage is often proportional to the number of pests or pathogens present, in which case the more that are present, the greater the damage done. It is, however, sometimes difficult to directly relate the extent of injury to crop plants with the damage that will be caused, as the injury occurs in a dynamic system in which biotic and abiotic factors can intervene. For example, it is well known that many crop plants can tolerate and compensate for a degree of injury without loss of yield, and some varieties are bred for this characteristic. On the other hand, some herbivores are capable of transmitting pathogens, which will have an effect over and above that of the feeding damage (which may be literally a pin prick). In all cases there will be a pest or pathogen presence or population level above which measurable damage will occur to the crop. This injury threshold level will vary considerably between crops and circumstances: for instance, the injury threshold for damage is usually much lower where the pest or pathogen attacks the parts of the plant that are harvested and sold (such as caterpillar damage on cabbage heads) as compared to when they attack other plant parts (such as aphid damage on carrot leaves). Some injury (for example cutworm attack) is proportionally more serious because the whole (in this case small) plant is killed and no compensation is possible.

Economic Damage

Although biological injury thresholds such as this can be useful, most crops are in fact marketed and/or have a market value, and so the important measure becomes the 'economic injury threshold'. This is based on the market value of the crop as compared to the damage resulting from pest or disease attack, and is defined as the pest density at which the damage or loss caused by the pest or pathogen equals the value of any possible control measures against it. The economic injury threshold arises from both the price and the market standard demanded of the produce, and is directly linked to them. For instance vegetables are often sold washed and packaged, and consumer tolerance of blemishes or feeding marks is very low, leading to very high market standards. In this case, even though pest damage – for example, carrot fly feeding tunnels – may be insignificant in terms of actual lost yield weight, the actual loss may be total because the crop will be rejected for marketing if (say) more than 5 per cent of carrots have any feeding tunnels.

Much research work has gone into defining economic thresholds for triggering control measures in crops against pests and diseases. This is normally defined as the level at which control measures need to be taken to prevent the pest or pathogen reaching economically injurious levels. Unfortunately economic thresholds, although a rational basis for pest and disease management decisions, are not easy to define and use, especially in organic farm situations. In fact they have been best defined for insecticide and fungicide application in conventional farming systems where the costs associated with control are discreet and easily calculable and the short-term effects measurable. Even in this situation they only apply to one, or at best a few, pests and/or diseases in any particular crop, where the costs of damage have been defined with some degree of accuracy.

The use of thresholds in organic systems is much more problematic for various reasons, but mainly due to the complexity inherent in these systems. Possibly the most serious shortcoming for organic farmers and growers is

Leaf injury by caterpillars – will it cause economic damage to yield?

that it is much more difficult to measure the direct costs of pest and disease control, which are often a property of the organic farming system as a whole. As will be seen, this makes it difficult to apportion costs to pest and/or disease management on which the calculation of economic thresholds depends, as costs are spread over the whole rotation and between crops. A further limitation to calculating economic thresholds is that they are normally only practically calculable for a single pest or disease attacking the crop, and become very difficult to estimate when a pest or disease complex is present, as the effects of multiple pests and diseases tend to be interactive rather than additive. Some of these interactions are likely to be more complex on organic farms and once again vary over the rotation.

Organic growers are also more likely to face a wider range of market conditions, and market standards and prices will vary with marketing arrangements. For instance box scheme customers are likely to be more tolerant of some damage than supermarket customers, meaning that thresholds have to be defined for different market systems.

Finally, crop injury and damage will vary depending on environmental conditions and may not be the same, differing, for example, between seasons, on differing soil types, and with the intensity of rainfall. All these factors are likely to be more influential on organic farms.

Pest and Disease Status

Notwithstanding the difficulties in defining meaningful economic action thresholds for pests and diseases in organic systems, it is possible to categorize pests so as to make decisions about their importance, and prioritize management decisions. Such categorization is a broad brush approach and needs to be tempered with the experience of the farmer and/or grower on a particular farm, and the use of the various monitoring and forecasting tools discussed later, so as to ensure that the risk of damage is kept low whilst the costs of control measures are also kept in bounds. Farm records on pest and disease attack and management will also be indispensable in the long run to home in on the best and most cost-effective strategies in any particular situation.

To this end, pests and diseases can be broadly classified in terms of the risk of economic damage to the crop as follows:

Non-economic: Not strictly speaking pests or diseases as they do not cause any economic damage to the crop at any time, although they may cause some injury. Their presence may need to be monitored to ensure that they do not become pests and diseases.

Occasional: Pests or diseases that occasionally cause economic damage, in some seasons, and which may in some cases be predictable (for example downy mildew in onions or thrips in leeks). In this case management or control measures can be taken when outbreaks are expected, and direct control measures are likely to be economically effective against these pests. Routine direct control measures will probably not be economic unless very cheap methods are available, or the cost of damage is very high.

Regular: Those pests and diseases that regularly cause economic damage to a crop, in most seasons. Cultural control methods will generally be the most cost-effective measures, reinforced by direct control measures where necessary. In this case it is often necessary to think about management methods that prevent the pest or disease reaching the economic threshold, and to use them routinely or when models or experience indicate levels of high risk (for example cabbage caterpillars or lettuce aphids). Moderate to high management costs would be expected to still be profitable as long as the crop is of high value, or crops can be sold in markets where standards can be relaxed.

Severe: Pests and diseases that always cause economic damage in every season will need to be managed at all times if a crop is to be grown profitably (for example late blight in potato). High management costs can be expected, which may make the crop marginal in terms of profitability. Cultural management techniques are likely to be necessary to reduce costs, as direct intervention at all times is likely to be costly so the value of the crop will need to be very high to maintain profitability. In some cases organic farmers and growers may not have sufficient tools or funds available to control severe pests or diseases, and there are recorded cases of them avoiding growing such crops as management costs mount up.

Whilst there are no easy answers to categorizing pest or disease risk to organic farm systems, there is an increasing amount of information available about them, coupled to sophisticated systems of predicting and monitoring their presence. When combined with knowledge about particular farm systems, these provide farmers and growers with a great deal of information about the best pest and disease management techniques to use.

MONITORING AND SAMPLING

Monitoring

Effective management of pests and diseases in organic systems depends on effective monitoring programmes and taking time to implement them, even under admittedly busy farm working conditions. Monitoring is the process of observing and sampling pests and diseases in crops within the farm system over the season and between seasons. Normally such a process would involve setting a schedule for walking crops and recording information about – in this case – pests and diseases, through some sort of sampling procedure. The point of the monitoring process is to enable management decisions to be made: either to treat a pest or disease in some manner, or to take no action until the next crop walk or sample. In some cases sampling may indicate that no action is necessary but that the crop needs to be sampled more frequently in order to keep an eye on a developing situation. When carried on between seasons, and with suitable evaluation, a picture can be built up indicating which are the most severe pests and diseases, under which conditions they are likely to become damaging, and what action might be effective at any particular part of the cropping cycle (*see* above).

A monitoring programme or schedule is only as good as the sampling methods used to observe and detect pests and diseases within the crop. These methods should satisfy three criteria:

- they should detect the pest or disease population sufficiently accurately;
- they should be easy to carry out (taking as short a time as possible);
- it should be possible to relate the result of sampling to the risk of damage.

Many sampling regimes fall short of this ideal, as all this information is not always available. The third requirement is particularly difficult to meet in that a lot of research work is often needed to relate population samples to actual damage done, which is in itself dependent on the socio-economic situation and management objectives of the farmer or grower in a specific context.

The relationship between pest and disease infestation and the damage done is often complex, crucially depending on crop, time and place as well as a range of other factors. Often damage at the vegetative stage and the beginning of the season is less likely to affect yield later in the season where the harvested part is grain or tubers. On the other hand if the leaves are to be harvested, early attack may be serious. Most plants recover from low or moderate pest attack. Even disease is often more prevalent on older leaves or later in the season, and trimming may suffice to remove direct damage. Exceptions include virus transmission, where even low numbers of infected and mobile pests can cause disproportionate damage. In any case, the complexity of the situation underlines the importance of a systematic monitoring programme which can be used to build up a picture of the pattern of pest and disease attack in the farm system, and can inform future management decisions.

Butterfly eggs found on cabbage during monitoring and sampling.

Sampling

Sampling for pests and diseases has been the subject of a great deal of research, much of it outside the scope of this book. Suffice it to say that most practical field sampling methods will rely on assessing crop plants *in situ* and will involve either directly counting the number of pests and/or diseases, or assessing the injury due to them. The former relies on direct counts of pest and disease individuals, or more often, colonies – for example aphid colonies on leaves, or necrotic (disease) spots on leaves. The latter relies on an estimate of the amount of damage caused to the plant by the pest or disease (such as defoliation due to caterpillars, or leaf area rotted by disease). In some cases keys have been produced which allow an accurate assessment of damaged areas by comparison with standard charts or pictures, although these are sometimes difficult to come by. They are particularly useful for use in assessing disease incidence in crops, and standardizing estimates of disease incidence between sampling dates.

It is generally better to take many samples divided into easily distinguishable categories, rather than a few samples with all individuals counted – for instance the number of aphid colonies per plant, rather than the number of aphids per plant. In this respect counts of pests may need to be made on plants, in plants or in soil around plants. It is even possible to mark and recapture individuals to get an estimate of a pest population, but farmers are unlikely to have the time to do this. When it comes to damage, symptoms might need to be assessed, such as plant stunting (for example in cabbage root fly attack to young brassicas) or degree of damage (for instance slight, moderate or severe tunnelling by carrot fly).

In each case the sample unit should be defined so that the individuals or damage is assessed in a defined sample. The sample may be small (say, a leaf), intermediate (a whole-crop plant) or large (say, 5m of the crop row). The sample unit and the number of samples needed is generally chosen to reflect the amount of work needed to get a meaningful answer for the crop and/or pest/disease combination in question, and in particular for the precision and accuracy required. For instance, counting the aphid colonies of ten plants in a small field may be sufficient to get a good idea of the infestation across the field, whereas fifty plants may need to be sampled in a large field. For practical purposes it may also be expedient to sample for all pests and pathogens at the same time so as to avoid excessive duplication of work. Sampling may also involve destruction of the crop plant, for instance if it needs to be uprooted to evaluate root growth and any damage. This is obviously more acceptable where a large number of plants are present.

Increasingly, new sampling methods are being developed which are aimed at reducing the amount of time a farmer or grower needs to devote to sampling. Most of these involve trapping pest insects, and these can be used in and around the crop at various times. Traps include light traps, baited traps, sticky traps, pheromone traps and suction traps. Some traps can even be used to capture fungal spores, but are unlikely to be practical at farm level. The disadvantage of such methods is that they are often fairly specific to the pest (or where they are not, they catch everything!) so miss others in the crop, and it is sometimes difficult to relate the actual catches to the number of pests in the crop. Although they are unlikely to replace crop walking, they are extremely useful in providing information about when pests are moving around the crop, and from this point of view demand serious attention. Crop walking may also provide supplementary information on field conditions or natural enemies, which might not be available with traps.

It is important to take care to avoid bias in sampling. People are very good at noticing anomalies and picking out differences without realizing it, and worry can lead to people taking a conservative approach and over-anticipating problems. There is a tendency to notice the novel and ignore the common, so that people search harder for pests or diseases when they are rare or more interesting. Similarly there is a tendency to search more for damage when it is rare, and to become accustomed to it and underrate it when it is familiar. Bias can therefore cause a pest or disease infestation to appear much worse than it is, or vice versa so that inappropriate action is taken.

The normal method of overcoming bias is to take random samples. Strictly speaking, all the sampling units should be divided up and chosen at random using a random number table, but this is impractical in field situations. Instead, sampling procedures normally involve either taking a random number of steps on a predetermined path across a field (normally a W zigzag across the field), where the number of steps can be determined from the random function of many cheap calculators; or throwing a quadrate blind (by shutting the eyes, turning round a few times, and throwing it over the shoulder) to determine the next sampling station (although both these methods are open to subtle biases).

Yellow sticky trap used for sampling carrot flies (among other pests).

Damage Assessment

Once samples are obtained it is necessary to interpret the observations to make an assessment of the damage. Even if actual individuals have been counted in samples, some evaluation will need to be made as to how severe an infestation is, or how much damage is being caused. This might need a simple analysis and categorization of samples. For instance, if aphid colonies have been counted, an average number of infested plants and an average number of colonies per plant can be calculated. Further categorization may be possible at this point, with aphids being (say) categorized as not present, some colonies in the field, many colonies, or all plants with colonies. In such circumstances some colonies may indicate a watchful eye needed, many colonies might indicate rapid direct action needed, whereas if all plants are heavily infested no further action can meaningfully be taken. In a similar manner disease severity can be evaluated across the crop, with leaf area and number of plants infected summed up across the field.

It is best to assess damage at the plant stage when action can be taken or effective decisions can be made. This will to some extent be dependent on the stage of the plant at which attack is occurring, and when it is likely to be more serious. For instance, flea beetle attack is usually more serious (and visible) on small brassica plants, whilst wireworm damage in potatoes is more serious and obvious close to harvest. In the latter case it would be more prudent to try and sample wireworm damage by burying pieces of potato before planting the crop, because there is little that can be done at the point of harvest. In some cases laboratory counts might be necessary, but field counts will always be more practical and useful to farmers and growers.

It should be borne in mind that infestations are often localized in patches or concentrated at the edges of fields – for instance on the upwind side of fields in the case of wind-borne infestations – and this should be taken into account when assessing any likely damage to a crop. Also infestation is likely to vary over time, depending on conditions: for example, disease epidemics may develop, spread and get worse in wet, humid conditions, but could slow down or even disappear in drier, less humid conditions. In some cases if natural enemies have also been counted, this can modify the type of action that may be taken, such as avoiding the indiscriminate application of soft soap, which might harm them.

Forecasting

Forecasting can provide a valuable adjunct to monitoring crops for pests and diseases. Forecasting services aim to make predictions about pest or disease activity at any point in the season based on inputs of climate data and real-time pest or disease monitoring. Most rely on mathematical models of pest and pathogen life cycles, which at best can predict when peaks of activity of population are likely to occur. To be locally relevant they usually need to incorporate local weather information; their accuracy is then improved by taking into account the results of trapping or other observational information, and obviously any on-farm monitoring will be useful in this respect.

Although it is possible to run forecasting models on the types of computer that most farmers and growers would normally have access to, they are not usually user friendly, and most forecasting services are normally run by centralized research or extension services. Forecasting services of this type exist in many countries. In the UK these are available for a range of pest and diseases, relying on regional weather information and a national trapping network to produce a regional forecast. The best developed cover a number of pests including currant lettuce aphid (*Nasanovia ribisnigri*), potato aphid (*Macrosiphum euphorbia*) and peach potato aphid (*Myzus persicae*). Other pests covered by the forecasting include cabbage butterflies, flea beetles as well as cabbage root fly (*Delia radicum*) and carrot fly (*Psila rosae*).

The predictions (*see* Table 5 for an example for a typical season) allow growers to target appropriate management methods with some confidence, for instance avoiding sowing at peak infestation periods, or applying crop covers. The models are most developed in the case of the root flies, which rely only on air and soil temperature to make a prediction. In the case of aphids the information is extensively backed up by trap data from a national suction trap network, which provides additional confidence. This national service also records the presence of a large number of other aphid pest species on a (near) real-time basis.

Forecasting services for plant disease are less well developed, in the UK at least, although in many cases the predictive models have been well developed by research. This has not translated into practical forecasts for farm use except in the case of potato late blight. In this case 'Smith periods' – periods favourable to blight sporulation and spread – are monitored on a regional basis, and farmers and growers can be alerted when they occur. Coupled with real-time monitoring reports from potato growers, this allows the progress of the disease to be charted during the season, and may allow organic growers to take precautionary measures, or at least assess the likelihood of their crop being infested. Other disease forecasts include the brassica leaf diseases ringspot (*Mycosphaerella brassicicola*) and dark leaf spot (*Alternaria brassicae*), and onion diseases, but it is not clear how useful these are to organic growers.

Table 5: Predictive Models for Brassica Infection with Various Pests During a Twelve-Month Period (darker shade = higher risk)

	J	F	M	A	M	J	J	A	S	O	N	D
Cabbage root fly												
Carrot fly												
Cabbage aphid												
Peach potato aphid												
Flea beetle												
Large cabbage white												
Small cabbage white												

COSTS OF PEST AND DISEASE MANAGEMENT

Once a farmer or grower has identified a pest or disease and has determined that it is likely to cause economic damage, they need to identify the relevant control measures to take, and examine the costs of taking them. The kinds of methods and options available to organic farmers and growers have been extensively discussed in the previous chapters, and in this section we will discuss the costing of these options. It is obviously difficult to cover all costs, but the intention is to give farmers and growers a pointer to the types of cost that might be involved, and then to discuss the kinds of decision that may need to be made in the light of this.

Types of Pest and Disease Management Costs

The costs encountered by farmers and growers in managing pests and diseases can be categorized into either the direct costs of the control technique or the indirect costs. Broadly speaking the former are those costs directly linked to taking management action against a pest or disease, whilst the latter are not, although it can sometimes be hard to distinguish between the two.

Direct Costs

Direct costs are the easiest costs to understand and calculate as they refer to the costs of an easily definable pest or disease control treatment such as applying a fleece to protect brassicas, anti-bird netting and/or soft soap against aphids. Prominent among the direct costs are the costs of the material used, for example a fleece or a soap treatment. In addition to the material costs there are likely to be handling and/or application costs for the treatment or material, which can also be directly attributed to the control action. These could be machinery costs such as the use of tractors or sprayers, or labour costs such as manual labour to install or apply the control treatment. In many cases direct costs will be due to a combination of both machinery and manual labour.

Flea beetle damage with (left) and without (right) mesh – a case where it is relatively easy to calculate direct costs.

Treatments are not usually a 'one off' procedure and isolated from other farm factors, and this often complicates their calculation. This is because it is necessary to allocate a portion of machinery or labour costs to the treatment, as the treatment only occupies a portion of the total labour and equipment input of the farm. In this case the cost should also include some element of maintenance or repair costs. In the case of equipment, these costs are normally described by calculating the proportion of the equipment's working time spent on the control measure, and then allocating this proportion of its total running costs to the measure. In the case of machinery the annual running cost is calculated from the original purchase price of the equipment, taking into account any depreciation, and the ongoing maintenance and repair costs. Equipment purchase and depreciation costs are often subject to many complex tax and accountancy rules, and in some cases it might be best to seek help from a farm accountant and/or adviser to best allocate these costs.

Similar reasoning obviously applies to the labour costs as well, which need to be allocated on the basis of a proportion of the working time spent on a task. When labour costs are considered, the time devoted to monitoring, sampling, crop walking and listening to forecasts, and the time spent making and discussing management decisions on pest and disease control, should not be underestimated. Although these factors are often not included in the cost calculation as they are not recorded and are difficult to allocate from the general management time, they can in some cases be considerable. Even the costs of attending open days or farm walks could sensibly be allocated, at least in part, to these labour costs, and any 'paid for' advice should also be costed. In these cases it is likely that different people will do these tasks, in which case costs have to be allocated from different salary or consultancy budget lines. In fact the labour costs may well be the highest cost in many pest and disease control programmes, and it is always worth trying to calculate the proportion of time spent on tasks and the cost to the farm business.

Although the costs of equipment and labour are often not easy to calculate, even a rough estimate is better than a wild guess or, worse still, ignoring them completely, when trying to judge the value of a pest or disease control measure to the farm business. In many cases they will be higher than perhaps expected, and this will set the context for taking future pest and disease management measures. On the whole it is better to find out that they are in fact trivial, as perhaps originally suspected, than to be faced with an unexpected expense or deficit at the end of the year. Cases in which these costs can be small include, for example, when equipment has a long working but low maintenance lifetime, and/or the treatment is very quick and easy to apply as part of the farm routine. On the other hand costs are likely to be high when, for example, a treatment needs to be applied many times by hand, or constantly adjusted, and/or involves a large amount of bought-in material. In some cases specialist services such as biological controls will also incur relatively high costs. In all cases, understanding the costs enables more realistic decisions to be made about pest and disease management on the holding, and is the key to reducing them in future seasons.

Many larger farms sidestep some of these issues in that they contract in discreet pest and disease management services; these can be simply allocated to the farm budget as contracted services involving a 'one off' payment or service agreement. If available, these services can be a way of reducing costs as, for example, labour and equipment can be provided by the contractor and does not need to appear as such on the farm books. Similarly paying for advice from an adviser is a 'one off' payment, and the adviser is responsible for many of the tasks, such as collecting and processing information – although in this case, a farmer or grower would be well advised to do some of this themselves.

Conventional farmers also often have access to 'free' advice through agronomists who take a commission on pesticides and fertilizers sold to the farm, but this route is not generally available to organic farmers and growers who are much more likely to need to pay for advice. Whilst there is some scope for contracted pest and disease management services on larger organic farms selling into wholesale or supermarket chains, medium- to small-scale enterprises are much more likely to be faced with taking on these costs themselves, as contractors will not be able to factor in the economies of scale that they themselves need in order to operate their businesses. This may change in future if cooperative working arrangements and machinery rings become more popular with organic farmers and growers.

The separation of direct costs into equipment, repair, machinery and labour is also useful when assessing both social and ecological sustainability on the farm. For instance, calculating carbon footprints is akin to economic accounting, and costs can often be carried between the two forms of accounting. For example, equipment and machinery have mainly fossil fuel-based inputs, while labour has little or none, and low repair and maintenance ratios favour a smaller carbon footprint – and having already separated out these factors for costing purposes, the calculation of the footprint is made easier.

Indirect Costs

The indirect costs are more difficult to grasp and to calculate, but once again it is important to recognize that they do exist and they need to be allocated, within reason and where possible. This is not always easy, as many of the preventative pest and disease management approaches described in the previous chapters are in many ways a property of the whole organic farm system. Moreover indirect costs may not, strictly speaking, appear on the budget sheet of the farm business, but need to be taken into account in order to judge the value of any particular pest and disease management approach. Some examples will serve to underline the difficulties of calculating and using these costs in organic farm management.

Many indirect costs in organic farming systems are fundamentally preventative costs – that is, the costs associated with having a particular rotation and growing certain varieties among others. Organic standards require wide and balanced rotations for various other good reasons beyond pest and disease control: fertility building, reducing nutrient leaching, and balancing the economic risk of narrow cropping. Only part of these costs will be due to the necessities of pest and disease management, and they will be almost impossible to allocate on the budget sheet. For example, it is difficult to cost the limit to crop choice in rotation design aimed at preventing severe and regular pests and diseases from causing damage. It is in any case questionable to account for the forgone profits for not growing a certain crop in a narrower rotation, as the alternatives are unclear or indefinable in the sense that, in some ways, the system you have is the system you have.

Further indirect costs might also be borne in the variety choice. For instance, a disease-tolerant or resistant variety may be more difficult to source as organically certified seed. Also, due to the fact that it is propagated on a smaller scale the variety may be more expensive and may not yield as highly as a more popular variety under little pest and disease pressure, or the market price of the alternative variety may not be so high. However, once again it is questionable or even impossible to account for the foregone profits. In both these cases although the costs may not appear on the balance sheet at the end of the day, they are hidden costs that organic farmers and growers have to bear, and are one of the reasons that higher premiums are required on organic produce, as they increase the cost of production.

Examples of indirect costs that may be more readily costed, but are still likely to be problematic, include, for instance, damage due to the treatment such as compaction in wheelings or the temporary reduction in the presence of beneficial species. Once again, although it might be difficult to put an exact figure on some of these effects, it is important to put a figure to them so as to account at least for the magnitude of the effect, even if it cannot be exactly measured. For instance compaction is well known to reduce yield in some cases, although wheelings may also open up the crop canopy to light and air, thereby increasing yield.

Similarly, while indirect costs for cover crops which can serve to control nematodes, or mustards which can be used as a natural bio-fumigant to control diseases such as *Verticillium* wilt in strawberry, can be more clearly accounted for as pest and disease control interventions, even they have many other benefits such as soil cover, weed suppression, better microclimate and protecting nitrogen from being leached away. Therefore it is difficult to quantify the effects exactly and then allocate the proportion to a disease control intervention.

This underlines the point that some indirect costs may also be beneficial and that they may, in the end, level each other out. In fact, lacking detailed information, it is often assumed that costs (seeds and cultivation) and overall benefits for the rotation at least level themselves out, and such interventions can be considered as cost neutral. Although this might be a reasonable assumption, and these costs may be difficult or next to impossible to allocate to the farm balance sheet, exploring the possibilities and weighing the effects will at least help to place pest and disease management decisions in context, and may help to prevent unexpected cost increases. In other words forewarned is forearmed. For example, if a treatment were to remove the natural biological controls on a key pest in a crop, how much would it cost to replace this with bought-in expertise and materials?

Variable Costs

Taken together, direct and/or indirect pest and disease control costs are classed as variable costs, because they vary with the degree and size of treatment. Other important variable costs in organic farming and growing are, for example, weed control, seed and transplants, and harvesting costs. In calculating the total costs of running the general farm business the variable costs are added to the fixed and overhead costs. The value of the variable costs is that it allows individual management categories or activities to be compared, and also allows the proportion of the business costs allocated to that particular activity to be estimated.

Table 6 provides an example of the variable costs of an average of more than twenty of the most common horticultural crops grown in the UK (including beetroot, calabrese, different cabbages, carrots, cauliflowers and so on) taken from data presented in the published farm management handbooks for organic and non-organic production. Although there is obviously some variation from crop to crop, it is generally low and it is interesting to note that, on average, the variable cost of pest and disease control of field-scale vegetable crops is only 1 per cent of the other variable costs in organic production. Even together with weed control (which is mainly mechanical in organic systems), they are only 3 per cent in comparison to 8 per cent in non-organic, conventional cropping of the same crops.

This perhaps surprising data reflects the fact that organic growers use a lot less direct pest and disease control than conventional farmers, and that the indirect pest and disease control costs are probably much higher in

Table 6: Average Variable Costs from Twenty Field-Scale Horticultural Crops, and the Percentage of Pest and Disease Control of Total Variable Costs for Organic and Non-organic Production

Different Variable Costs as Percentage of Total Variable Costs	**Organic**	**Non-organic**
Seed, transplants	20%	12%
Fertilizers	1%	11%
Direct pest and disease control	1%	8%
Weed control	2%	in P&D control
Packing, drying and transport	27%	23%
Market commission	17%	13%
Casual labour	31%	33%

organic systems. This is not really unexpected because many direct interventions are not allowed under the organic standards and because indirect methods such as wider crop rotations, resistant varieties, and accepting lower yields are commonly used in organic production and also serve a wider pest and disease management role (as discussed in previous sections and chapters).

The relatively high proportion of seed and transplant costs in organic production at 20 per cent, as compared to 12 per cent in non-organic, confirms the importance of variety choice and quality in organic production as well as the popularity of raising transplants for organic crops, rather than planting directly, as a means of avoiding weed competition and early pest and disease pressure. In terms of pest and disease control it is also interesting to note from this comparison table that casual labour costs are very similar. This is caused by the relative high cost of hand planting and harvesting of specialist horticultural crops. The fact that organic pest and disease control (and weed control, for that matter) may require higher casual labour costs is not very significant as these additional labour costs are small in comparison to costs incurred at planting and harvesting.

An overall conclusion is that because direct pest and disease control costs are so low in comparison to other variable costs, few savings can be made by reducing them further. From a farm business point of view, it is therefore likely to be more beneficial to concentrate on reducing costs in other farm operations. In economic terms, expenses in pest and disease control are like an 'insurance policy' because regular and severe pest and disease incidents can have such a devastating effect on yield, so it is better to err on the side of caution when implementing direct pest and disease management methods. In effect it is better to incur a small cost to prevent a pest or disease problem and to avoid a potentially large cost – that is, complete yield loss due to pest or disease.

However, in ecological and sustainability terms, it is not possible to completely dismiss the costs associated with pest and disease management. After all many, especially the indirect, do not fully appear on the farm account sheet, but still play a significant part in pest and disease management on the holding and actually serve to hold down the direct costs in this area. For instance excessive use of soft soap, although apparently economically rational, is likely to lead to a drop off in natural enemies and natural biological control, with the consequent need to either spray more or reintroduce biological controls, both costly.

Similarly there is still every point in maintaining as broad a rotation as possible, and using resistant varieties where available, as these have the potential to reduce longer term costs by building an ecologically sustainable system, whose main pest and disease management costs are indeed indirect, but effective, whatever the short-term economic imperative to maximize yield with short rotations. Shortening the rotation is likely to increase direct pest control costs down the line, as soil-borne diseases or other problems increase, necessitating costs in rectifying the situation; using susceptible varieties is likely to have the same effect. In fact this is one of the main challenges of organic and agroecological farming: balancing the short-term financial problems of running a viable business with the longer-term necessity of having a truly ecologically and socially sustainable farming system.

Growing Costs: Fleece and Polytunnels

One area of rising costs in temperate organic farming systems is the use of soil covers such as fleece and mesh, or completely protected cropping as exemplified by polytunnels. They serve as a good example of the complexities of costing pest and disease control practices in organic farming systems because although, in some cases, they can be considered a direct pest and disease control cost, they also have several other beneficial effects such as improved microclimate, protection from wind and flood erosion, and provision of a more controlled harvesting and general plant growth environment, amongst others.

In calculating the pest and disease control costs it is necessary to allocate a certain percentage of the total fleecing and/or polytunnel costs to pest and disease management. In order to allocate or proportion the costs it is, firstly, important to calculate and understand the total cost of the fleece or polytunnel to avoid missing any big cost contributions. The costs of fleece and polytunnels consist of, as described above, the direct costs of the material, and as they can be reused, the repair and maintenance ratio. The third major cost factor is hand labour for laying and removing fleece or constructing the tunnel. Other costs are likely to include labour for supervision during their seasonal use, and the time and resources used to store them off season, mainly in winter.

An example calculation has been made for fleece from information derived from long-term research work carried out by HDRA/Garden Organic on commercial organic farms. The calculation illustrated in Table 7 assumes that one hectare is used, of which 90 per cent of the area is fleece and the rest headlands or pathways. In this case the fleece is used for two years (at a material cost of 3.5p per m^2), but only for one crop per cropping year. In this example the fleece was removed manually twice for weeding operations, and then finally prior to harvest.

Table 7: Example Calculation of Different Variable Costs for Fleece

Fleece				**£/ha**
Fleece (18g/m^2)	£315	2 years	depreciation	£158
Fleece repair ratio	10% of annual costs			£16
Fleece laying (labour)	*20 h/ha*		*7 £/h*	*£140*
Fleece maintenance (labour)	*10 h/ha*		*7 £/h*	*£70*
Fleece removing 3-times (labour)	*40 h/ha*		*7 £/h*	*£280*
Fleece storage (labour)	*5 h/ha*		*7 £/h*	*£35*
Fleece storage (space)			*7 £/h*	£50
Sum (£ per ha)				**£748**
	Labour costs of total costs:			*70%*

Labour costs, as might be expected, are the bulk of costs incurred, and in this case make up a not unrealistic 70 per cent of the total costs. Using machinery to lay fleece may reduce those costs, however other handling costs for maintenance and removing cannot be avoided easily. In fact in ongoing work it has been observed that protection costs for fleece can vary widely from £100 to £1,000 per hectare, depending on the time the crop is covered as well as the area covered. Fleece can be moved from one field to another, and may be used on several crops in a season, and this can help mitigate costs as the depreciation, repair and storage costs can obviously be 'shared' by the different crops. As a final consideration, expressing the costs in terms of cost per hectare will also help in comparisons of costs between different techniques and across different management methods.

In this case, the proportion of the cost allocated to pest and disease management would depend entirely on the circumstances. For instance, a fleece laid to prevent cabbage root flies or cabbage white butterflies accessing the crop might be put down entirely to crop protection. In contrast, a fleece applied early in the spring to protect and warm carrots but which also prevented carrot flies attacking the crop through the spring might have a quarter or even a half of its costs allocated to pest control, depending on the intention of laying the fleece and the practical experience of the grower.

Understanding these costs raises the possibilities for exploring other options to reduce them. In this case other possibilities could include using mesh, which is initially more expensive then fleece, but lasts longer (five to ten years). Mesh has a lesser effect on crop advancement or frost protection and therefore can be considered mainly as a pest and disease expense. The mesh size is chosen depending on the pest targeted: for example 1.3mm excludes cabbage root fly, carrot fly, aphids and many caterpillars; smaller sizes (0.8mm) will exclude flea beetle or thrips (0.27mm). In this

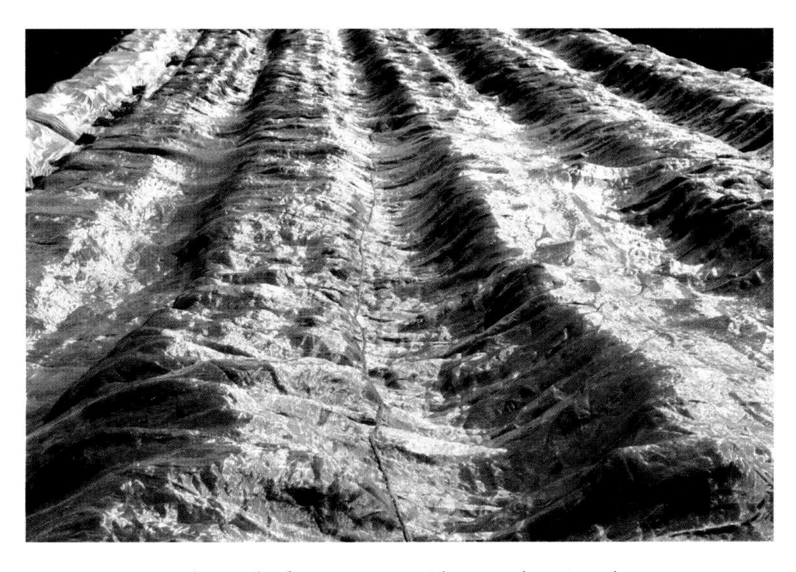

Fleece on carrots – an increasingly common cost in organic enterprises.

case it can be argued that a more general size mesh is more cost effective as it can be used on more crops, but specialist size meshes to control one pest per year can also pay off. As with fleece, most of the cost of using mesh is likely to be in the labour of laying and removing the cover, and other costs are likely to be comparatively small.

In Table 8 we give an example of a polytunnel costed for organic strawberry production, but in principle the calculation will be similar for any horticultural crop. The data are per hectare (ha), assuming tunnels cover 85 per cent of the area. The remainder is left for headlands and biodiversity. Data are only given as an example to show the type of calculation to be made, and the details are likely to vary in individual cases.

For polytunnels a depreciation period of fifteen years is used, and for the polythene itself a three-year period. The labour time (hours per hectare) and casual labour costs (£ per hour) are calculated on the assumption that the tunnels have to be taken down each season to avoid the negative impact on the landscape, and on tourist and amenity value, although this is less likely to be done in vegetable production systems. The various labour inputs are for construction, maintenance, venting, dismantling and storage, and it is important to record all the labour involved to assess if the whole enterprise can actually pay back all its cost from increased yields and improved harvesting conditions. In our example labour costs are almost half of the total cost and therefore important to monitor closely.

From a pest and disease management point of view the proportion of the costs allocated to pest and disease management also depends on the circumstances. In this case it is likely that this allocation would only be a very small proportion of the total tunnel cost, as the overwhelming driver for such protected cropping is to extend the season or provide suitable (hot) conditions for certain crops, rather than protect the crops against pest and disease. Once again, the proportion actually put down to pest and

Table 8: Example Calculation of Variable Costs for (Spanish) Polytunnels

Polytunnels				**£/ha**
Tunnel frame	£20,000	15 years	depreciation	£1,333
Polythene replacement (15% of tunnel)	£3,000	3 years	depreciation	£1,000
Tunnel repair ratio	10% of annual costs			£133
Tunnel construction (labour)	*150 h/ha*		*7 £/h*	*£1,050*
Tunnel maintenance (labour)	*20 h/ha*		*7 £/h*	*£138*
Tunnel venting (labour)	*50 h/ha*		*7 £/h*	*£350*
Tunnel dismantling (labour)	*120 h/ha*		*7 £/h*	*£840*
Tunnel storage (labour)	*10 h/ha*		*7 £/h*	*£70*
Tunnel storage (labour)				£100
Sum (£ per ha)				**£5,015**
			Labour costs of total costs:	*49%*

What proportion of costs, if any, is attributable to a polytunnel? (Courtesy Maggie Haynes)

disease management would depend on the intention and experience of the grower. It is interesting to note that at certain times of year and with certain crops it might be necessary to introduce biological controls into the tunnel, and/or plant them with some flowers to encourage natural enemies – both increasing the costs for this method of production, and which would have to be added to the costs of the enterprise in question.

Environmental and Other Subsidies

The difficulty of defining or even calculating the indirect costs of pest and disease management raises the interesting possibility of counting them against income gained from environmental subsidies. In this case this is the effective use of government environmental subsidies to, in effect, pay for the indirect and difficult-to-cost aspects of pest and disease management. In a sense this is a payment made by the government as a social good to address a market failure – that is, to protect biodiversity, which has no adequate market cost.

Many of these recoverable subsidies have been discussed under conservation biological control in Chapter 4 and will not be discussed in detail here, except to say that the details for such schemes are likely to vary between countries. In the UK the Environmental Stewardship Schemes have already been mentioned, as have the possibilities for payment under the Organic Entry Level Scheme and Higher Level Schemes. Under these schemes many elements of function diversity are eligible for points under the agreements to pay for stewardship. As already mentioned, features beneficial to pest and disease management include beetle banks, nectar flower mixes and margin management schemes, all of which are eligible for points.

Apart from environmental subsidies there is also the possibility to use crops and other plants for both a yield and a pest and disease management purpose, normally as a natural enemy attractant, where the yield obtained subsidizes the pest management function of the plant. For example, coriander can be cropped until it goes to flower, at which point it can be left to attract natural enemies. Similar uses could be found for crops such as nasturtium, calendula and borage, all flowers that can be included in salad bags.

ECOLOGICAL AND CARBON FOOTPRINTING

Ecological and carbon footprinting are becoming alternative methods of describing environmental sustainability in farming systems, including organic ones. In the former case the footprint measures the amount of resources that a given farm operation or process requires, and the latter measures the amount of CO_2 or climate change gas equivalent that a given farm operation or process emits. Both methods are in some sense an attempt to make sustainable practice more definable and accountable, and also to counterbalance the persuasive influence of balance sheet economics over sustainable farm practice. However they are calculated, and definitions and standards are still evolving, they at least provide a measure of performance and, like costs, can be used to benchmark and share best practice between farms.

It is not the place of this book to rigorously define or describe ecological and carbon footprinting except to say that pest and disease management practices are likely to have a footprint attached to them. Obviously a farm operation should aim to reduce and live within its footprint if it is to become ecologically sustainable, and pest and disease management options should also be examined from this point of view as they are being costed. Organic farms are in many ways already better placed to reduce their footprints in this respect as they do not use synthetic pesticides which are produced using a large fossil fuel input. However, other farm practices used for pest and disease control, such as tillage and netting, are usually no better than conventional systems. Fleece and mesh, for example, are produced from fossil fuels using similar energy-intensive processes to pesticides, and their use should not be adopted without a critical analysis.

For a guide to reducing footprints the ecological mantra of 'reduce, reuse and recycle' is likely to be the best guide to farm practice, and many good sources of information now exist. For a low ecological and carbon footprint it is better to try and avoid the use of material produced in a patently unsustainable way. However, this is likely to be hard to achieve given the current supply chains, and it is more likely to be a case of reducing dependence on bought-in products where possible. Reuse is also important when some use is unavoidable, and in this sense it pays to buy good quality material and equipment. For example good quality fleece, mesh or plastic can be used for longer periods (for example five years for fleece or twenty years for polytunnels) and their costs can also be spread over a longer period of time. In many cases the main costs and inputs to crop protection practices are provided by manual labour, in which case the potential for reducing environmental footprint is unlikely to be as low as for other farm operations which require a greater input of fossil fuel or other resources.

In some circumstances it may also be possible to intensify production and in effect reduce ecological or carbon footprints as resources are used much more efficiently in smaller areas. For example, in a carbon footprinting calculation the achieved higher yields per unit of land can easily compensate for the carbon emissions of the crop protection practices, and avoiding the transport of alternative crops from other climates can compensate for the use of crop covers and polytunnels. In the end, many of the issues are still to be defined by farmers, growers, consumers and academics, and it will be some time before clear principles can be laid down and applied – although this does not preclude farmers and growers taking the initiative with the best information currently available.

DECISION MAKING

At some point a farmer or grower will need to make a decision about the control measures it is necessary to take against a specific pest or disease. In an ideal world, they will have determined what the pest or disease is, what damage it is likely to do, what control measures are likely to be effective at that stage of the crop cycle, how much they are likely to cost, and most importantly if the control method is likely to reward the costs incurred by higher yields or better quality. Making an active decision to do nothing (and let biodiversity and predators do their job) might not always be the wrong decision.

Up to this point this book has described these factors in general terms to give a broad overview of the decision-making process. The subsequent chapters (Chapters 7 to 9) describe the key pests and diseases likely to be encountered in temperate European organic agricultural systems, along with some of the key management options for them. The final chapter (Chapter 10) synthesizes this knowledge and discusses the practical implementation of organic pest and disease management programmes.

7

Pest Insect Management Strategies in Crops and Systems

Pest insects comprise the largest single group of pest species, and the damage they do to organic crops is only rivalled, as a group, by the diseases (*see* Chapter 9). Although crops can potentially be attacked by a wide range of pest insects, many crops are only attacked by a few key insect pest species, and a few years of observation will probably enable a farmer or grower to become acquainted with the most common and the most damaging. This chapter provides an overview of the insect pests most likely to be found in organic field crops, together with pointers as to best practice for their management in organic systems. The severity of pest attack and the response will depend on many factors, which have been widely discussed in this book to this point, and this chapter is, to some extent, a synthesis of this knowledge as concerns dealing with insect pests.

Some insect pests are specific to certain crops or crop families, whilst others attack a wider range of crops. In this chapter, insect pest management is discussed in broad outline with reference to the main crop types. Key pests and management strategies are discussed in the general text, which should be read in tandem with the tables (Tables 9 to 17) that give specific details for the more common and damaging pests in the most widely grown crops (or crop types) in temperate organic systems. Some of the less common pest insects might not be covered, and more specialist texts should be consulted if this is suspected. In any case it should be borne in mind that pest attack and damage can be very specific to location and situation, and that farmers and growers should be prepared to obtain as much information as they can about pests from as many different sources as they can. This is likely to enable them to tailor their management strategies to their particular circumstances. Other pest groups such as nematodes and molluscs can potentially do considerable damage, and these are discussed in the following chapter (Chapter 8).

INSECT PESTS OF CROPS

Arable and vegetable crops are drawn from a range of plant families and species, and as might be expected, can potentially also support a wide range of different insect herbivores, some of which have the potential to be pests. Some pest insects confine their attention to a restricted group of crop species or families, the so-called specialist pests, whilst others are capable of attacking a broader range of crops; these are termed generalist pests.

Insect pest damage is generally more serious when the harvestable part of the crop plant is attacked and consumed, and less serious where other parts of the plant are attacked. In the latter case some compensation in yield can be expected, and the plant might recover to yield well, or even be more nutritious. In the former it is much less likely that marketable yield will be recovered. Consumer tolerance for insects or insect damage is also low where the harvested plant parts are eaten directly with minimum preparation, as in salads; it is likely to be more lax if the food is to be peeled and prepared for cooking. Injury to crops such as cereals that are processed before being sold may not come to the attention of the consumer at all. Tolerance for pest insect damage is also low where they transmit disease or allow entry of rots, which can cause damage disproportionate to the actual feeding injury.

Due to the potentially large number of different pest and crop combinations, the discussion of insect pest management is naturally limited in a book of this size. Although we (hopefully) provide a good overview of the issues in the text and tables, it is likely to be necessary to obtain further information on more detailed aspects of insect life cycles and management in field vegetables and arable crops from other sources, and to supplement this with on-farm observation in order to be able to better manage insect pests in any specific farm situation. We would encourage farmers and growers to take the time to do this research and make these observations, which will represent a valuable investment in their pest (and disease) management strategy.

INSECT PEST GROUPS AND SPECIES

Insects comprise the largest group of land-dwelling organisms by far, represented by well over a million different species in thirty-two family groups or orders – though fortunately the number of pest species is nowhere near this. The main pests of insects are found in six or so orders and are restricted to certain herbivorous or omnivorous families within these orders, and any farmer or grower can easily become familiar with them. These pest groups and families are described below, but a wealth of readily accessible information exists in books and on the internet (*see* Sources of Information), and these should be consulted to build a broader knowledge.

Orthoptera (Grasshoppers and Crickets)

Although many species are generalist herbivores, this order is not usually of great importance in temperate organic farming systems, although members of the order (for example locusts) are certainly well known in tropical and sub-tropical agriculture. Where they occur the life cycle is more or less simple, with the nymphs resembling the adult and gradually acquiring adult characteristics and sexual maturity with successive moults.

Dermaptera (Earwigs)

Earwigs are also not generally regarded as agricultural pests although they can be locally troublesome, especially in protected crops where they eat holes in leaves and young plants. On the other hand they are also valuable biological control agents, and consume other pests. Like grasshoppers, the nymphs develop into adults with successive moults, although the adults may show some degree of parental care for the newly hatched nymphs before they disperse.

Hemiptera (True Bugs)

This order contains many important pest families, and many are plant feeders as both nymphs and adults. They have highly adapted mouthparts (stylets) to suck plant juices. Their life cycles are more or less straightforward, with the nymph gradually turning into an adult through successive moults. There are numerous families, and they are divided into four suborders (of which the Heteroptera, Sternorrhyncha and Auchenorrhyncha contain important pest families and species). Many families and species have also evolved parthenogenesis and the females are able to give birth directly to young nymphs without fertilization; this means they can rapidly increase in numbers in favourable conditions. Many have also evolved the ability to exploit different host plants at different parts of their life cycle, with complex life strategies for dispersal and fertilization, thus avoiding natural enemies; this allows them to build up their numbers very rapidly on crops at certain times of year, until the natural enemies 'catch up'. The most important families and super families are as follows:

Miridae (capsid bugs): A group of small bugs, of which only a few species are likely to be encountered as pests, and some of which are important natural enemies.

Aphidoidea (aphids): These are ubiquitous on plants in temperate zones, and are probably the single most troublesome insect pest group. All crops are attacked by one or more aphid species (*see* Tables 9 to 17), which often form conspicuous colonies covered with sticky honeydew, frequently tended by ants. They are important vectors of plant viruses. A great deal of research work has been done on aphid behaviour and ecology, much of which cannot be detailed here but can be consulted for more information.

Capsid bug presence and damage on carrot.

Psylliodea (jumping plant lice): Small, plant-sucking insects that tend to be specialist pests on fruit or soft fruit, although some are found on vegetables or other crops.

Aleyrodioeae (white flies): Very small, waxy insects, noticeable for rising in clouds when crop plants are disturbed. They can be particularly troublesome in protected cropping, and are important virus vectors.

Coccoidea (scale insects and mealy bugs): Another family of insects covered in a waxy coating, which lead a rather sedentary life. They occasionally reach pest status in temperate areas, but rarely in field crops.

Cicadelloidea and Cercopoidea (leafhoppers and spittle bugs): These are abundant in grass and not normally considered pests. Some species produce copious froth or 'cuckoo spit', noticeable on weeds or plants in field margins.

Fulgoroidea (planthoppers): Feed on plant sap, causing damage when present in high numbers. They can also be vectors of plant viruses.

Thysanoptera (thrips)

Very small but abundant insects that can be found widely on vegetation and are especially noticeable in flowers. They suck the contents of plant cells with modified mouthparts or stylets, often leaving empty cells; this gives plant foliage a characteristic sheen, especially noticeable on leeks. The life cycle is more complex than the orders previously described, with the nymphs initially active and feeding but becoming quiescent in the third and fourth instars (stages), these often comprising the overwintering stage, either in soil, leaf litter or in suitable crevices in vegetation. The final instar is the adult with feathery wings; these can disperse over wide distances on the wind, often becoming concentrated in 'storm fronts' in summer; in this form they become annoying as they land everywhere and crawl into the smallest of spaces, earning themselves the nickname 'thunder flies'. The most common pest species includes onion thrips (*Thrips tabaci*) (*see* Table 15) and grain thrips (*Limothrips* spp.) (*see* Table 17).

Coleoptera (Beetles)

This is another large and diverse group of insects, many families or species of which are herbivorous and crop pests. It should be remembered that many of them are also effective generalist predators that need to be conserved in crops and fields. The pest species, as might be expected, form a diverse group of insects with different habits; the main pest families are as follows:

Carabidae (ground beetles): Mainly predacious beetles, and thus to be encouraged in organic farming systems (*see* Chapters 2 and 3). There are a few omnivorous and herbivorous species, but these are not generally regarded as agricultural pests although they have been recorded as damaging grasses such as cereals and soft fruits. In general the benefits of encouraging ground beetles will far outweigh any risk of pest damage.

Elateridae (click beetles/wireworms): Click beetle larvae are known as wireworms and can be very problematic locally in organic farming systems, especially in the cropping period after grass leys. Because of their importance a lot is known about their life cycles, behaviour and management, and this is discussed in more detail below.

Hydrophilidae (mud beetles): Drab brown beetles that often look muddy or dirty; they can be readily found in arable fields. They are not generally regarded as serious agricultural pests, but larvae have been recorded as causing damage in cereals, especially after grass leys, and in some species both adults and larvae attack turnips and other vegetable crops. In organic systems a good rotation and good husbandry practice would normally prevent any pest damage due to these beetles.

Nitidulidae (pollen beetles): Small insects that can easily be found in flowers during the hot summer months, especially those of brassica plants. They have been recorded as damaging seed set in brassica crops, but unless present in abnormally large numbers they probably do little damage in organic farm systems as long as plants are well established and growing vigorously in periods when the beetles are active.

Tenebrionidae (mealworms) and Bruchidae (seed beetles): Both these families are generally pests of stored products, and as such are just noted as potential pests. Infestations of stored products, especially of bruchids, often start in the field because the females lay eggs on the mature seeds (for example, bean seed beetles), which are then brought into store where the eggs hatch. Hygiene of both storage spaces and seed is usually the best means of protecting against damage from these pests.

Chrysomelidae (flea beetles, leaf beetles and relatives): This is a large family of often colourful beetles, which contains many herbivorous species. Both adults and larvae generally feed by chewing, and different species can be found specializing on different plant parts. Many feed on leaves, some mine in stems and leaves, and others feed below ground on roots. This family contains the famous Colorado potato beetle (*Leptinotarsa decemlineata*) as well as the flea beetles (*Phyllotreta* spp. and others), which can be very troublesome when establishing brassica crops (*see* Table 11). Many of the leaf beetles probably do little harm to vigorously growing crops, although some – such as the asparagus beetle (*Crioceris asparagi*) and mustard beetles (*Phaedon* spp.) – have been recorded as causing damage in some circumstances.

Curculionidae (weevils): A large group of species, many of which are pests as larvae and/or adults. Larvae often feed inside plant roots, stems or other tissues, and some adults specialize in feeding on seeds. They can be serious pests of peas (Table 14), clover (Table 14) and brassicas (Table 11).

Scarabidae (chafer beetles): Not normally regarded as pests unless they occur in grassland in very high numbers where the larvae feed on the roots of (mainly) grasses.

Lepidoptera (Butterflies and Moths)

A group of insects containing many crop pests, mainly in the larval or caterpillar stage of the life cycle. The adults often feed on nectar and actively seek out sites to lay their eggs, normally food plants. The eggs hatch to the larval stage, which is recognizable as caterpillars on many crops. The caterpillars pupate to become adults. Caterpillars are voracious feeders capable of defoliating entire plants, with their mouthparts adapted to biting and chewing. The families containing important pest species include:

Hepialidae (swift moths): Occasionally and locally important when they attack roots and other underground parts of plants.

Tineoidea (tineid moths): Inconspicuous moths with many important pests among them including the diamond back moth (*see* Table 11) and leek moth (*see* Table 15), which can be very significant when present in high populations.

Tortorticoidea (tortrix moths): More normally regarded as pests of fruit, although the pea moth is a member of this family.

Cabbage white adults mating on brassicas.

Papilionoidea (butterflies): Not normally regarded as pests, although the cabbage white butterflies are capable of serious damage to brassica crops (*see* Table 11).

Noctuoidea (noctuid moths): A large family of night-flying moths, with many pest species represented including cutworms (*see* below), cabbage white caterpillars (*see* Table 11) and stem-boring moths.

Diptera (True Flies)

An advanced order of easily recognized insects with two wings; they are generally strong, manoeuvrable fliers with well developed sense organs. The mouthparts are normally modified for piercing and/or sucking plant juices, liquid organic matter or blood. They have a complex life cycle, and larvae can often be found in very different habitats to adults, often recognizable as 'maggots', and often feeding off liquid or semi-liquids in habitats where organic matter accumulates. Some larvae have mouthparts adapted for tunnelling and feeding in plants. Although some species are pests, others are predators or parasites on pests and are thus valuable natural enemies; for instance the larvae of hoverflies eat aphids. Some of the better known pest families include the following:

Cecidomyiidae (gall midges): Attack all plant parts, feeding internally and causing a wide range of symptoms such as leaf curl, leaf distortion, galls, deformed flowers and distorted pods. Grass seed midges damage grass seed crops, but midges also exist on legumes (pea midge), cereals (hessian fly) and brassicas (swede midge). One or two feed on other insects and are natural enemies of aphids.

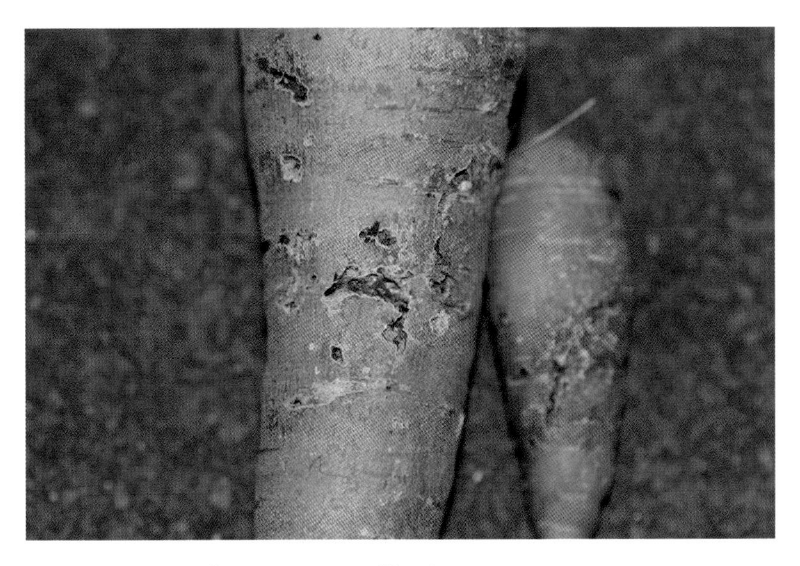

Injury caused by carrot fly maggots tunnelling in carrot.

Tipulidae (daddy long legs): The adults of leatherjackets that feed on grass roots including cereal crops (*see* p. 195 for management methods).

Chloropidae (frit and gout flies): Pests of grasses and cereals (*see* Table 17).

Psilidae (carrot flies): Well known pest of carrots and related crops (*see* Table 9) that burrows in roots and reduces the quality of the marketable produce.

Trypetidae (celery fly and relatives): Leaf miners in celery and related crops (*see* Table 9); often form conspicuous mines in leaves, although the damage caused is normally limited except when they are present in large numbers.

Agromyzidae (leaf and stem miners): Include a number of pest species that mine in legumes and brassicas, although some are parasitic on other insects.

Anthomyiidae (root flies and relatives): Contain a number of important pests that attack crop plants, feeding inside the stem at or below ground level. Crops affected include cereals, legumes, alliums and brassicas. Wheat bulb fly (*see* Table 17) and cabbage root fly (*see* Table 11), among others, belong to this group of flies.

Hymenoptera (Ants, Bees and Wasps)

In some ways one of the most advanced insect orders, containing some pest species but also many parasitoids and predators. Many are social with complex communal life histories, but all share a similar life cycle, passing through larval, pupal and adult stages, with the adults provisioning the larvae with food in many families (the wasps and bees). Pest families include sawflies in the families Cephidae and Tenthredinidae, which either consume foliage or tunnel in stems and/or other plant parts. Some wasps form galls on crop plants, mainly species in the families Chalcidoidea and Cynipoidea, but many more members of the former are in fact parasites on other insects.

INSECT PEST LIFE CYCLES

Entomology is a popular science, and insect herbivores a convenient model system that can be used to explore wider ecological theory. For that reason we know a great deal about insect life cycles, and this knowledge alone is now vast, encompassing many functional levels from the genetic, to physiological and behavioural, and up to ecological. The task of this book is not to provide onerous details about pest life cycles, but merely to indicate that this information is available in many forms and can be obtained by farmers and growers with some basic research. Much of it is free on the internet (*see* Sources of Information). In this book we illustrate the life cycles of key pests of crop plants likely to be encountered in temperate organic agriculture in the assumption that this will be useful in designing effective management programmes.

Insect pests adopt a huge range of life styles or life strategies that have evolved to exploit specific ecological niches. In the case of pests these often amount to adaptations to survive in the absence of the host plant, and to locate and colonize the host plant when it is present. Many have also evolved life cycles that allow them to escape from and evade their predators and parasites, at least temporarily during part of the season. The ecological adaptations underlying pest ecology have been described in Chapter 2, and this understanding has been used to develop strategies that undermine them and serve to manage pest populations as discussed in subsequent chapters (Chapters 3 to 5). The key features of insect life cycles have been summarized below to assist this understanding.

Insects can be roughly divided into two groups: those that follow a similar life strategy as nymphs and adults (hemimetabolous insects with incomplete metamorphosis), and those that pursue different or discontinous life strategies, first as larvae and then as adults with an intermediary pupal stage (holometabolous insects with complete metamorphosis). In the former case the nymphs and adults normally display broadly similar feeding habits and thus cause similar types of damage, just more of it as they get bigger. These insects can be generalist feeders with chewing mouthparts (for example grasshoppers), or specialist feeders with sucking mouthparts (for example aphids). It is interesting to note that they may pursue different life strategies at different times of the season and in the absence of the host plant. For instance, aphids often overwinter on alternative hosts and follow a sexual life cycle, as compared to feeding on crop plants and following an asexual life cycle in summer (for example the lettuce currant aphid in Table 10).

In contrast, holometabolous insects often display a discontinuity between the feeding habits of the larvae and the adult. For instance, lepidoptera larvae, caterpillars, are often herbivorous foliage feeders with mouthparts adapted to cut and chew leaves, whilst the adults are the familiar butterflies and moths with tubular mouthparts adapted to suck nectar from flowers. In this case there is less need for different life strategies at different times of the year, and normally one stage in the life cycle – for example pupa or egg – is adapted to overwinter in periods of absence of the host. The larvae are often the 'pest' stage of the life cycle, actively consuming large amounts of plant material, whilst the adult is adapted to find a mate and locate the host plant in order to place the egg where the larva has the best chance of survival. Larvae can be more, or less active: contrast the active movement of caterpillars with the sedentary habit of weevil larvae in stems. In some cases both adults and larvae are herbivorous, causing pest damage – for example the Colorado potato beetle (*see* Table 16).

In all cases the habits and behaviour of the insect pests will influence the types of control and/or management methods that are likely to be effective in any given situation. In fact many of the strategies, techniques and methods discussed in the previous chapters (especially Chapters 3 and 4) are aimed at disrupting or interfering with insect life cycles. It is important to note that what works with one species at one time might not work with another, or even the same species at a different time. Tables 9 to 17 summarize the life cycles of the key pests of field crops, although for many of them alternative texts will need to be consulted for a more detailed account (*see* Chapter 10 and Sources of Information).

PEST INSECT SYMPTOMS AND DAMAGE

It is safe to say that, between them, pest insects attack all plant parts including roots, stems, leaves, flowers, fruit and seed. However, any given species is likely to concentrate its attack on specific plant parts, tissues or organs, and its behaviour, mouthparts and body form are likely to be adapted for this. Any cursory examination of insect behaviour on crop or weed plants will reveal a number of plant feeding habits or 'guilds'. Chewing insects (such as grasshoppers and caterpillars) normally feed externally on plants, macerating plant tissue and creating the familiar holes in leaves, although they can also attack other plant parts such as stems, roots and seeds. Other groups of insects (such as aphids or Hemipteran bugs) siphon plant juices through tubular mouthparts while remaining outside on stems or the underside of leaves. Others live inside plants eating the plant tissue from the inside. Leaf miners, for example, tunnel in leaves and stems, whilst gall formers actually cause plants to swell and form structures that not only protect, but also provide nutrients to, their insect hosts.

The main symptoms of key insect pests are given in Tables 9 to 17, but the symptoms displayed will be a result of the pest following one of the following main feeding strategies: chewing or biting, sucking, or gall forming.

Chewing or Biting

Many insect larvae and adults have hard chewing mouthparts that are adapted to cut plant material and macerate it. Most chewing insects feed externally on leaves causing the familiar holes, although others feed inside leaves (leaf miners), and yet others in stems and roots. Symptoms are usually feeding holes, skeletonization of leaves, windows in leaves, hollow stems with exit holes and/or mines in leaves and stems. In extreme cases the whole leaf can be devoured and the plant appears defoliated. Chewers include lepidoptera larvae, beetles (adults and larvae) as well as grasshoppers feeding on leaves, leaf mining flies feeding inside leaves, and weevil larvae feeding inside roots and stems.

Sucking

Other insects siphon plant fluids through tubular mouthparts. They may specialize on different fluids: for instance, aphids generally feed on phloem sap, members of the cicadoidea generally feed on xylem sap, and whiteflies feed on sap in leaf mesophyll. Physical injury to the plant from sap feeders may be limited, but they often need to extract large amounts of sap to gain sufficient nutrition, especially amino acids (or nitrogen), and they often excrete large amounts of sugary solution on to plant surfaces, causing black moulds to develop. This can give plants a ragged or blackened look, especially if dust and other debris sticks to the sugar solution as it dries. Ants in particular also value this honeydew and may protect sucking pests to obtain it. Plants infested with sucking insects may appear stunted, foliage may curl up, and heavily infested plants will generally not thrive. Other symptoms can include silvering, mottling or streaks on leaves or other plant parts. In some cases the injection of toxic saliva will cause necrotic (dead) spots on the leaves in a similar manner to plant diseases.

Flea beetle and feeding damage on brassica.

Gall Forming

Gall-making insects have taken the manipulation of the plant to a more sophisticated level. They influence the plant to produce a shelter which also provides them with nutrients. The so-called galls are visible as hard swellings on the plant tissue, and are often caused by flies or wasps although thrips and aphids can also be responsible. Gall formers are not often serious pests of field crops.

MONITORING AND RISK ASSESSMENT OF INSECT PESTS

The art of organic insect pest management is in monitoring insect presence and in evaluating risk to the crop before taking action. It is an art because it depends on juggling and balancing the likely damage to the crop with specific farm circumstances, which include the resources available to manage the problem and the income and demands of intended markets. As explained in earlier chapters it is possible to tip the balance in the crop's favour by learning about the likely key pests on any crop, the range of effective management measures available, and by systematically monitoring the crop for their presence and their damage. This is discussed in more detail in Chapters 6 and 10. Here we merely note that monitoring pests requires active observations of their presence, and that this can be greatly aided by regular field walks, trapping and pest forecasts.

The tables on key pests (Tables 9 to 17) indicate monitoring techniques for specific pests where these have been developed and are available. These tables also indicate the conditions in which risk to the crop is likely to be higher (where this information is available). Ultimately judgement of risk depends on the farmer or grower, but we would urge that any risk should be assessed in the light of an estimate of the economic damage likely from an observed level of pests in any situation, and that this be placed in the context of organic farming principles and best farming practice (*see* Chapter 6 and Chapter 1 respectively). The tables (Tables 9 to 17) give an indication of the main indicative risk factors for each pest where these are known. Farmers and growers are likely to be able to add detail to these risk factors as part of their long-term monitoring programmes.

MANAGEMENT OF INSECT PESTS

Given the wide range of herbivorous behaviours exhibited by insect pests, it is difficult to generalize about the management approaches that should be adopted. It is also obviously impractical to describe management tactics for each and every potential pest insect. In the remaining sections of this chapter we have therefore opted to look at broad groups or guilds of pests characterized by the plant parts attacked, as this allows some brevity at the same time as allowing a commonality of approach with the following chapters and the tables (Tables 9 to 17). We have categorized the management of pests by the plant organs attacked: soil, root and seedling pests, stem and bulb pests, foliage pests, and flower, grain and seed pests. Storage pests are not covered in detail in this book but their economic effect can be considerable and more specialist guides should be consulted in this respect. The text should be read while keeping an eye on the specific tables that detail the key pests likely to be encountered in any particular crop or crop type, together with some key management techniques.

MANAGING SOIL, ROOT AND SEEDLING INSECT PESTS

Soil pests are among the most intractable and difficult of insect pests to manage in organic systems. The injury they cause to crop plants ranges from subtle effects on vigour to complete destruction. Diagnosis is often also difficult, as by their nature they are hidden from view and difficult to spot, even when a plant is pulled up for inspection, as they often remain in the soil. It is also difficult to judge whether root systems have been 'grazed' off as it is difficult to pull plants with completely intact root systems. Feeding damage to roots is often also obscured and hidden by soil clinging to the roots.

Seedling pests can be frustrating as they prevent crops establishing, a prerequisite for successful organic growing. Obviously once a seedling is dead no yield can be expected from the plant, though if damage is only sporadic neighbouring plants may be able to compensate for missing plants by producing higher yields. However, where pests leave large bare patches yield *is* likely to be affected, and such damage can necessitate complicated gap-filling or resowing operations, increased lodging, as well as causing slippage in field and harvesting operations.

Many insect soil pests have a long and slow life cycle, and so are persistent across seasons. In addition, many of the most important pests, especially wireworms and leatherjackets (*see* Managing Specific Problem Soil Pests p. 199), are normally pests of grasses and so are encouraged by periods of fertility-building ley. They are often problematic when moving to cropping after longer ley periods. Management can only be practised as part of the longer-term soil and rotation management plan for the holding as a whole, and short-term solutions are not generally available to organic farmers.

Susceptible Crops

Many, if not most, crops are susceptible to damage by soil and root pests, but especially cereals (Table 17), brassicas (Table 11), carrots (Table 9), lettuce (Table 10) and potato (Table 16). Persistent pests in organic systems

Cabbage root fly damage in cauliflowers.

include leatherjackets, wireworms and cutworms, and these are discussed in more detail below as problem pests. Root flies are particularly damaging in brassicas and umbellifers, especially where they damage young plants and/or damage the harvestable root. Nematodes and slugs, as soil-dwelling pests, also share some similarities with these pests, but are discussed in the next chapter (Chapter 8).

Pests of seeds and seedlings also normally spend at least part of their life cycle in the soil or close to it. The most important seedling pests in organic systems are likely to be cutworms, which attack a wide range of crops and are discussed in more detail below as a problem pest. Flea beetles, whose larvae live in the soil, are frequently problematic in brassicas (*see* Table 11). A number of crops are susceptible to damage caused by other seedling pests, including onions (Table 15) and beans (Table 14), which are attacked by seed flies that can mine in the germinating seed and cotyledon leaves, leading to poor establishment.

Management Strategies

Cultural Management

As soil-dwelling organisms that live part, or even most, of their life cycle underground, soil pests are profoundly affected by soil ecology and soil cultivation techniques. These have been discussed in detail in Chapter 3 as a basis for preventative pest and disease management, and in Chapter 4 from the perspective of a more adaptive response to managing pests and diseases, and these chapters should be consulted for more details. Suffice it to say that promoting an active soil biology is likely to encourage the natural enemies of soil pests, which are, in turn, more likely to work to suppress the pests when present in adequate numbers. Tillage operations are also part of good husbandry technique, and carried out appropriately will work to help plants establish well, and grow away rapidly, thus indirectly helping to manage pest problems, but also more directly to bury or expose soil pests. On the negative side they may adversely affect natural enemies or disrupt soil food webs.

Rotation can also be used to help manage soil pests. Some crop types are not as susceptible to specific soil pests, and these may be used as break crops to reduce pest populations, as can green manures. Susceptible crop types and/or varieties should not be planted in the rotation at points where soil-borne pests are likely to be more prolific (for example, after breaking a ley). In some cases tolerant or resistant varieties are available, as for instance Resistafly, a carrot variety partially resistant to carrot root fly (*see* Table 9), and these can be used as an aid to management in problem areas. Some green manures, such as 'caliente' mustard varieties, are also reputed to have a soil fumigant effect when turned in, although their effect on soil insects might be variable.

A range of adaptive cultural techniques is likely to be useful in managing various soil and root pests. Adequate and timely soil preparation is a key component in establishing all crops and in promoting vigorous growth that will allow plants to grow quickly through susceptible stages. Some crops could be raised as modules and protected from infection in the case of seedling pests (for example bean seed fly, *see* Table 14). Other adaptive techniques are also likely to be useful. Cabbage root flies (*see* Table 11) can be managed by a combination of tactics such as judicious timing of sowing (using forecasts), crop covers (to exclude flies), undersowing (to confuse egg-laying females), and earthing up (to encourage secondary root growth and compensation for damage). Similarly carrot flies (*see* Table 9) can be managed by timing sowing to avoid peak periods of infestation, and combining this with other tactics such as the use of fleece, withholding irrigation at key periods of the life cycle, and even vertical barriers in some circumstances.

The management of companion weeds can be important in some situations. For example, adult cutworm moths lay eggs on weeds, and the caterpillars then move down into the soil – so removing the weeds hinders this phase. In some cases – for example flea beetles (*see* Table 11) and carrot flies (*see* Table 9) – trap cropping may be effective, although this will require more management input.

Biological Control

Biological control of soil-borne pests is normally by preserving and conserving natural enemies *in situ*, and many farm practices should be aimed at conservation biological control, especially those that promote ground predators and soil biological activity. Techniques such as mulching and composting are likely to be positive in this respect, and both have shown the capacity to increase natural enemy activity under various circumstances. Research has demonstrated the potential for augmentation of biological control by the addition of biological control agents (for example nematodes to control carrot root fly, or predators to control cabbage root fly (*see* Tables 11 and 9)), but these have not been generally effective for commercial field application to date. In contrast, entomopathogenic nematodes are available to control vine weevils in container growing, and are used in commercial systems.

Direct Control

Direct control measures are hampered by the fact that soil pests are normally hidden from view in the soil; there is also the potential for adverse effects on other soil flora and fauna. Consequently direct control measures, apart from the effects of soil tillage and tillage operations, are extremely limited. Garlic granules have been shown to be effective against pests such as cabbage root fly (*see* Table 11) and carrot fly (*see* Table 9) *in vitro*, but are not normally effective in field conditions where placement and timing have to be very accurate to have an effect in repelling the fly and/or killing the larvae.

MANAGING SPECIFIC PROBLEM SOIL PESTS

Leatherjackets

Damage and Risk Assessment

Leatherjackets are the larvae of (Tipulid) crane flies (*Tipula* spp., often *Tipula paludosa*), more commonly known as 'daddy long legs', which are often seen flying on warm and humid late summer or autumn evenings. The larvae live in the soil and are, as their name implies, tough with leathery-looking skin, not unlike large maggots. They feed on the roots of plants, including crop plants, and when present in high numbers cause damage by preventing adequate root development.

Leatherjackets are normally found in grassland and the population will tend to build up with time, so that swards of longer duration have larger populations. They will reduce yields of grass as they consume the roots, but it is only when they actually destroy the sward that the damage tends to be noticed. Other crops sown after grass are most at risk from leatherjackets, and this obviously creates a higher risk in organic rotations which often incorporate grass leys as fertility-building periods. The longer the grass has been down, the more likely there will be a problem with leatherjackets, especially if a ley has been down for more than two years.

Leatherjackets can cause damage in a wide range of crops, and those where it is difficult to adjust sowing rates to compensate for damage will be more sensitive to economic damage. Crops at most risk include spring-sown grasses such as spring barley and sweetcorn (Table 17), but also vegetables such as brassicas (Table 11), leeks (Table 15) and lettuce (Table 10). Damage can also be seen on a wide range of crops such as beans, peas (Table 14), potatoes (Table 17), swedes and turnips (Table 11), although it is usually less of a problem.

Life Cycle

Adult crane flies fly from midsummer to late autumn. The females mate soon after emerging from their pupae in the soil, and tend to lay most of

Adult crane fly resting in vegetation.

their eggs close to the site where they emerge as they are weak fliers. The eggs hatch in two to three weeks, and the larvae, or leatherjackets, feed slowly through the winter. It is in the latter stages, when they are large, that the leatherjackets can consume considerable quantities of plant material, after which they 'rest' in the soil until they pupate to repeat the cycle. The larvae are susceptible to desiccation, especially when small, and drought in late summer can cause high mortality.

Monitoring

Leatherjackets can be assessed in grassland before it is turned in. One method recommends that the assessment is made between November and March by hammering 30cm lengths of 100mm of open pipe into the grassland where leatherjackets are suspected. The pipe should then be filled with saturated brine (salt water) and the number of leatherjackets that come to the surface counted after 15 minutes. Pipes should be hammered in at random in the field or on a W transect as with other monitoring methods, and the more pipes that can be used the better the evaluation of the infestation. A minimum of ten sampling points will be needed in a hectare to get reliable results. The pipes will need to be removed before tillage operations. If particular areas of infestation are suspected, these can be targeted for sampling, although this should be accounted for in the final assessment of the infestation across the whole field.

Alternatively monitoring services that give general forecasts are available (in the UK through levy bodies and research organizations), which advise when the risk of damage is likely to be high over larger areas. They are usually based on models that use weather conditions to make predictions about how populations will develop (warmer winter weather favours leatherjackets while dry autumn conditions do not).

Management

Cultural Management Manage rotations to avoid planting susceptible crops after a grass ley, especially if this has been down for more than two years. Pasture and leys can be cut or grazed from July to September to disrupt adult egg laying, although the effects are likely to be variable. Delaying planting (until July) to avoid the peak spring feeding period might be feasible in some crops. If transplanting vegetable crops use good, sturdy transplants, with a good root system.

Regular cultivation and tillage in preparing seed beds (and consistent with good soil husbandry) can expose the leatherjackets to predation and desiccation, and this is likely to be the mainstay of any cultural management. Early cultivations (starting in January or February) will have a greater effect on leatherjackets. Tillage in late summer can expose small leatherjackets when they are vulnerable to desiccation, and this is useful when planting winter cereals.

There is conflicting evidence that green manures, especially mustard, grown over the summer and then turned in can reduce leatherjacket populations, but it is probable that the effect is due to the tillage operations, rather than the mustard. This accounts for the variable results reported, in that removal of vegetation where the females prefer to place their eggs is likely to cause disruption. Tillage operations to establish the green manure will also expose resting larvae and pupae before they emerge.

Biological Control There are currently no reliable biological controls for leatherjackets, but turning the soil will obviously expose them to predation by birds, small mammals and other predators. Chickens might also possibly be used to remove leatherjackets. Biopesticides, based on microbial pathogens, are a possible future development, but as of yet are unreliable on a field scale, although nematodes developed for use against vine weevil can be effective.

Direct Control Crop cultivations may help to reduce damage to the growing crop, but are only probably economic where weed management is also needed. On small areas, for example seed nurseries, barriers of sacking, tarpaulin, cardboard or grass clippings may help to prevent egg laying, and can be used to 'collect' leatherjackets and crane flies, although this is likely to be uneconomic and ineffective on a large scale.

Wireworms

Damage and Risk Assessment

Wireworms, contrary to their name, are the larvae of click beetles (mainly *Agriotes* spp.). There are three common species in the UK – *Agriotes lineatus*, *A. obscurus* and *A. sputator* – and these are responsible for the vast majority of attacks on crops. They all have a similar capacity to damage crops, and attack a wide range of crops including cereals (Table 17), sugar

Wireworms around the base of a leek plant.

beet and vegetables. In the latter case they generally do most economic damage to potatoes (Table 16), but can also cause considerable damage in shallow-rooted crops such as lettuce (Table 10).

High populations of wireworms are normally associated with long-term grass leys, and therefore crops are most at risk in the first few seasons after ploughing in grass. However, there is an increasing trend for damage being reported irrespective of cropping history. Damage often appears patchy across fields or sites, but areas of high infestation tend to persist for a number of seasons and can be troublesome to manage. Wireworms are normally found in the top 20cm of soil, but high soil temperatures and low soil moisture in the upper soil layers will drive them lower in the soil profile, and they will try and move away from dry soil as lack of moisture reduces their chances of survival.

In vegetable and other crops they can attack and damage the stems and roots of seedlings, leading to loss of vigour, wilting or even death of the plants. They are attracted to the carbon dioxide (CO_2) given off by the seedlings in the soil. Subsequently they curtail the root development of vegetable and other crops, leading to plants showing a lack of vigour. Potato growers are particularly badly affected in some years, as the small holes made in the tubers cause them to be rejected due to the increasingly stringent demands for high quality in retail marketing chains (that tend to sell clean, washed potatoes). The small feeding holes and narrow tunnels leading into the tuber also allow other opportunistic pests and diseases to enter the tuber, decreasing marketability and storage longevity.

Life Cycle

Adult click beetles are common in the UK, and are active between April and August with a peak of activity in May. Females lay eggs just below the surface of the soil either singly or in small batches, usually under grass or on weedy ground. Small white wireworms hatch after about four to six weeks, and feed on soil organic matter. Mortality of both eggs and small larvae can be high in dry periods due to desiccation.

As the larvae mature they pass through one to three instars, or moults, each year, and feed increasingly on living plant material. The peak feeding periods are March to May and September to October, though the larvae may spend long periods without feeding, especially around moults. As a consequence they grow slowly and may take four to five years to mature. When they reach this point they burrow down into the soil (to 5 to 30cm) and pupate. After three to four weeks the adult beetles are ready to emerge, although they may remain over winter in their pupation cell before emerging.

Monitoring

Any management strategy should aim to monitor the presence of wireworms and gauge the risk. A simple risk assessment will take into account the agronomic factors such as the presence of a grass ley in the rotation, the previous history of infestation, and weather and soil conditions. High risk factors could include long grass leys and moister soil types (for example, alluvial soils), and some work has indicated that south-facing slopes are more likely to be infested.

Wireworm populations can also be directly sampled in various ways including soil sampling, bait traps and adult trapping.

Soil Sampling The standard UK method uses twenty soil cores to a depth of 15cm on a W-shaped transect across a field (normally 4–10ha) and then washes the larvae from the soil using standard soil laboratory equipment. This method is very labour-intensive and prone to large sampling errors, especially when the infestation is patchy or low. However, it can demonstrate the presence of wireworms when they occur at high populations and the relationship between the numbers of larvae found and the damage to the potato is (with a deal of uncertainty) known from accumulated experience.

Bait Traps These traps can be used simply to assess the presence or absence of wireworms. Most simply, potatoes can be cut in half or quarters and buried in the soil, and then dug up four days to a week later. Any wireworms in the soil will be attracted to the potato bait and can be counted. A more sophisticated bait trap consists of a plastic pot with wireworm-sized holes into which a bait bag filled with pre-soaked and germinated seed is placed. The trap is buried and dug up sometime later and examined for wireworms. Wireworms may also be present in the soil around the trap.

Adult Trapping Adult click beetles are in principle easier to catch, but it is much harder to relate the number of adults caught to any likely level of damage to the crop. Trapping should be carried out in the periods when the adults are most active, and normally takes the form of either plastic

Pheromone trap for adult click beetles.

sheet traps or pheromone traps. Plastic trap sheeting is unreliable and involves laying a ½m^2 sheet of plastic on bare soil and covering it with freshly cut rye grass. After three days the trap is inspected for beetles under the grass.

Purpose-designed pheromone traps are also available that have been designed for use with female sex attractants. The sex attractants are species-specific, so a set of traps for each of the three common species is normally necessary in any situation. The traps are fixed into the ground, normally sited well away from field margins, with at least 40m between traps. Relationships have been established between the number of beetles caught and the number of wireworms in the soil; however, at some sites this is sometimes inaccurate.

Management

Cultural Management Careful crop rotation should avoid sowing susceptible crops following grass leys or weed fallows. Wireworm populations should decrease more or less quickly (three to four seasons) under arable cultivation (*see* Direct Control below). Crops such as peas and beans are also considered tolerant of wireworm, although this might be due to husbandry practices associated with this crop rather than any true resistance *per se*. Green manures such as mustard that contain glucosinolates are reputed to control soil-borne pests such as wireworm, but these effects have been hard to demonstrate in trials over and above the beneficial effects normally attributed to green manures. Care to reduce perennial grass weeds in crops (couch or bent) might also help to manage these pests.

Potatoes at risk (see above) can be lifted early before peak feeding activity in autumn. Some varieties of potato – specifically those with a high glycoalkaloid content – are reputed to be less susceptible to wireworm damage, but this is difficult to verify as the glycoalkaloid content depends on environmental factors to some extent, and high levels may not be palatable to consumers.

Biological Control The only form of biological control is natural predation by birds and other predatory insects, though this is probably not effective in isolation. Similarly many natural diseases have been observed but not developed as introduced biological controls. Sensitive management of the farm system will maintain these natural predators at high background levels.

Direct Control Shallow cultivation of the soil in March to April and/or interrow hoeing in the summer can expose the larvae to predation and desiccation.

Cutworms

Damage and Risk Assessment

Cutworms are a frustrating pest for vegetable growers because of their habit of cutting plant stems at the soil surface and killing them. They are soil-dwelling caterpillars of noctuid moths (*Agrotis* spp., *Euxoa* spp.). The caterpillars are quite large and dull in colour, and by the time they are noticeable, about 3–5cm long. They often curl up into a C-shape when dug up from around the base of plants that have been cut.

Larger cutworms cut plants off at or just below soil level and can cut several seedlings in a night. Damage can often be seen along a row, and the cutworm can normally be found easily just under the soil nearby. Where populations are high this may leave bare patches in crop stands and reduce yield accordingly, especially in row crops such as lettuce, tomatoes or celery where neighbouring plants are unable to compensate. Where populations are high and plants more developed, cosmetic damage is also likely to be found, and this can be serious on vegetable crops such as carrots (Table 9), potatoes (Table 16), swedes and turnips (Table 11). Damage often occurs in the same area year on year, and high soil populations of caterpillars are a good indication of high risk to crops.

Life Cycle

Female moths lay eggs on plants, often crop weeds, close to the ground on stems or leaves, but some may also lay eggs in leaf litter. They are usually in flight from late spring through to late summer. The first instar larvae hatch and often feed on these above-ground parts, but during the second or third instar, drop to the soil where they complete their development feeding on plant material. At this stage feeding is normally nocturnal. Fully fed caterpillars overwinter in the top layers (5cm) of soil as caterpillars or pre-pupae, and then pupate in the spring (April to May). Adult moths emerge from pupae and mate to complete the cycle. Wet weather in June and July can cause high mortality in the early instar larvae of some species of cutworm, and this can be simulated to some extent by irrigation at this time.

Monitoring

Adult moths can be monitored by trapping, using either pheromones or light traps. The latter are less efficient, especially when the moon is bright, but both suffer from the difficulty in predicting actual crop damage from the number of adults caught, although it will indicate when adults are on the wing and likely to be looking for egg-laying sites. Crop and weed monitoring is likely to give a better estimate of the potential for crop damage. Adults lay eggs on 'weed' plants, and these can be monitored in and around crops. Larvae can be observed by digging around the base of plants that have obviously been cut overnight. Unfortunately management options may be limited at this stage in organic systems (*see* below). Alternatively cutworm risk forecasting may be available (in the UK from the HDC Pest Bulletin run by Warwick HRI), which uses a model based on temperature combined with light trap catches to forecast risk to crops over larger regions.

Management

Cultural Management Adult females generally lay eggs on 'weed' plants, and the field should be kept weed free at peak egg-laying times and especially late summer. Cultivation to remove weeds as potential laying and feeding sites at least ten days before planting will help reduce populations on subsequent crops. It will also expose caterpillars and pupae to predation and desiccation. The extent to which field margins can be kept weed free, and the effect that this has in preventing increased attack to all but crops planted very close to the margins, is a moot point. Within the rotation try to avoid planting susceptible crops close to areas likely to be attractive to moths, such as grass or dense cover crops.

Turning the soil in tillage operations will help to expose larvae and pupae and will bury crop or weed residues to help in reducing populations, but will probably only be of economic value as part of weed management or seed-bed preparations. More frequent tillage is likely to be more effective but might have undesirable side effects such as disturbing soil structure or increasing nutrient leaching. Thorough preparation of seedbeds will also help expose caterpillars and pupae, and will also promote good establishment and help the crop to pass rapidly through the susceptible seedling stage where damage is most serious. Sowing at higher rates will allow for some seedling loss where cutworms are anticipated. Where transplants are used these should be vigorous, and hopefully large enough, to be less susceptible to being cut. Irrigation can serve a dual function, not only in helping the plants establish quickly but also in increasing mortality of the first and second instar larvae, which significantly reduces the future cutworm population.

Biological Control Noctuid larvae suffer from a range of parasites and parasitoids; the latter can easily be observed emerging from collected larvae kept in a jar. This natural background parasitism helps to suppress populations, but given the nature of the damage, is usually insufficient to prevent economic losses. Large larvae are also susceptible to predation by birds, small mammals and other insectivores. Chickens can turn over the earth and will readily eat the caterpillars and pupae.

Direct Control In small areas searching for and destroying caterpillars around cut plants can be very effective if done systematically; similarly card collars may also be used to help prevent access to plants. They can be made of recycled card, and should circle the plant stem and be pushed a few centimetres into the seed bed to prevent access. Traditional remedies have also included bran baits, which cause the caterpillars to swell up, stop feeding and die, and molasses which hardens once ingested and prevents feeding; however, their efficacy is unproven or variable, depending on the conditions in which they are applied.

Moth caterpillars are susceptible to Bt, which stops them feeding if they eat it. Although it can be sprayed on the ground near affected stems, this is not always effective, presumably because caterpillars do not eat enough of the microorganism to cause death. It has been suggested that presenting the Bt in a bran bait mixed with molasses is a more effective treatment and might serve to make the two approaches more effective.

MANAGING STEM AND BULB PESTS

Stem and bulb pests can directly affect yields in the case where this is the harvestable part (for instance celery, onion or leeks), and can indirectly reduce yield by interfering with the translocation of nutrients within the plant which leads to thrifty plants. This is obviously more serious where there is one main (or a few) yield-bearing stems, as in cereals or beans. Many of these pests are maggots (fly larvae) or lepidoptera (caterpillars), which eat holes in the stems or tunnel inside them and may be difficult to diagnose as they are hidden inside the plant. Some pests found on stems also suck plant sap (for example aphids or plant hoppers), and in this case many also attack the foliage (*see* below).

Susceptible Crops

The crops most susceptible to stem and bulb pests include (obviously) onions and other alliums (Table 9), cereals (Table 17) and legumes (Table 14). Some crops may be less susceptible, and in some cases resistant varieties may exist. If so, they should be used.

Management Strategies

Cultural Management

Good crop husbandry is important as a basis for managing these pest insects. In particular the resting stages (large larva, pupa) are likely to be found in crop residues, and these should be managed, for instance by ploughing them in, as is recommended for leek moth and onion fly (Table 15). Management of tillage and timing of sowing can be important in reducing attacks by cereal pests such as gout fly and frit fly (Table 17). Many of the shorter-term tactical methods useful for controlling these pests will be similar to those used for foliage pests (*see* below).

Biological Control

Conservation biological control should be used to promote generalist predators such as ground beetles that are likely to predate eggs, larvae and pupae close to the ground. For example ground beetles are likely to be especially useful in this regard, and beetle banks have been shown to increase their presence in crops. There are no biological controls generally and commercially available that can be used to augment the background level of biological control by inundation.

Direct Control

For those pests that feed externally on the plant many of the direct control methods used to tackle foliage pests are also likely to be useful (*see* below). However, many of these pests are likely to be hidden close to the ground or within plants, and are thus markedly more difficult to reach whatever

Ground-dwelling predators should be encouraged to manage stem and bulb pests. (Courtesy *Colin Newsham)*

method is chosen. For example, Bt could be used for caterpillar control but is unlikely to be effective where caterpillars are feeding internally in plants or hidden within stems at ground level, as the spray will not reach them.

MANAGING FOLIAGE PESTS

Insect leaf pests are in some ways the 'classical' crop pest. They are ubiquitous and easy to spot in crops. Foliage pests feed by chewing holes in the leaf or sucking sap from the leaf (and other plant parts). Paradoxically they are also the pest that the crop is most likely to be able to compensate for, as plants can often tolerate a degree of defoliation (up to 50 per cent) or sap extraction with no apparent reduction in yield. They are not tolerable, however, where the leaf itself is harvested and eaten. Furthermore modern consumers also have a very low tolerance for insect damage or the presence of insects, especially on salad leaves.

Crops Susceptible

All crops are susceptible to some degree to leaf insects, which may also frequently be found on stems or other plant parts such as flowers or stems. Crops most affected include cereals (Table 17), brassicas (Table 11), lettuce (Table 10) and potatoes (Table 16). Problem pests are likely to be aphids and caterpillars (the larvae of lepidoptera), whose management is described in more detail below.

For some pests and crop combinations resistant varieties are available. Where possible resistant and/or tolerant varieties should be used to manage pest attack, and in some cases it may be possible to mix varieties of different resistances to 'protect' susceptible crops and increase the range of varieties grown.

Management Strategies

Cultural Management A wide array of cultural control methods exists which can be brought to bear on foliage pests. These have been extensively discussed in Chapters 3 and 4, and only the most useful are summarized here. The first line of defence will always be preparing the cropping area in an appropriate way and sowing the crop to give it the best possible conditions for rapid establishment and growth. Vigorous crops are more likely to resist attack, and pass quickly through susceptible stages. The timing of sowing may be useful to avoid times of peak infestation, but it may also have detrimental effects on crop development and growth. Other factors that are likely to strongly affect pest insects are the state of crop nutrition (for instance excess nitrogen will attract many pests), and crop combinations that may in some cases serve to confuse insects by hiding 'susceptible' plants among non hosts or resistant varieties.

Other cultural approaches will also be important in reducing or mitigating insect attack. Crop hygiene can be important in preventing pests bridging seasons in crop debris, which should be ploughed in (although this will also affect natural enemies) or removed (and thoroughly composted or destroyed). Irrigation at key times will also reduce crop stress and will help promote the plant's ability to defend itself. Prompt harvesting is also likely to reduce exposure of crops to pest attack.

Other more direct measures include using crop covers to exclude pests, although these should be used sensibly so that pests are not trapped under the cover with the crop which might exacerbate the problem, especially if natural enemies are excluded by the cover – for example, covers are used to exclude carrot fly (Table 9). Other methods such as trap cropping and mulching can be used in certain circumstances to reduce or to mitigate the effect of pest attack.

Biological Control

Conservation biological control is the cornerstone of managing foliar pests, and this is described at length in Chapter 3. Farmers and growers should appreciate that almost all the pest control occurring in a crop will be the result of endemic predators and parasites, and that every possible effort should be made to promote them within and around crops; the methods for doing this are described in Chapter 4.

Inundative biological control measures exist and some of these are widely and commercially available, especially for protected cropping systems. These include insect predators (such as ladybirds), insect parasitoids (mainly small wasps) and even insect diseases (for instance entomopathogenic fungi and nematodes). These have been extensively discussed in previous chapters (*see* Chapters 4 and 5).

Direct Control

Possibly due to the fact that foliage pests are very visible in crops, and they, together with leaves, present obvious and direct targets for control measures, a range of direct methods are available for controlling foliage pests. These measures will also be applicable to insect pests found externally on other plant parts such as stems and grain. Many of these methods have been described in detail in Chapters 4 and 5 so will not be exhaustively discussed again; however, they are summarized below.

In small-scale systems direct removal of pest insects can be effective if carried out systematically and regularly. Potentially this could also have a minimal impact on natural enemies. As the crop grows it is likely to be less effective and more costly.

More cost effectively, sprays can be used to target pests feeding externally on crops. Organic farmers should always use selective sprays where possible to avoid damaging any natural biological control that is taking place. In this respect sprays containing biological control agents active against a limited range of species (preferably the pest alone) are better than broader spectrum biocides. In the former case sprays such as Bt are active against caterpillars (and other selected groups), as are entomopathogenic nematodes, and these should be preferred to plant-derived pesticides such as pyrethrum or microbe-derived spinosad, which have a much broader spectrum action.

Insect pests can also be controlled by a range of sprays with a physical action (soaps, oils, dusts), which can be effective if targeted towards the insects themselves. These are likely to be intermediate in their effect on natural enemies and/or beneficial insects, and should be used at the times of day when these are not likely to be active where possible.

MANAGING PROBLEM FOLIAGE PESTS

Aphids

Damage and Risk Assessment

There are numerous species of aphid worldwide, and aphids are a major pest of temperate crops because of their complex life strategies for overwintering and reproducing. Although easily overlooked until they have reached serious numbers in a crop, they are also prey and hosts for numerous predators and parasites, and these hold the key to their successful control in organic farming systems. They cause considerable damage due to their habit of sucking plant sap, but this is only detrimental once they reach large numbers on the plant, at which point they begin to extract significant amounts of sap and nutrients causing the plant to lose productivity. Of equal consequence is the fact that they are major transmitters of viral diseases. Due to their prolific production of honeydew they can also cause sooty moulds to develop on plants, giving these an unsightly, blackened, sticky look, and these moulds can, in themselves, reduce the photosynthetic efficiency of the leaves.

All major crop types suffer aphid attack, but possibly the most seriously affected are brassicas (Table 11), cereals (Table 17), legumes (Table 14), lettuce (Table 10) and potatoes (Table 16). In some cases (for example brassicas and salads) the presence of aphids is itself a major problem as they both stunt the growth of plants and contaminate the harvestable product, whilst in other crops (for example potatoes) they are the primary agent responsible for the transmission of viruses.

Life Cycle

Aphids are ubiquitous and successful in part due to their complex life cycle. The two features that make them successful on (especially) annual crops are, firstly, the ability to alternate between hosts during the different phases of their life cycle, and secondly, the ability to reproduce parthenogenetically and/or viviparously during a part of their life cycle. Whilst more specialist texts should be consulted for the exact details of aphid life cycles, which obviously vary in detail between species, they generally show a variation on a similar pattern.

Most species are capable of overwintering on a perennial (and woody) host as either eggs or aphids. Winged forms arise from this host in the

Cabbage aphid.

spring, and these migrate in search of herbaceous summer hosts, which include many crop plants. Aphids are capable of parthenogenetic and viviparous reproduction on the herbaceous hosts, which allows them to build up their numbers on these hosts very quickly. Secondary cycles at this time allow winged forms to seek and find new herbaceous hosts, and thus epidemics can rapidly build up. Towards autumn the winged forms increasingly seek out the perennial and woody alternative hosts on which they complete the winter, often as eggs after sexual reproduction. Some aphid species are capable of feeding on a wide range of herbaceous plants – for example black bean aphid (Table 14) – which also allows them to increase rapidly in numbers on a range of crop types, whilst others appear to have a more restricted range – for example currant lettuce aphid (Table 10).

However, many variations on this basic strategy exist. For example some aphid species only appear to feed on herbaceous species – as the cabbage aphid (Table 11) – and do not alter between types of host. Some aphids appear to have no sexual phase (at least in the UK), and many overwinter not as eggs but as aphids at various stages of development. In addition many aphid species – for example peach potato aphid (Table 16) – are increasingly capable of overwintering on summer or herbaceous host plants as the winter weather has tended to become warmer, a trend likely to increase with global warming.

This basic life-cycle strategy allows aphids to avoid both agricultural operations and natural enemies. In the latter case the alternative hosts are usually wild plants scattered in the landscape; they are thus effectively beyond management by farmers at this point as these alternative hosts are

Aphids in cereal.

valued within the landscape for all sorts of reasons over and above their ability to overwinter aphids. It is in any case impossible to locate all host individuals and to destroy or control aphids on them. Secondly, because they are able to multiply rapidly on herbaceous hosts (including crop plants) their numbers rapidly outpace the ability of their natural enemies (principally parasitic wasps and insect predators) to consume them. This results in a characteristic peak of aphid numbers in early summer (around mid July in the UK) followed by a crash in population in late summer (August) as the natural enemies catch up and begin to exert their controlling effect. Unfortunately this can lead to a crash in predator and parasite populations, allowing a resurgence of aphids in the autumn.

Monitoring

Aphids are monitored on a national scale in the UK using a network of suction traps, and similar systems may exist in other regions. These trap catches are used to monitor which species are flying, and can be important for developing forecasts (*see* Chapter 6) of likely aphid attack later in the year. Some of these forecasts can also be supplemented by information from mathematical models, which can also make predictions based on day degree temperatures or other factors that also predict the probability of population increases. In all cases these forecasts can serve as general warnings for farmers and growers, and can indicate times at which control measures such as crop covers might be most effectively applied.

Aphids in crops can best be monitored by crop walking or by the use of sticky traps against which they are (mainly passively) blown. Plants selected at random should be sampled for the presence of aphid colonies, or the numbers counted on sticky traps. The advantage of sampling crop plants is that it gives a direct indication of infestation in the crop, and allows a prediction of future damage to be directly assessed. It also allows an estimate to be made of the level of natural enemies in the crop by observing (for example) syrphid larvae, ladybirds and 'mummified' aphids that have been parasitized. This allows a judgement to be made about the impact of any direct control treatments on natural enemies, or even whether or not they are likely to bring the aphid population under control. It should be recognized that infestation might vary across a crop – for instance aphids are often more prevalent on the edges of crops where they are 'filtered' out of the wind – and sampling should take care to be representative of the crop as a whole rather than by just a few plants round the edges.

Management Strategies

Cultural Management A range of cultural methods has been developed that are of use in managing aphids. It should be borne in mind that those used should be as compatible as possible with conserving the natural enemies of aphids, as these predators are likely to be the most effective control measure the farmer or grower has. From that point of view, and for some crops and aphid species, resistant varieties are available and these should be used where possible (*see* lettuce, Table 10, for example).

Good crop husbandry is crucial in minimizing the effects of aphid attack. Raising crops as transplants where appropriate can prevent infections at the seedling stage which are likely to develop into more serious infestations in the field, and where affected plants will mature more slowly and unevenly. Good seed-bed preparation will also aid plants to establish and grow quickly and vigorously, avoiding long periods of high susceptibility to damage. Seed-bed preparation can also be important in eliminating cracks in the soil where root aphids can enter and attack plant roots (Table 10). Time of planting may also help avoid the worst periods of attack, although this should be balanced against the effect on the development of the crop. It may also be possible to minimize damage by selectively harvesting and discarding those parts of the plant (for example the outer leaves) most heavily infested.

Irrigation might also be important, as drought aggravates the water loss caused by aphid attack, and wilting is more likely unless the water is replaced. Watering can be particularly important where root aphids are present, because these prevent efficient uptake of water by the plant (Table 10). In some cases irrigation can wash aphids off foliage or increase aphid fungal diseases, but this might be at some risk of increasing fungal diseases within the crop as these are also favoured by high humidity.

More direct cultural control measures include the use of crop covers such as fleece or mesh during periods of high risk as indicated by forecasting tools or experience (normally June to August in the UK). Care should be taken not to trap large numbers of aphids under the cover, which might exacerbate problems if natural enemies are excluded.

Biological Control The mainstay of aphid management is conservation biological control (Chapter 3), which should aim to preserve natural enemies in and around the crop. Many of the techniques for conservation biological control have already been discussed (Chapter 4), but one of the most useful for aphid management is flower strips with nectar sources for parasitoids. Flower strips can be interplanted with field vegetable crops to attract parasitoids into the crop, and strips of green manures might likewise provide shelter for generalist predators of aphids. Field margins can also be used to encourage and maintain predators and parasites, with sensitive sowing of both nectar sources and plants on which aphids can survive along with their natural enemies. In some cases it is possible to augment the background level of biological control by increasing the resources available for natural enemies artificially (*see* Chapter 4): for example, some research has indicated that spraying sugar solution or other baits on to a crop can increase the activity of some parasitoids.

Inundative biological controls are available for protected crops using parasitic wasps and/or predators, and these can be extremely effective, lasting for a whole season once applied (*see* Chapter 4). They tend to be much more effective in protected cropping, and much less effective under field conditions which are open and subject to the vagaries of the weather. Entomopathogenic fungi have been shown to be promising as potential biological control agents in the field, but the exacting requirements they have for infecting aphids means that practical application methods are still not available.

Direct Control A number of direct control measures are available for use against aphid pests. Most of these take the form of sprays of one

Hoverfly larva devouring aphids, one of the important natural enemies of aphids.

sort or another, although on very small areas direct removal or squashing might be a useful stopgap. It should be stressed that the main control of aphid populations is natural enemies, and when using sprays, those that are damaging to these predators should be avoided, and should be applied at times that avoid their peaks of activity. In addition many of these products are unlikely to work unless targeted directly at the aphids, and in some cases even this will not be effective, as aphids are protected by waxy secretions that repel water – for example, cabbage aphids (Table 11).

The main spray used against aphids is soft soap solution, which has a physical effect on the insect cuticle (and can have the same effect on natural enemies). The spray should be applied directly on to the aphids to have an effect, and is more likely to be effective early in the infestation phase, before vast numbers have built up. A range of plant extracts and botanical pesticides has also been shown to be effective against aphids, but their use needs to be balanced against damage to natural enemies (*see* Chapter 5).

Caterpillars

Damage and Risk Assessment

Caterpillars are the larvae of Lepidoptera (moths and butterflies) which can be observed in all cropping systems, both on crop and non-crop plants. They have tough generalist mouthparts for biting and chewing. As crop pests they often feed on foliage, and are most conspicuous when actively feeding as they often have cryptic coloration and/or shapes that conceal them at rest. Some take a defensive position when disturbed, either aimed at startling predators or hiding the caterpillar in the foliage. Caterpillars generally chew holes in leaves, stems and/or other plant parts. Characteristically they cause holes in leaves, but they may also only eat the lower surface leaving windows, or leave behind skeletons of leaves, or even defoliate the plants completely. They may also live inside larger structures such as stems (so called 'stem borers').

The most affected crop types are those where caterpillars directly eat the part of the plant that is to be sold, such as brassicas (Table 11), and/or where they mine in stems or other parts, such as leek moth in alliums (Table 15). Arguably cutworms (discussed above), a type of caterpillar that lives in the soil, can be more troublesome.

Life Cycle

The general life cycle of butterflies and moths is the same. The adults, which feed mainly on nectar from flowers, actively seek out the host plants on which to lay their eggs during their brief life. The eggs hatch into small caterpillars which get larger as they actively feed, passing through five or six stages before pupating. The pupal stage, which may be attached to the plant in some way, or may be in the soil, transforms into the adult insect to complete the cycle. Different species often specialize in feeding on different host plants as caterpillars, although others

Moth (Garden Pebble) caterpillar on cabbage.

have a wider host range. Adults can take nectar from a range of flowering plants.

Monitoring

Caterpillars are best monitored by crop walking and inspecting plants for eggs, caterpillars, and/or damage. These can be supplemented by pest forecasting services which are issued on a nationwide basis in the UK based on trapping and/or modelling. Localized trapping (light or pheromone traps) can also provide local estimates of likely population trends, and observation of pest butterflies by day is likely to indicate periods in which adults are active.

Management

Cultural Management Adult moths and butterflies often actively seek out crops on which to lay eggs, and are stronger fliers than other pest insects (despite appearances). In such cases, although good husbandry can help a crop compensate for damage, it is unlikely to prevent attack completely. Although direct mechanical control is possible – that is, picking and squashing – in crops of limited area and size, it is unlikely to be completely effective as caterpillars will already be large and doing conspicuous damage. In some cases it might be possible to search out the egg masses and destroy these instead, but this is likely to be onerous on large areas.

Other common and more adaptive management methods include the use of crop covers, which can exclude adult butterflies and moths, thus preventing egg laying. As with all pests, it is important to ensure the cover is on before the pest is present, because if eggs have already been laid on the crop, the cover could exacerbate the problem because natural enemies would be excluded. In some cases adults can lay eggs through mesh or netting so it may have to be lifted off the crop. There is some evidence that intercropping and companion planting can deter adults from laying eggs, but the effects are not generally well understood and this method is difficult to use with consistency on a large scale.

Biological Control Natural enemies normally exert considerable control of caterpillars. This can be inferred by a comparison of the often large number of eggs laid by adults as compared to the one or two caterpillars that manage to pupate. The natural enemies of caterpillars include parasites, birds and diseases. Parasitic wasps include Brachonids, Chalcids and Ichneumonids. Ichneumon wasps (5–10mm) in particular prey on the caterpillars of butterflies and moths very effectively. Chalcid wasps are amongst the smallest of wasps (less than 3mm) and parasitize the eggs and larvae of cabbage caterpillars, whereas the tiny Brachonid wasps attack not only caterpillars but a wide range of other insects (although some species specialize in caterpillars). Tachinid flies also parasitize a wide range of insects including caterpillars such as cabbage moth, whilst ladybirds, lacewings and larval hoverflies may all also consume small caterpillars.

Birds are also generalist predators of caterpillars. Starlings and many other small bird species such as tits will take caterpillars to feed their chicks. However, some caterpillars are more palatable than others: for instance, the caterpillars of the large white butterfly actively sequester a glucosinilate called sinigrin (itself a plant defence chemical) and store it in

Caterpillar parasitized by parasitic wasp (Cortesia *spp.*). (Courtesy *Colin Newsham*)

their bodies, making them toxic to predators. Frogs and toads will also eat caterpillars if they are within reach.

All these natural enemies can be promoted by conservation biological control techniques as discussed in Chapter 3, which can be further enhanced within the crop by those mentioned in Chapter 4. In addition to promoting background biological control some inundative biological control agents have also been developed and commercialized. For instance, proprietary products are available containing the wasp *Trichogramma brassicae,* a wasp parasitoid of caterpillar eggs, provided within parasitized eggs and supplied attached to cards for distribution within the crop. Adult wasps emerge from the parasitized eggs to seek out pest eggs, and have been shown to reduce crop damage by between 40–80 per cent under some conditions. In this case timing is critical, as the parasites will not have any impact when the caterpillars have hatched. Where several generations of the pest occur within a given crop early releases, made when pest pressure is low and adults are still laying eggs, are more likely to be effective.

Direct Control Once again a range of direct control measures can be taken against caterpillar pests that are present at a level likely to cause considerable economic damage. These are most cost effectively applied as sprays. The most selective should be used for preference, and in some cases it might be necessary to mitigate their effects on non-target insect species, including natural enemies. Permission might also need to be sought before using them in organic systems. Possibly the most effective and selective biological control agent is the bacterium *Bacillus thuringiensis* (Bt), a highly selective biological insecticide which acts on the digestive system of the caterpillar after it has eaten it on the treated leaf, and whose action is described in Chapter 5. The microbial insecticide Spinosad has been approved for use in the EU under organic regulation, although its action is non-specific, so it should be used with caution. Entomopathogenic nematode treatments are also available which are effective in attacking foliar caterpillars, but are likely to require repeated application. Their use is likely to be expensive in large scale field crops.

MANAGING FLOWER, GRAIN, FRUIT AND SEED PESTS

Flower, grain, fruit and seed pests are not normally regarded as serious except in the case of cereals (Table 17) and legumes (Table 14) in UK organic production, but are nevertheless capable of affecting the marketable yield of many crops. Fruit and seed pests are commonly carried from the field into storage, where they can cause significant and serious losses, although their management is not discussed in any detail in this book. Many foliage pests in particular can also be found on flowers and fruits at various times, and the main pest species are likely to be the same, including aphids and caterpillars, although beetles can also be problematic under some conditions, and especially in storage. Pollen and blossom beetles, for example, are pests under some circumstances, as for example when they contaminate broccoli heads before harvest. Thrips are common in and around blossoms, and plant bugs (Hemiptera), weevils (Coleoptera) and pod borers (Lepidoptera) attack grain and seeds.

Susceptible Crops

Various insect pests can be problematic on flowers and developing seeds and fruits in brassicas (Table 11), cereals (Table 17), legumes (Table 15) and squashes (Table 13). Many foliage pests can also be found devouring fruits and seeds, depending on the circumstances.

Management Strategies

Cultural Management

Many of these insect pests arrive late in the crop cycle when plants are already large and diverting their resources to flowering, fruiting and setting seed. As such, many cultural control measures will be difficult and/or awkward to implement, and many of the measures effective against foliage pests might now be impractical. In addition crop plants may tolerate less interference at this stage, which is crucial for ultimate yield in many cases. In some vegetable crops the harvestable yield may well have been taken before this point, so these pests are largely irrelevant anyway.

Prompt harvesting is likely to be one of the main methods of preventing insect damage at this stage. After harvest it may be possible to take some corrective action if damage or contamination is found by washing or trimming. In the case of pollen beetles infesting broccoli, the produce can be stored in a dark room with a (small) open window or door. The beetles are attracted to the light and leave the produce, although this technique is unlikely to be completely effective.

In some cases resistant varieties are also available, and these are resistant to, or are not attractive to these pests – for instance orange blossom midge in wheat (Table 15). Appropriate treatment of crop residues will help to reduce the reservoir of overwintering pests in some cases.

Many pests that are problematic in storage can infect grains and seeds in the field before harvest, and in these cases seed cleaning and drying is likely to be very important in reducing their impact in storage. Specialist texts on this topic should be consulted.

Biological Control

The main biological control on these pests is likely to be exerted through background biological control, and in common with other pests this should be promoted through various conservation biological control methods applied at farm level and the level of the crop (Chapters 3 and 4). Parasitic wasps in particular are likely to be specialized in seeking out these pests, which would otherwise be hidden in floral structures, pods or grains.

Direct Control

Direct control methods against these insects are problematic because they are normally hidden within plant structures. Broad spectrum insecticides are also likely to affect beneficial organisms that frequently visit flowers to collect both pollen and nectar. Any treatment should therefore be carried out with an aim to targeting the pest and minimizing impact on beneficial insects.

Table 9: Key Pests of Apiacae (umbellifer crops including celery, celeriac, carrot, fennel, parsnip and parsley)

Crop	Pest	Symptoms/Damage	Life Cycle	Monitoring	Management
Carrots Celery	Cutworms (noctuid moth larvae and/or swift moths)	Cutworms may reduce yield on late-drilled crops by severing seedling plants from their taproots but they also affect quality caused by cutworm larvae mining into maturing roots	*See* Chapter 7, section on managing soil, root and seedling insect pests, p. 205	Economic damage is mostly on light soils and in dry seasons	*See* Chapter 7, section on managing specific problem soil pests, p. 205
Carrots Celery Fennel	Slugs (various species)	Slugs can be a severe problem of fennel and celery, damaging leaves and affecting marketability. They can also attack carrot seedlings and damage the roots of carrots, especially crops stored in the ground	*See* Chapter 8, section on the management of slugs and snails, p. 285	*See* Chapter 8, section on the management of slugs and snails, p. 285	*See* Chapter 8, section on the management of slugs and snails, p. 285
Carrots Parsnip Celery Celeriac Parsley	Carrot fly (*Psila rosae*)	A key pest of organic carrots. Young seedlings can be killed. Growth of older plants may be stunted and leaves reddened although often there are no foliage symptoms. The maggots mine into roots and cause dirty tunnels that are	Adults are minute flies (8mm). There are usually two distinct generations of carrot fly each season, although in some areas there may be	Root mining damage is most extensive around the edge of fields. The first generation is normally the highest risk period but this depends on crop and harvest period, with longer	Carrot fly is not a strong flier and so windy exposed sites and large fields suffer less damage. Isolating umbellifer crops as part of a rotation plan (damage

	prominent under the surface of washed carrots. Carrots left in the ground until January or later often show the worst damage as some larvae continue to feed right through the winter and can move between plants. It can also be a pest of parsnips, celery, celeriac, parsley and other umbelliferous herbs and can survive in umbelliferous weeds	three. The first generation emerges from pupae from April to early May to lay eggs around young host plants. These hatch after about seven days and the maggots burrow into the ground to feed on and in the roots. The maggots' larvae (8–10mm long) are creamy-white in colour. The second-generation flies lay their eggs in July or August. Some of these will develop into adult flies by autumn,	crops likely to suffer more damage. Celery and parsley plants are at risk when they are small as a few larvae can cause wilting and yellowing. Larger plants can withstand higher populations of larvae without visible symptoms. For parsnips and celeriac late-lifted crops are most at risk, with severity of damage increasing from November onwards. Forecasting models based on accumulated temperature data are used to predict timing of generations (rather than the number of pests) and so (sticky)	drops off after 500m), together with a combination of crop covers and drilling times, can be used to manage infestation. Maincrop carrots sown in late May or early June can miss the first generation but may be affected by later generations. Holding off irrigation at key stages in the pest life cycle can be beneficial as maggots are killed by drought. Early harvest can avoid damage from late generations.

(Continued)

Table 9 (Continued)

Crop	Pest	Symptoms/Damage	Life Cycle	Monitoring	Management
			others survive as larvae through the winter in host roots. The third generation of flies can appear in October and November	carrot fly traps should be used to measure actual pest populations at any location. Traps need to be changed weekly and can be assessed by an experienced eye or sent away	Partially resistant varieties are available (e.g. Fly Away and Resistafly) but efficacy is variable. The use of vertical barriers (1.7m, with a 0.4m external overhang) can be effective on a small scale. Sacrificial trap crops of a susceptible variety or species around the edge of the field has been tried, as has intercropping (e.g. with onions) with variable effect. Destruction of crop remains, or feeding remains to animals, can help break the pest life cycle

Celery Parsnips Parsley	Celery leaf miner (*Euleia heraclei*) (also known as celery fly)	This pest attacks celery, celeriac and parsnip leaves, but also less commonly parsley and carrots. The larvae mine the tissues between the upper and lower epidermis of the leaves, causing large blotch mines, which may cause withering and death of the leaves	Tiny (5mm) fly lays about 100 eggs inserted singly into leaves in spring and early summer. Eggs hatch in 1–2 weeks, and maggots produce characteristic mines or blisters or hollowed out areas after 2–3 weeks burrowing in the leaves. Later larvae emerge, drop to the ground and pupate in the soil. The second generation of adults emerges 3–4 weeks later,	While infestations would have to be huge to have any marked effect on yield, marketable quality can be reduced, particularly where harvesting leaves (e.g. celery leaf, parsley)	Not normally serious enough to merit specific treatment. On a small scale picking off and destroying affected leaves can help. Crop covers can be used, or sprays such as potassium soft soap, if used early in an infestation, may have an effect. In protected cropping leaf miners can be controlled by introducing the parasitic wasp *Diglyphus isaea*. This wasp thrives in hot conditions, and can

(Continued)

Table 9 (Continued)

Crop	Pest	Symptoms/Damage	Life Cycle	Monitoring	Management
			and this can be followed with a third generation in late summer. It overwinters as pupae in plant debris or in the soil. As the second and third generations overlap, attacks can occur at any time during the summer		quickly eliminate the pest
Carrots Parsnip Celery Parsley	Willow carrot aphid (*Cavariella aegopodii*)	Pest of umbelliferous crops, including carrots, celery, parsnips and parsley. Feeding activity can cause stunting of growth, but it is principally important as it transmits carrot motley dwarf virus, which is a	This aphid overwinters as eggs on the bark of willow trees, and hatches out in a winged form to infest umbellifers during May. Aphids can be	Often present late spring to early summer. Aphid monitoring is available through the HDC pest bulletins	Not normally serious enough to warrant specific treatment unless viruses are expected to be a problem, otherwise treatment as for other aphids (*see* p. 214)

		particular problem, and parsnip yellow fleck virus	found all over the leaf but commonly on the main veins. Peak infestation is in early June but continues until early July when another generation emerges to reinfest willows and hedgerow umbellifers		
Parsnips	Stem nematodes (*Ditylenchus* spp.)	Cause rotting crowns, which dry and split. Difficult to distinguish from other diseases which cause similar symptoms	*See* Chapter 8, section on managing nematodes, p. 271	*See* Chapter 8, section on managing nematodes, p. 271	*See* Chapter 8, section on managing nematodes, p. 271

Table 10: Key Pests of Asteraceae (mainly salad crops including endive, chicory, cardoon, artichoke, lettuce, salsify)

Crop	Pest	Symptoms/ Damage	Life Cycle	Monitoring	Management
Lettuce Chicory Endive	Cutworms (noctuid moth larvae and/ or swift moths)	Plants collapsed and cut through at stem	*See* Chapter 7, section on managing soil, root and seedling insect pests, p. 205	Walk crops and inspect soil around cut plants to unearth caterpillars. Especially serious when crops are recently transplanted as will remove plant	*See* Chapter 7, section on managing soil, root and seedling insect pests, p. 205
Lettuce	Wireworms (larvae of click beetles *Agriotes spp.* etc.)	Plants unthrifty with poor growth. Roots undeveloped and stunted	*See* Chapter 7, section on managing soil, root and seedling insect pests, p. 201	Most damage in spring and autumn, but *see* text	*See* Chapter 7, section on managing soil, root and seedling insect pests, p. 201
Lettuce and others	Slugs	Can be a major pest of lettuce, devouring leaves and leaving slime trails	*See* Chapter 8, section on the management of slugs and snails, p. 285	*See* Chapter 8, section on the management of slugs and snails, p. 285	*See* Chapter 8, section on the management of slugs and snails, p. 285
Lettuce	Leaf aphid including peach potato aphid (*Myzus persicae*) and potato aphid	Abundant and widespread pests which can be present throughout the	Aphids commonly overwinter on brassicas, beets, weeds and on protected lettuce and in	Damage is principally caused by their ability to transmit lettuce mosaic and other viruses.	*See* Chapter 7 on controlling problem foliage pests, aphids (p. 211). Commonly excluded by the use of crop covers or by spraying with insecticidal soap.

	(*Macrosiphum euphorbiae*)	season feeding on the underside of leaves. Aphid damage is often negligible or cosmetic, but aphids feed on many hosts and transmit many important viruses including cucumber mosaic virus to lettuce	greenhouses, generally. Also overwinter on alternative hosts as eggs (e.g. peaches, roses). Migration of winged forms begins in May and June, peaking in mid- to late July. Two or three migrations may occur in a season, and aphids move between plants within a crop when crowded	Degree of intervention depends on tolerance levels of the intended market, as aphid presence may be unacceptable. The HDC pest bulletin can provide forecasting and details of populations	Natural enemies, given sufficient time, will catch up, and will also reduce the numbers of aphids. Biological controls are available for control on both glasshouse and outdoor crops, but are more effective in protected crops
Lettuce Chicory	Currant-lettuce aphid (*Nasonovia ribisngri*)	Widespread and serious both in protected lettuce and outdoor crops. Colonies of green aphids on leaves can develop rapidly in warm weather and render plants stunted and	Overwinters as eggs on currant and gooseberry, hatching in spring. Can also overwinter on protected lettuce crops and weeds in mild winters. Migration by winged forms starts in May, with numbers peaking in June and	*See* above and Chapter 7, section on managing problem foliage pests, aphids, p. 211	*See* Chapter 7 on controlling problem foliage pests, aphids, p. 211. Resistant varieties are available. Soft soap sprays if applied early and directly on to colonies may check infestations, as can natural enemies. Biological controls are available for control on both glasshouse

(Continued)

Table 10 (Continued)

Crop	Pest	Symptoms/ Damage	Life Cycle	Monitoring	Management
		unpalatable. Also a pest of other compositae (hawkweeds), gooseberries and redcurrants	July. Return migration takes place in September/October		and outdoor crops, but are more effective in protected crops
Lettuce Endive Chicory	Lettuce root aphid (*Pemphigus bursarius*)	First noticed as plants wilt. Developing heads remain soft, fail to develop properly, and yields are reduced. Affected plants may mature more slowly and unevenly with smaller heads. Colonies, covered by white powdery wax, can be found on the roots of uprooted plants. Ants can also	Overwinters on poplar trees, especially Lombardy poplars. In June the winged forms migrate to crop hosts where they initiate colonies on the roots. Migration occurs over a period of about 4–5 weeks but can be prolonged during hot settled weather. Tiny nymphs can crawl short distances to infect neighbouring plants. The return migration occurs in September and	This can be a very serious pest, resulting in large crop losses, particularly in hot dry seasons. The start of the aphid migration from poplar to lettuce occurs when about 672 air day-degrees (above a base temperature of 4.4°C) have been accumulated from 1 February. Lettuces sown or planted between mid-April	*See* Chapter 7 on controlling problem foliage pests, aphids, p. 211. For root aphids prepare a fine seed bed avoiding cracks in soil allowing access to aphids. Resistant varieties are available. Irrigation can mitigate water loss caused by root damage. Crops raised in irrigated blocks can tolerate small populations. Adult aphids can be excluded effectively by covers (put in place before migrating aphids arrive at crop). Use mesh (rather than fleece) during

		indicate root aphid presence. Pest of other compositae, including weed species such as sow thistles	October, when aphids can be seen congregating around the stems of infected plants. They can also overwinter in the soil without host plants	and the end of June are at most risk. The HDC pest bulletin can provide forecasting and details of populations	hot weather to avoid leaf scorch, rots and misshapen heads. Biological controls are available for protected crops using parasitic wasps. Garlic sprays or granules have been claimed to reduce damage from lettuce root aphids
Lettuce	Birds, rabbits	Leaf margins pecked and or eaten. Hearts damaged and/or eaten	See Chapter 8, sections on management of rabbits, p. 306 and bird management, p. 294	*See* Chapter 8, sections on management of rabbits, p. 306 and bird management, p.294	*See* Chapter 8, sections on management of rabbits, p. 306 and bird management, p. 294

Table 11: Key Pests of Brassicaceae (brassicas including cabbages, cauliflower, broccoli, Brussels sprouts, kohl rabi, swedes, turnip, radish, cress)

Crop	Pest	Symptoms/Damage	Life Cycle	Risk/Monitoring	Management
Brassicas (all crops)	Cutworms (*Agrotis* spp. etc.) and swift moths (*Hepialus* spp.)	Plants collapsed and cut through at stem and/or roots with holes eaten out of them	Caterpillars live in or near surface of soil feeding on leaf material and/or roots. *See also* Chapter 7, section on managing soil, root and seedling insect pests, p. 205	Walk crops and inspect soil around cut plants to unearth caterpillars	*See* Chapter 7, section on managing soil, root and seedling insect pests
Brassicas	Slugs and snails	Irregular holes in leaves accompanied by slime trails or other evidence of slugs	*See* Chapter 8, section on the management of slugs and snails, p. 285	*See* Chapter 8, section on the management of slugs and snails, p. 285	*See* Chapter 8, section on the management of slugs and snails, p. 285
Brassicas (including oilseed rape)	Flea beetle (*Phyllotreta* spp. including small striped flea beetle (*P. undulata*), the large striped flea beetle (*P. nemorum*), turnip flea beetles (*P. atra*, *P. cruciferae* *P. nigripes*))	Pests of brassicas, radish and mustard. Particularly damaging to young seedlings, especially during dry weather. Beetles eat small holes and pits out of young leaves and stems, and a severe attack will check growth and can kill young plants.	Flea beetles are small (<3mm), shiny beetles. Adults pass the winter as adults, hibernating in tussocky grass, in debris under hedges and similar situations. In spring they move out on to young plants to start feeding. They may	Small glossy beetles, which jump when disturbed, will be seen on and around plants. Damage is always more severe in hot, dry weather and when crops are establishing (small plant size). Crops that are harvested for their leaves, such as	Management approaches will vary between crops depending on market. Prepare the soil well and choose appropriate sowing times to encourage rapid and vigorous growth of young plants. Irrigate regularly if necessary. Fleece or mesh (<0.8mm) will keep flea beetles off

		Large plants relatively unaffected unless leaves sold as salads	fly for a kilometre or more to find suitable food. Female beetles lay their eggs in the soil near suitable plants in May/June. The larvae feed in 'mines' in leaves or on plant roots before pupating in the soil to hatch as adults in autumn. Adults feed for a few weeks before hibernating to repeat cycle	salad rocket, oriental mustards and ironically radishes and turnips for bunching, are also at risk for longer periods. Most damage happens in April and May. There may also be an influx of beetles in summer, after oilseed rape crops have been cut	crop if put in place immediately after sowing, and some (Chinese cabbage, radish or rocket) may need to be covered throughout their cropping life; others will only need protection when young. Trap crops such as a sacrificial row or two of radishes around crop may help to protect other young brassicas. Flea beetle trolleys can be used to brush plants and stimulate beetles to jump on to grease traps
Brassicas	Cabbage aphid, (*Brevicoryne brassicae*)	The mealy cabbage aphid (*Brevicoryne brassicae*) is common on all plants in the Brassica family (especially cabbages, Brussels sprouts,	The aphids are grey-green in colour and covered with a whitish-grey, mealy wax which repels water. They form dense colonies, and	Examine young plants regularly during June–September and use forecasting services to predict periods of risk	*See* Chapter 7 on controlling problem foliage pests, aphids p 211. Techniques that have been used against cabbage aphid include removing brassica plants

(Continued)

Table 11 (Continued)

Crop	Pest	Symptoms/Damage	Life Cycle	Risk/Monitoring	Management
		cauliflowers and swedes). Dense colonies occur on the lower leaves of brassica plants from July onwards and continue increasing in numbers until the winter. They can also occur in the heads of Brussels sprouts plants. Leaves become distorted and turn yellow as a result of the aphids feeding. Severe infestations can check growth and kill young plants; the effect is less damaging on mature plants. This aphid is an important carrier of cauliflower and turnip mosaic viruses	winged forms move from one brassica crop to another throughout the year. Although they can survive the winter as adults in milder areas, they usually overwinter as eggs on the stems and leaves of brassica plants. Young aphids hatch out in April and begin feeding on leaves, moving to flower buds and stalks as the season progresses. During May to July winged aphids fly off and establish colonies on younger plants. The build-up of aphid numbers is linked to temperature – years		after cropping (especially in successional plantings during the season), sowing wildflower strips to encourage beneficials, and leaving some crop plants to overwinter as a source of natural enemies. Undersowing (with clover) has been shown to reduce the numbers but may cause management problems and have yield penalties in commercial systems. Soft soap applied directly to the colonies may have some effect if carried out regularly with good quality spraying equipment

			with mild winters see a much quicker increase in infestations. Predator numbers also increase in these conditions		
Brassicas	Cabbage root fly (*Delia radicum*)	The maggots tunnel in the roots and stems of young plants, weakening them. Plants may turn bluish or red and may wilt or die. The damage is most serious in young transplants. In older plants, damage to the root system can reduce yield, especially in dry conditions. In moist soil conditions new roots can form and the crop can recover.	The adults lay eggs at the base of the plant after overwintering in the ground as puparia, emerging from spring onwards (coinciding with the period when cow parsley is in flower). Two to three generations can take place between April and September. The females lay their eggs, which are 1mm long and white, in cracks and	The first (spring) generation tends to be larger and do more damage, but summer-planted crops such as overwinter cauliflower are also at risk from the later generations. A third generation can be severe in areas where oilseed rape is grown. A forecasting system has been developed at Warwick HRI,	Many cultural control methods can be combined into a management strategy. Cultivation and good soil preparation may help manage populations. Earthing up plants will encourage secondary root formation, and this is aided by irrigation. Fleece or mesh (<1.3mm) applied immediately after planting can exclude adult flies (but may trap adults emerging from puparia in soil in

(Continued)

Table 11 (Continued)

Crop	Pest	Symptoms/Damage	Life Cycle	Risk/Monitoring	Management
		For root crops (swedes, turnips, radishes etc.) the tunnelling can render the crop unmarketable or reduce yield by necessitating trimming. Tunnelling allows pathogens to gain access to the root, leading to soft rots	gaps in the soil around the stem of the host plants. The white maggot-like larvae hatch after about six days and enter the stem, provided it is not too woody. The maggots pupate after 3–4 weeks in the soil close to the host plant. The red/brown pupae are 5mm long and oval shaped. New cabbage root flies emerge after 2–3 weeks	which predicts the timing and duration of populations based on statistical information and weather data. This is available through the HDC pest bulletin. Populations can be monitored using trap systems	successional or double crops). Timing of application can be judged using forecasts. Undersowing companion plants (in modules or on bed) can decrease pest infestations in some situations (e.g. birdsfoot trefoil (*Lotus corniculatus*)). Planting through a mulch can reduce or prevent egg-laying, but costs will not be justifiable in most situations. Predatory beetles and other generalist predators can be encouraged to prey on larvae and pupae. Parasitoid rove beetle (*Aleochara bilineata*) have been evaluated but are not especially effective. Spinosad can

					be used as a pre-planting drench on block-and module-raised plants. Garlic sprays or granules can check root fly larvae when used as a plant stimulant
Brassicas	Caterpillars (large cabbage white butterfly (*Pieris brassicae*), small cabbage white butterfly (*Pieris rapae*), green-veined white butterfly (*Pieris napi*), cabbage moth (*Mamestra brassicae*), diamond-back moth (*Plutella xylostella*), the garden pebble moth (*Evergestis forficalis*)	Brassicas are all attacked by a range of caterpillars. Damage will depend on the species, type and season. Caterpillars of cabbage moth and small white butterfly can be more damaging as they bore into the hearts of cabbages, whereas others feed on the outer or lower leaves which can be discarded. In some cases fouling with frass is a problem. Caterpillars can	There are two to three generations a year in summer and autumn, depending on temperature and species. The first generations normally occur in May and June, and the second generation starts to emerge and lay eggs from early August onwards. This generation is usually more damaging. The behaviour of species can differ: large whites often feed in	Inspect plants regularly by walking crops and checking for eggs, caterpillars and damage. It is also important to note any natural enemies present. Pheromone traps are available to monitor males of the diamond-back moth. The HDC pest bulletins will also give updates of population movements. Damage often occurs in the headlands or in sheltered places	*See* Chapter 7 section on managing problem foliage pests, caterpillars, p. 216. Commonly used management methods include fleece or mesh. Bt (*Bacillus thuringiensis*) applied directly on to areas where caterpillars are feeding can be effective. Parasitic wasps (Brachonids, Chalcids and Ichneumonids), flies (Tachinids) and generalist predators can be encouraged by flower strips

(Continued)

Table 11 (Continued)

Crop	Pest	Symptoms/Damage	Life Cycle	Risk/Monitoring	Management
		damage the crowns or growing points of young plants, severely stunting growth. They may also bore into heads of calabrese or cauliflower, or in the flower buds of stalks, causing direct damage	conspicuous colonies whilst small whites are solitary. Diamond-back moths feed under silken webs which protect them to some extent and can sometimes produce four generations per year	and can therefore be worse in small plots. Damage is normally more severe from summer onwards, and worst during hot, dry summers	
Brassicas	Cabbage whitefly (*Aleyrodes brassicae*)	Common on many brassica types (especially Brussels sprouts, broccoli, kale and cabbage). Small white-winged adults live on the underside of leaves of brassica plants, and fly up in clouds when disturbed. The young whitefly, known as 'scales', remain on the leaves. The whitefly themselves do not often cause severe	Adults are small, white, up to 2mm long, with two pairs of wings folded over the back of the abdomen when at rest. Immature whiteflies are seen as creamy yellow-coloured scales found on the underside of leaves. Adults overwinter on brassicas and move on in the spring to lay eggs from mid-May onwards in	Whitefly can be active at most times of the year once temperatures are sufficient	Increasing generalist and specialist predation and parasitism is the best way of managing whiteflies. Flowers such as coriander, fennel, cow parsley, lovage and other members of the Apiaceae (Umbelliferae) and Asteraceae (Compositae) families will attract parasitic wasps such as *Aphelinus*. Lacewings, predators which feed on whitefly, are attracted by members of the

	damage, but the presence of the scales can make leaves unappetizing, as can the sticky honeydew that covers them, exuded by the feeding insects. Sooty or black moulds often grow on the honeydew, spoiling leaves and flower buds (e.g. broccoli). The covering of mould will reduce the amount of sunlight reaching the leaves, impairing their ability to photosynthesize, thus reducing cropping	semi-circular groups on the underside of leaves. The tiny nymphs which emerge are mobile for a short while, but then settle down to feed. They insert their needle-like mouthparts into the plant tissues and suck the juices. The nymphs become immobile and develop a scale-like protective covering, which darkens with age and in which the larvae develop and pupate. The life cycle from egg to adult lasts about four weeks in the summer, and will continue through to the autumn, when both eggs and adults will overwinter on brassica plants		Asteraceae (Compositae) family such as dandelions, yarrow and shasta daisy. As a last resort potassium soft soap can be used, but care should be taken to target the adults directly on the underside of leaves and is more effective when adults are not so active (e.g. early morning when temperatures are low). Spray under the leaves, using a good quality sprayer. Spraying once a week for 3–4 weeks may be necessary to see a significant effect

(Continued)

Table 11 (Continued)

Crop	Pest	Symptoms/Damage	Life Cycle	Risk/Monitoring	Management
Brassicas (including oilseed rape)	Thrips (a number of different thrip species attack brassicas, but *Thrips tabaci* is the most common)	Feeding damage by adults and larvae can result in wart-like spots (oedema) and silvering or bronze-like stripes, particularly in headed cabbage. If the head of a cabbage continues to form after thrips have started feeding, damage can continue during storage, making the cabbage unsaleable. It is also a pest of leeks, onions, carrots, winter cereals and clover	Thrips survive the winter as eggs, larvae or as wingless adults. Infestation can come from neighbouring plots due to the large number of host plants. Development from egg to adult insect takes between 14 and 30 days, depending on temperature. Winged adults fly to colonize brassica crops in May or June. Eggs are laid on plants and the nymphs feed for a week before pupating. Populations can explode in hot and dry weather (they are often called thunderflies), badly affecting the storage quality of cabbages	Because of their small size (0.8–1.2mm) and the fact that they prefer to hide within the crop, monitoring can be difficult. Adults can also be monitored by the use of yellow sticky traps	Good husbandry practice and avoiding crop stress will help to suppress thrip damage. Some white cabbage varieties are moderately resistant to thrip damage, particularly to the development of oedema. Either fleece or a very fine mesh covering (*see* thrips as a pest of alliums p. 256) can be used. It is difficult to control once thrips are present and the cabbages begin to heart. Spinosad is licensed for leafy brassica crops (including rocket) grown for baby leaf production (i.e. crops harvested up to eight true leaf stage)

Brassicas	Swede midge (*Contarinia nasturti*)	All brassica crops can be affected, particularly at the seedling stage. It occasionally causes severe localized damage in the growing points of young plants, resulting in the death of the plant. In some types of cabbage, cauliflower and broccoli it can cause 'whiptail': after the midge has destroyed the growing point the leaf starts to twist in the growing point, causing abnormal growth	The swede midge overwinters in the ground, the first of three generations appearing in the second half of May and the beginning of June. Eggs are laid in batches of 15–20 on young leaves, and the larvae hatch out after a few days and feed mainly on the growing points	Midge numbers build up rapidly in periods of high humidity; drought slows up or stops development	Crop rotation is important. Also avoid cultivating on plots close to plots that were affected by swede midges in the previous year. Crop covers should work, provided there is no previous history of infection on that plot. Potassium soft soap, sprayed when the larvae emerge from the eggs, can be effective
Brassicas (including oilseed rape)	Pollen beetles (*Meligethes aeneus, M. viridescens*)	Also known as blossom beetles, most damage is caused by adults, which can damage the developing florets of calabrese and	Adult beetles are greenish-black, 1.5–3mm long and with clubbed antennae. They overwinter in field margins or	Beetles can be seen on flowers in crop	Try to avoid neighbouring oilseed rape crops (in the case of field vegetables). Fleece and fine mesh will exclude these beetles. If they are

(Continued)

Table 11 (Continued)

Crop	Pest	Symptoms/Damage	Life Cycle	Risk/Monitoring	Management
		cauliflower. They can result in greater losses to seed crops by totally destroying flowers. More common because of the increasing area of oilseed rape being grown	other sheltered sites, emerging in the spring. After about four weeks the adults fly to brassica crops, feeding on buds and flowers. One to three eggs are laid in flower buds, which hatch about a week later. The larvae feed on pollen for 25–30 days before dropping to the soil and pupating. Adult beetles emerge 2–3 weeks later		present in cut calabrese or broccoli, boxes should be left in a dark place with open access to a lit area: the beetles will be attracted to the light, away from the vegetables
Brassicas	Birds (especially pigeons)	Rooks can remove transplants as they search for grubs and food underneath, necessitating replanting. Pigeons peck and tear leaves, sometimes reducing them to a skeleton	*See* Chapter 8, sections on bird management, p. 294	The risk is higher on smaller scale plantings or near colonies or roosting sites. Overwintered crops are at risk when there are few other sources of food around	*See* Chapter 8, section on bird management (p. 294), but scaring devices can have some deterrent effect depending on the availability of alternative food sources

Table 12: Key pests of Chenopodiaceae (salad and root beets, chard, spinach)

Crop	Pest	Symptoms/Damage/ Period	Life Cycle	Monitoring/Risk	Management
Beet Chard Spinach	Mice	Attack seeds at germination causing gappy stands and non-emergence	*See* Chapter 8, section on management of rodents and other small vertebrates, p. 303	Checking along the drill for signs of mouse activity and/or discarded seed remains	*See* Chapter 8, section on management of rodents and other small vertebrates, p. 303
Beet Chard Spinach	Black bean aphid (*Aphis and/ or fabae*)	Dense colonies of black aphids on young shoots and/or leaves (can also be found on beans and many other plants). May cause distortion of shoots and leaves and stunting of plants. Vector of viruses (especially bean yellow mosaic)	Wingless females or more often eggs overwinter on *Viburnum* and *Euonymus* to give rise to colonies in early spring, which originate winged adults that migrate to crop hosts in summer. Adults return to winter hosts in autumn to continue cycle	Capable of rapid increase in populations in favourable weather, especially following mild winters. Colonies can be monitored by crop walking and observation	*See* Chapter 7 on controlling problem foliage pests, aphids, p. 211. Infestations can be pinched out in small scale production. Treatment with soft soap early in colonization phase may reduce infestation. Natural background controls should be encouraged
Beet Chard Spinach	Beet leaf miner or mangold fly (*Pegomya hyoscyami*)	Maggots tunnel in leaves making large blotch mines. Severe attacks may affect marketable yield of leaves and check growth, but older plants generally unaffected	Adults emerge from puparia in soil in spring. Eggs laid on underside of leaves and larvae eat into leaves and feed before dropping to soil to pupate. Two to three generations are possible per season	Main period of risk is early summer when the potential for damage is highest. Not generally regarded as a serious pest	Tillage and cultivation can reduce puparia in soil. Establish plants quickly under favourable growing conditions to move crop through vulnerable phase quickly

Table 13: Key Pests of Cucurbitaceae (cucumber, courgette, squashes, marrow, pumpkin, melon)

Crop	Pest	Symptoms/Damage	Life Cycle	Monitoring	Management
Squashes Cucurbits	Slugs (various species)	Pests of new transplants especially under plastic mulch. Can cause damage to stems, flowers and buds as well as soft-skinned fruit such as courgettes	*See* Chapter 8, section on managing slugs and snails, p. 285	Moist soils with high organic matter content are favourable to slugs and they can collect under mulches. *See* Chapter 8, section on managing slugs and snails, p. 285	*See* Chapter 8, section on managing slugs for general management tactics, p. 289
Squashes Cucurbits	Pollen beetles (*Meligethes* spp.)	Small shiny beetles in flowers, often in large numbers. Flowers may be destroyed	Adults overwinter in soil and emerge in late spring (April–May) to lay eggs on unopened flower buds. Larvae hatch and feed on pollen before dropping to soil to pupate	Can be troublesome in areas where oilseed rape is commonly grown, as large populations build up	Not generally necessary or possible to control these pests. Direct control methods likely to be problematic as they feed in the flowers. Wasp parasites should be encouraged by conservation biological control
Squashes Cucurbits	Whitefly (in glasshouse, *Trialeurodes vaporarium*)	Large infestations reduce plant vigour, and secondary infections of sooty	Mainly a pest of glasshouses in the UK. Parthenogenic females lay eggs on underside of leaves.	Monitor (glasshouse) plants regularly and search for adults and nymphs under leaves. Yellow sticky traps	Soft soap may be applied to the underside of leaves before the infestation becomes fully

		mould may reduce photosynthesis	Hatching larvae are initially mobile but settle to become mobile scales before pupating to give rise to adults	can be used to monitor adults	established, and be repeated regularly. Biological controls are available for glasshouse use, especially the parasitic wasp *Encarsia formosa*, and should be deployed under specified conditions for best effect
Squashes Cucurbits	Root knot nematodes (*Meloidogyne* spp.)	Galls develop on roots and may coalesce to give roots a knotted appearance	Soil-dwelling organisms. *See* Chapter 8, section on managing nematodes, p. 271	*See* Chapter 8, section on managing nematodes, p. 271	Hygiene to prevent spread of nematodes. *See* Chapter 8, section on managing nematodes for other measures, p. 276
Squashes	Rodents and badgers	Eat fruit leaving holes and teeth marks, or move fruit looking for prey. Damage prone to secondary rots	*See* Chapter 8, section on managing vertebrate pests, p. 303, 310	Mainly visual monitoring of the field, and in the case of badgers, damaged fences	*See* Chapter 8, section on managing vertebrate pests for general exclusion methods, p. 305, 313

Table 14: Key Pests of Fabaceae (legumes including runner beans, French beans, field beans, peas, leguminous green manures)

Crop	Pest	Symptoms/Damage	Life Cycle	Monitoring	Management
Peas	Mice	Seeds fail to germinate or seedlings grow poorly and appear damaged. Many missing or die	*See* Chapter 8, section on management of rodents and other small vertebrates, p. 303	Check along the drill for signs of mouse activity and/or discarded seed remains	*See* Chapter 8, section on management of rodents and other small vertebrates, p. 303
Peas	Slugs, millipedes	Poor germination and growth. Cotyledons with pieces eaten out of them	*See* Chapter 8, section on the management of slugs and snails and other arthropods, p. 285	Can be monitored by trapping. *See* Chapter 8, section on the management of slugs and snails, p. 285	*See* Chapter 8, section on the management of slugs and snails, p. 285
Broad beans Runner beans French beans Peas Clovers	Bean seed flies (*Delia platura, Delia floriga*)	Larvae (small white maggots) attack the growing point and seed leaves of germinating peas and beans. The seeds are then vulnerable to fungal rots, and in many cases, the seedlings fail to emerge. 'Snake-head' condition in runner beans is caused when growing points are destroyed. Also	Adults are small flies that emerge in the spring to feed on nectar and pollen, laying eggs near germinating seedlings. The eggs hatch after 2–4 days, and the larvae move down into the soil to feed on seedlings. After about three weeks, the larvae pupate in soil close to the host plant.	Females prefer to lay eggs on freshly disturbed soil and attacks are less likely to be serious where seeds have been sown into a stale seed bed. More common where there is high soil organic matter and/or the peas and beans are struggling to germinate in cold soil. Usually only the	Raise plants under glass as blocks or modules for transplanting. Delay sowing outside until the ground has warmed up. Sow into well prepared seed beds, and ensure adequate water. Plastic mulches can prevent egg laying, and covering with fleece or mesh immediately after

		attack other crops (especially onion and leeks)	There can be as many as four generations a year	first generation that causes major losses (due to large numbers coinciding with vulnerable stage of the crop). Often occur in local hot-spots	sowing are other options (but probably of limited effectiveness)
Legumes (all crops, but especially broad bean)	Black bean aphid (*Aphis fabae*)	Common on many other crops such as vetches, beets, spinach and common weeds such as fat hen and docks. Black bean aphids cluster together and form colonies on the undersides of leaves as well as on new growth. Feeding activity reduces the plants' vigour, leading to distorted growth and failure of shoot and flower	Adult aphids are up to 2mm long and elliptical in shape. Generally black in colour but variable from brownish-black to olive green or purple. They overwinter as eggs on the bark of spindle (*Euonymus eurpaeus*) and snowball trees (*Viburnum opulus*). The primary migration from overwintering hosts to bean crops is	Crops that are flowering during the period from mid-May to mid-June are at most risk. Late-sown crops of broad and field beans are generally at greater risk. Monitoring of aphid numbers on overwintering hosts can alert growers, through warnings, of the scale and timings of likely infestations. Aphid populations initiated after	*See* Chapter 7 on controlling problem foliage pests, aphids, p. 211. Encouragement of natural predators is recommended as part of a management plan. Fleeces or mesh can be used for higher value crops. Growing of partial resistant varieties of field beans, or the intercropping of field

(Continued)

Table 14 (Continued)

Crop	Pest	Symptoms/Damage	Life Cycle	Monitoring	Management
		development. These aphids can be very damaging if colonies develop just prior to flowering. As they feed, the aphids secrete honeydew, which can become infected with sooty mould. They can also transmit several viruses	from mid-May to mid-June. The life cycle takes from 8–5 days, depending on temperatures. In autumn, winged adults migrate to the overwintering hosts. Wingless adults can sometimes overwinter on some herbaceous plants or weeds	flowering cause little damage	beans with spring cereals separately, has been shown to reduce the infestation of black bean aphids. As a last resort there are organically acceptable sprays, such as potassium soft soap, that can be used with permission from your control body
Pea Clover	Pea aphid (*Acyrthosiphon pisum*)	Mainly affects peas, but can also be found on beans, clover, other legumes and weeds. Pea aphid feeds on the young growing points causing stunting, distortion and	A relatively large aphid living all year on leguminous plants, and which often drop to the ground when disturbed. Eggs and adults overwinter on clovers, lucerne, sainfoin and trefoils.	Crops are most susceptible as they start to flower, particularly in late June/early July when the population peaks. Conventional growers use a population threshold	No specific measures, just as for other aphids, but *see* Chapter 7 on controlling problem foliage pests, aphids, p. 211

		yellowing of leaves and pods, but rarely causing major damage. Heavy infestation can significantly reduce yields. Pea aphid can also transmit viruses (including pea leaf roll virus, pea mosaic virus and bean leaf roll virus)	Eggs hatch in February to March to give rise to winged forms during May, which then migrate to peas and other legumes, where they remain for the summer months. Numbers can build up rapidly, peaking in late June and early July, and populations remain on successive sowings of peas through to early autumn. There is a small autumn migration in late September back to the overwintering sites	of 5 to 10 per cent of growing tips of peas infested with colonies to guide growers in control programmes	
Peas Broad beans Clover	Pea and bean weevils (*Sitona lineatus*)	These weevils eat u-shaped notches out of the edges of the leaves of peas, broad	Adults (small, brown/grey in colour and short-snouted) overwinter in piles of	Damage can appear in early spring and also in late summer/early autumn, but is	Encourage strong, fast growth by providing plants with the best possible growing

(Continued)

Table 14 (Continued)

Crop	Pest	Symptoms/Damage	Life Cycle	Monitoring	Management
		beans, clovers and vetches. Small, scalloped edges can be seen on the leaves of preferred plants. Heavy infestations of larvae can reduce yields, and reduce the protein content of peas and beans – but this is of little importance except on poor soils. Can occasionally transmit broad bean stain and bean true mosaic virus, causing leaf malformations and reducing yields	leaves, tufts of grass and other debris, or hibernate amongst leguminous plants such as vetches or overwintered broad beans. In early spring they begin feeding on pea and bean leaves, and the female lays eggs in soil around the plants. The larvae hatch two weeks later and feed on root nodules during late spring/early summer. They pupate in the soil and the adults emerge 2–3 weeks later, often after the peas and beans have been harvested, moving on to other hosts (in late July to early August) before moving to hibernation sites in autumn	normally only a problem when the weevils are extremely numerous. Severe infestations can cause seedling losses, especially in cold and wet conditions, but older, established plants are little affected. Early sown crops in a dry spring can be more severely affected by weevils. Later sown crops are less likely to suffer severe attacks. Monitoring is possible through commercial traps	conditions; avoid cloddy (preferred) seed beds. Leguminous overwinter cover crops may increase the numbers of adults that survive hibernation, and alternative (less preferred) legumes, such as clover, should be sown when problems are anticipated. Alternatively sow a non-leguminous green manure such as grazing rye, prior to pea or bean crops. A fleece or fine mesh cover can be used to protect young plants (if put in place immediately after sowing) but is unlikely to be justifiable

Pea Vetch	Pea moth (*Cydia nigricana*)	Pea moth can be a serious pest of field and garden peas and can also attack other legumes, including vetches. Caterpillars feed on peas/ seeds inside pods. Tolerances for damage are very low and infested crops can be rejected by processors. It is generally not a problem for mangetout types where the pods are harvested before the seeds have developed	The adult moth is inconspicuous, emerging from cocoons in late May to early June. Eggs are laid on the foliage of host plants, hatching after one week to bore into pods to feed on developing seeds. In late summer the caterpillar bores its way out of the pod and descends to the soil to form a tough silken cocoon where it overwinters. There is only one generation per year	Damage is usually most severe in summer (July/ August). Pheromone traps can be used for monitoring purposes	Cultivation can expose cocoons to predation. Early or late sown crops are less vulnerable to attack as they flower and form pods outside the egg-laying period (June/July). Mid-season crops can be covered with fleece or mesh during June to mid-August to stop pea moth females laying eggs on the plants (though practically this may be difficult). Crop rotation is important: avoid vetches prior to the pea crop, which might allow moths to emerge from underneath the crop (and any covers)

(Continued)

Table 14 (Continued)

Crop	Pest	Symptoms/Damage	Life Cycle	Monitoring	Management
Pea	Pea midge (*Contarinia pisi*)	The pea midge causes local and sporadic damage in Britain, particularly in areas of Lincolnshire and Yorkshire where it can be a serious pest of vining peas, causing yield loss in some seasons. Feeding activity of the larvae causes 'nettle heading' symptoms, distorting or killing the shoots and flowers if attack occurs at early crop stages	Adult midges emerge from the soil during June. The females emit a pheromone to attract males, and lay their eggs on shoots, flower buds and the young pods of host crops. Eggs hatch 4–5 days later, and the larvae feed for about 10 days on plant tissues. This can cause the plants to become sterile, and botrytis infection can also develop, which will further reduce yields. They then drop to the soil and pupate. Some emerge after 10–11 days as a second generation, whereas others overwinter to emerge in the following or subsequent years	Pea midge is not usually troublesome out of traditional pea-growing areas. Monitoring can be done with pheromone traps, and warnings are available to inform growers of the need to inspect their crops for signs of midge activity. Outbreaks are difficult to predict as a combination of weather, soil conditions and locations of previous midge infestations come into play	Varieties of peas with an extended flowering period can compensate for some loss of flowers and pods. Avoid growing crops in close proximity to a previous year's infested crops

Legumes	Bean seed beetle (*Bruchus rufimanus*)	The beetle, sometimes known as the bruchid beetle, can affect both winter and spring varieties of peas, beans and other legumes. It can cause serious losses in the field, and can destroy 20–35 per cent of seeds in some areas. Adults emerge from the seed leaving a circular hole. This can destroy the seed, or the seed may rot due to secondary infections. The beetles do not breed in grain stores, but damaged produce may not be accepted	The adult is about 3–5mm long and has black markings with white hair on the wing cases. Adults fly from their overwintering sites such as in hedgerows and verges to beans during flowering in late May or June. They feed on pollen and lay eggs on the surface of the developing pods. Larvae hatch and bore through the pod and into the seed, where they feed. As the seeds mature and dry, the fully grown larvae pupate. Adults	Numbers of the beetle are difficult to predict	NIAB have been working assembling many parent and exotic bean breeding lines to assess resistance characteristics as part of Defra's Sustainable Arable Link Programme. Crop hygiene is important, including efficient disposal of crop residues. Thorough inspection of seed for signs of damage is necessary, especially when home-saved seed is used

(Continued)

Table 14 (Continued)

Crop	Pest	Symptoms/Damage	Life Cycle	Monitoring	Management
		for human consumption	emerge through the seed coat leaving circular holes in store, or the insect may overwinter in the seed		
Peas Broad beans Red clover	Nematodes (various species including stem and cyst nematodes (*Heterodera* spp., etc.)	Poor, stunted and unthrifty plants are often apparent as patches of yellow plants within the field. Plants have cysts on their roots and may die	Soil-dwelling organisms. *See* Chapter 8, section on managing nematodes, p. 271	*See* Chapter 8, section on managing nematodes, p. 271	Field hygiene to prevent spread of nematodes. *See* Chapter 8, section on managing nematodes for other measures, p. 271
Peas	Birds	Seeds missing, seedlings pecked or pulled out of soil, killing plants. Later pods pecked and/or removed. Peas missing from damaged pods	*See* Chapter 8, section on bird management, p. 294	*See* Chapter 8, section on bird management, p. 294	*See* Chapter 8, section on bird management, p. 294, but scaring devices can have some deterrent effect depending on the availability of alternative food sources

Table 15: Key Pests of Lilaceae (allium crops including onions, shallots, leeks, garlic, chives, asparagus)

Crop	Pest	Symptoms/Damage	Life Cycle	Monitoring	Management
Onions Shallots Leeks Garlic Chives	Stem nematodes (*Ditylenchus* spp.)	Plants swell at the base early in the season, and leaves can be swollen and distorted (onion bloat). Bulbs prone to crack and rot	More or less typical life cycle of nematodes. *See* text in Chapter 8, p. 274	*See* text in Chapter 8, section on nematode management, p. 276, but mainly from symptoms on plants	*See* text in Chapter 8 on nematode management for general approach to control, p. 276
Onions	Onion fly (*Delia antiqua*)	Attacked leaves become yellow and wilt, and seedlings may disappear. When maggots have finished on one seedling they move on to an adjacent one, so seedlings tend to be killed in patches. In older plants they feed in the bulb, working upwards	Overwinter in the pupal stage in soil (of previous allium fields) to emerge in May. Flies are responsive to onion odours and lay their eggs on plants, or on soil nearby. Maggots emerge after three days to burrow into the bulbs or leaves. After three weeks they return to soil to pupate to emerge as adults in about 17 days, to repeat two or three generations (in	Check plants and monitor for bulb damage. Damage most likely in summer (June and July). Females prefer to lay eggs in freshly cultivated ground and soils high in organic matter, and these factors increase the probability of attack	Good sanitation, post-harvest ploughing, crop rotation and siting crops distant from past infestations will help manage the pest. Crop covers could also be used during the most vulnerable periods, but are unproven. Affected plants should be removed and burnt, if practicable

(Continued)

Table 15 (Continued)

Crop	Pest	Symptoms/Damage	Life Cycle	Monitoring	Management
			July, and August–September), with increasing numbers of pupae becoming dormant to overwinter		
Onions Leeks	Onion thrips (*Thrips tabaci*)	Thrips feed on the leaves, rupturing the surface cells so they appear silvery white from the reflection of air trapped inside the ruptured tissues; leaves may later become distorted or blasted, and can wither and die. More noticeable and serious on leeks, reducing marketable yield and necessitating extra trimming	Complete life cycle, adult insects are minute, 2mm long, slender and tapered with two pairs of wings. Thrips can complete their life cycle in 11 to 30 days, depending on temperature. Breeding is largely parthenogenic and males are rare. Eggs are laid singly in plant tissue and hatch in 5–10 days; larvae feed for several days on the youngest leaves at the top of the plant before	Thrips are easily detected by visual inspection of the crop by their characteristic damage and presence. Adults can also be monitored by the use of yellow sticky traps. Numerous in hot 'thundery' weather	Ploughing in crop debris can help reduce pupae numbers, as can separating allium crops by space and time (rotation). Thrips are not good flyers, but they move long distances on the wind. Early sown plots should be planted upwind (relative to the prevailing winds) of later plots to reduce migration between crops. Heavy rain or irrigation can help reduce symptoms.

			becoming pre-pupae. The pupal stage follows in soil or leaf litter. The adults may live up to twenty days, with one or two generations a year in the UK		Crop covers can also be used, either fleece or fine mesh (0.17 x 0.37mm mesh will exclude thrips), but as with other crops, the covers need to be in place before the pest arrives
Leeks Onions Chives Garlic Shallots	Leek moth (*Acrolepiopsis assectella*)	Leaves of leek and onion develop whitish-brown patches where caterpillars are feeding within the leaves. Young caterpillars tunnel into the leaves; older caterpillars tunnel down into the stem and bulb causing extensive damage. Secondary rotting occurs within areas of the stem, causing parts to become brown and slimy. As	Adult moths overwinter in plant debris. As temperatures rise in the spring, the moths start to become active and egg laying begins. Caterpillars hatch after a week or so and tunnel into the leaves to feed. They mine the leaves, leaving the outer skin intact, eating down through the outer leaves to feed on the inner leaves and growing points of stems. The	The first generation of caterpillars causes damage in May and June. As these become adults and lay more eggs, a second generation of caterpillars emerges to cause damage from August to October. Most damage is caused by the second generation, especially after warm, dry summers. Pheromone traps can be used for monitoring purposes	Crop hygiene (disposing of severely infested plants) and rotation is effective in reducing numbers. Cultivation will disturb overwintering adults and pupae and expose them to predators. Birds, bats, hedgehogs, frogs and beetles will eat adult moths, pupae and caterpillars. Keep leeks watered in summer to ensure strong growth as larger plants are

(Continued)

Table 15 (Continued)

Crop	Pest	Symptoms/Damage	Life Cycle	Monitoring	Management
		damage becomes more extensive, leaves start to turn yellow with brown patches. Silken cocoons on the leaves may contain brown pupae	caterpillars feed for about a month before crawling back up the leaves to pupate in cocoons. Two generations of caterpillars per year (in UK), one in spring and one in late summer. Adult moths and pupae of the later generations hibernate in plant debris to overwinter, emerging in spring to start the cycle over again	and forecasts are available. Check leek plants for damage in the spring. Caterpillars pupate in white silk cocoons found within the foliage. They contain the reddish-brown pupae	more tolerant of damage and can survive to produce usable crops. *Bacillus thurungiensis* can be used for caterpillar control, although caterpillars may be hidden in plants. Crop covers can be used from seedling stage onwards to prevent adult moths from laying eggs. Later plantings, after May, can avoid the first generation of caterpillars
Asparagus	Asparagus beetle (*Crioceris asparagi*)	Leaves and stems eaten, with many black larvae or colourful beetles present. Persistent infestations may defoliate and weaken plants. Can be serious on shoots	Adults overwinter in soil and plant debris, and emerge in May to feed on shoots and lay eggs on leaves. Larvae feed on leaves and drop to soil to pupate. 2–3 generations per year	Look for larvae, beetles and defoliation to plants	Remove manually or tolerate. Although highly visible, only occasionally a serious pest

Table 16: Key Pests of Solanaceae (potatoes, tomatoes, peppers, aubergine)

Crop	Pest	Symptoms/Damage/ Period	Life Cycle	Monitoring/Risk	Management
Potato	Wireworms (*Agriotes* spp. and others)	Thin brown larvae tunnel into tubers leaving small 'clean' holes that may, however, be enlarged by slugs or other pests	*See* Chapter 7, section about managing specific soil pests, wireworms for general life cycle, p. 201	Tend to be problematic after the ley phase. *See* Chapter 7, section about managing specific soil pests, p. 201	*See* Chapter 7, section about managing specific soil pests p. 201, but early lifting helps to prevent damage accumulating on tubers in wireworm-infested soil
Potato (and other Solanaceae)	Aphids (*Macrosiphum euphorbiae, Aulacorthum solani, Myzus persicae* but also many other species)	Medium to large green or pink aphids (depending on species) with a wide host range including potatoes and tomatoes and other crops. Physical damage due to aphids limited unless present in large numbers. Mainly important as virus vectors (e.g. potato leaf roll, potato virus X and virus A (crinkle))	Aphids overwinter on host plants in protected situations or (eggs) may overwinter on alternative hosts (e.g. roses, peaches). Capable of rapid increases in numbers as weather warms from spring onwards. Winged aphids spread virus infestations through crops	Aphid forecasts are available. Walking crops will confirm presence of individuals or colonies, but only very small numbers are necessary to transmit viruses	*See* Chapter 7, section on managing problem foliage pests, aphids, for general prevention techniques (p. 211), as treatment on a field scale unlikely to be economical. Use certified seed, as virus damage is often greater in saved seed

(Continued)

Table 16 (Continued)

Crop	Pest	Symptoms/Damage/ Period	Life Cycle	Monitoring/Risk	Management
Potato (and other Solanaceae)	Colorado beetle (*Leptinotarsa decemlineata*)	Large black and yellow beetles and red larvae feed on potato leaves and can defoliate plants to cause extensive yield reductions. Not present in the UK but extensive in other areas	Adult beetles hibernate in soil to emerge in spring or early summer to seek new crops. Females lay large number of eggs, which give rise to larvae that feed on leaves before dropping to soil to pupate. These give rise to adults that will produce a second generation, or hibernate	Crop walking to monitor conspicuous beetles and larvae	Isolation of crops in rotation may reduce infestation. Mechanical removal methods to dislodge beetles may be effective, or barrier methods to prevent them arriving on crop (but likely to be uneconomic on field scale potatoes). Flaming potatoes after emergence to kill feeding beetles has been used, but may damage the crop. Any suspected infestations in the UK must be reported to Defra
Tomatoes Peppers	Red spider mites (*Tetranychus urticae*)	Light speckling and yellow spotting on leaves. Later leaves become discoloured	*See* Chapter 8, section on management of mites, p. 281	Generally worse in hot, dry conditions but at any time in protected crops. *See* text p. 283	Good husbandry accompanied by physical methods and biological

		or bronzed and covered by webbing. Mites and eggs present on underside leaves under magnifying glass			controls. *See* Chapter 8, section on management of mites, p. 283
Tomatoes Peppers	Whitefly (*Trialeurodes vaporariorum* and others)	Clouds of small 'white flies' arise from underside of leaves when disturbed from sticky plants covered in honeydew. Severe infestations reduce vigour in plants and leaves might become discoloured	Adults lay eggs on underside of leaves, which hatch giving rise to nymphs, which settle to feed becoming immobile scales before pupating and giving rise to new adults. Females parthenogenic	Even small infestations can rapidly increase in protected crops. Yellow sticky traps can be used to monitor populations, although frequent visual inspection will be sufficient	Treat early stage infestations with insecticidal soap and then introduce biological control agents (e.g. *Encarsia*). Good husbandry and adequate ventilation to reduce plant stress may help manage symptoms
Potato Tomato (and other Solanaceae)	Cyst nematodes (*Globodera* spp.)	Poor unthrifty plant growth with reduced yields. Patches of yellowed plants in field. Plants may	*See* Chapter 8, section on managing nematodes for general life cycle, p. 274. Eggs contained in cysts which are	Land used to grow potato crops is always at risk of nematode build-up. Soil tests are available that can	*See* Chapter 8, section on managing nematodes for generic nematode management methods, p. 276.

(Continued)

Table 16 (Continued)

Crop	Pest	Symptoms/Damage/ Period	Life Cycle	Monitoring/Risk	Management
		yellow and die in severe attacks. White or yellow cysts present on roots (visible with a hand lens)	the dead bodies of female nematodes and may remain dormant in soil for many years. Hatch when host plants grow nearby due to root exudates and young nematodes locate and penetrate roots to feed. Females eventually burst through root epidermis although head remains attached. After fertilization develop into brown cysts which remain in soil	assess the number of cysts in soil and predict the likelihood of severe infestation	Most effective methods are rotation (potatoes not grown in the same place for at least four years and preferably longer), planting early, or using trap crops (e.g. *Solanum sisymbriifolium*). Some varieties show resistance to some strains or species of nematodes

Potato	Tuber nematode (*Ditylenchus destructor*)	Tubers cracked and wrinkled with internal pockets of nematode infestation. Infected seed potatoes fail to establish well and plants unthrifty. Eventually causes plant collapse and death	*See* Chapter 8, section on managing nematodes, p. 274, but this nematode can survive on weeds and soil fungi but enter through wounds or lenticels and feed on plant tissues. Dormant in dry conditions	*See* Chapter 8, section on managing nematodes for monitoring methods, p. 276	*See* Chapter 8, section on managing nematodes for generic nematode management methods, p. 276
Potato	Slugs (various spp.)	Large holes and cavities excavated into tubers, with slugs sometimes present. Other pests such as chafers, cutworms and swift moths may cause similar damage	*See* Chapter 8, section on managing slugs and snails, p. 285	Moist soils with high organic matter content are favourable to slugs. *See* Chapter 8, section on managing slugs and snails, p. 285	*See* Chapter 8, section on managing slugs (p. 285), although some varieties seem resistant or not preferred by slugs. Early lifting will prevent damage, and potatoes should not be left in infested soil

Table 17: Key Pests of Poaceae (cereals (Gramineae) including wheat, barley, oats, sweetcorn)

Crop	Pest	Symptoms/Damage/ Period	Life Cycle	Monitoring/Risk	Management
Cereals	Wireworm (various species)	Most active in early spring and late summer/ early autumn, causing damage to crops at or below ground level	*See* management of key soil pests in Chapter 7, p. 201	Heavier soils after grass leys, but *see* Chapter 7, section on managing soil problem pests, p. 201	*See* Chapter 7, section on managing soil problem pests, p. 201. Reducing the length of grass leys to below four years, and the proportion of grasses in leys will help manage this pest. Management of grass weeds such as couch will also help
Cereals	Leatherjackets (various species)	Larvae feed below surface, grazing on plant roots, stunting and slowing growth above ground. Spring-sown crops tend to suffer more damage as larvae are larger after developing through winter	*See* management of key soil pests in Chapter 7, p. 199	Highest risk after grass leys have been ploughed in. Can be counted in 30cm lengths of row excavated to below root depth and sorted on a tray to count larvae. 0.6 million or more per ha (two per 30cm row) are likely to cause damage to the crop after ploughing	*See* Chapter 7 p. 199, but early ploughing will reduce numbers. Omitting grass from clover leys will reduce number of egg-laying sites

Cereals	Slugs (various species)	Slugs hollow out newly sown grains preventing emergence, graze off young shoots. Worse when plants are small and slugs active (Oct–Nov, Apr–May)	*See* Chapter 8 for description of slug life cycle, p. 286	Likely to be worse on heavy soils. Place baits in field and/or traps and count slugs, observe plant germination	*See* Chapter 8 for general description of slug management methods, p. 289
Cereals Sweetcorn	Frit fly (*Oscinella frit*)	Larvae kill the central shoot (deadheart symptoms), most damaging Oct–Nov, and again in late spring (May–Jun) when deadheart plants and withered spikelets may become evident. May attack flowering oats in July and damage grain	Larvae overwinter in grasses and crops, arising from eggs laid on early sown winter crops, or can move on to crop after ploughing in leys. A new generation emerges in May–Jun to lay eggs on spring grasses and cereals. The subsequent generation emerges in late July to produce the generation of adults that lay eggs on winter cereals again	Pest of spring oats and winter-sown cereals. Eggs can be sampled in stubble or grass before ploughing	Plough in grass and leave fallow for as long a time as possible before sowing winter cereals (but *see* bulb fly below). Sow spring crops on time. Establish crops so that tillering (preferred laying site) is well advanced in the egg-laying period

(Continued)

Table 17 (Continued)

Crop	Pest	Symptoms/Damage/ Period	Life Cycle	Monitoring/Risk	Management
Cereals (not oats)	Gout fly (*Chlorops pumilionis*)	Larvae kill central shoot causing deadhearts, most damaging Oct–Nov, otherwise plants may appear swollen, 'gouty' and stunted. Spring crops may fail to produce ears, and in advanced crops grain is destroyed and yield reduced by feeding	Two generations per year. Adults lay eggs on upper surface plants in May–Jun, larvae burrow into and feed on plants before pupating and giving rise to second generation in Aug–Sept. Eggs laid on early sown winter crops (and grasses) cause damage, pupating in spring to repeat cycle	Highest risk on early sown winter cereals and late sown spring cereals. Crops can be monitored for egg laying in late Sept–late Oct	Avoid sowing in peak egg-laying periods. Garlic sprays are reputed to discourage flies from laying eggs and larvae from feeding, but this has not been verified
Cereals (not oats)	Wheat bulb fly (*Delia coarctata*)	Larvae feed on central shoot causing it to yellow and die (deadheart symptoms)	Adults lay eggs from mid Jul–Sept on bare soil, and eggs hatch from Jan onwards. Larvae feed on central shoot	Wheat sown after potatoes, peas, rape, field veg and fallows is particularly at risk, as are late sown crops that have not tillered before attack. Soil samples can be used to count eggs, and counts	Establish crops early to produce vigorous, well tillered plants by early spring, or sow wheat late in spring (from mid March). Rolling may help reduce damage. Keeping ground

				above 2.5 million/ha indicate high risk	covered in summer will reduce number of eggs laid or alternatively fallow fields destined for other crops as a trap. Tillage after egg laying will increase egg mortality. Significant numbers of eggs will be predated by ground beetles and other predators so these should be conserved (*see* Chapter 4, p. 118)
Cereals	Cereal aphids (various species including grain aphid (*Sitobion avenae*), bird-cherry aphid (*Rhopalosiphum padi*), rose grain aphid	The main concern is the transmission of viruses early in the season (Oct–Nov) and in summer after May when sheer numbers can affect yield. Grain aphid infects flag leaves and heads and	Complex life cycles with both sexual and asexual stages (*see* Chapter 7 p. 211). Migrant aphids arrive in the crop from overwintering hosts and sites in spring. Non-migrant aphids	Crop walking and inspecting, aphid forecasts from suction trap network. Early infestations are more likely to inflict more damage. Mild winters are more likely to lead to early and severe infestations	*See* Chapter 7, section on managing problem foliage pests, aphids, for general management techniques, p. 211. Avoid sowing winter cereals too early (before mid-Sept).

(Continued)

Table 17 (Continued)

Crop	Pest	Symptoms/Damage/ Period	Life Cycle	Monitoring/Risk	Management
	(*Metopolophium dirhodum*) and grass aphid (*Metopolophium festucae*)	is most likely to do direct damage to the crop. Other aphids generally inhabit lower leaves	might also be present on host plants in and around fields		Encourage predators and parasites with whole-farm management practices (*see* Chapter 4, p. 118). Generic aphid management practices also described in Chapter 7, p. 214
Wheat	Orange wheat blossom midge (*Sitodiplosis mosellana*)	Shrivelled grain and reduced grain quality in milling wheats. Main period of risk May and June	Adults lay eggs on ears in warm weather after emerging from pupae in the soil and mating. Eggs hatch in 5–7 days and larvae enter and feed in grain. Orange grubs fall to the soil and overwinter in soil in cocoons. When weather warms up they move to soil surface to pupate	Main risk in early summer and between ear emergence and onset of flowering when air temperatures exceed 15°C. Crop walking, midges may be observed in crop in early evening. Use of sticky yellow pheromone traps to trap adults. Counting larvae in heads. 6 million/ha pose a risk to crops	Resistant varieties are becoming available

Cereals	Thrips (*Haplothrips* spp., *Limothrips cerealium*)	Yield loss may occur if sufficient young thrips present, reducing sap flow by feeding on stems and flag leaf sheaths. Larvae may cause black spotting and possibly reduced flour quality if present on grain	Thrips fly to cereal crops in May and lay eggs in emerging ears. Larvae also feed under the glumes and on developing grain	Warm weather at the appropriate times of year	Good crop husbandry to reduce susceptible periods. Direct control measures are not usually justified and likely in any case to be ineffective
Cereals	Rabbits	Feed on plants, most damaging between Oct and Apr when plants are small	*See* Chapter 8 for general description of rabbit life cycle, p. 306	Observation of field margins, conspicuously grazed field margins, burrows and scrapes in hedgelines	*See* Chapter 8 for generic methods of rabbit control, p. 306
Cereals Sweetcorn	Birds (especially rooks, but occasionally pigeons, sparrows and geese)	Can remove seeds from soil on sowing, pull up transplants (sweetcorn) and strip grain heads from significant areas when nearly mature	*See* section on bird management in Chapter 8, p. 294	Rookeries near cereal fields, especially barley. Geese along riverbanks, especially close to urban areas or areas of open water	*See* section on bird management in Chapter 8, p. 294

8

Other Pest Management Strategies in Crops and Systems

Pest attacks are not limited to insects. Organic vegetable and arable crops can potentially suffer attack from a range of other pests including nematodes, mites, molluscs and various vertebrates. Whilst the best management tactics in any situation will obviously vary with the specific pest species and/or crop, it is often possible to take a broadly similar approach to managing many of these groups to that of managing pest insects as described in the previous chapter. This is because they share many of the characteristics of herbivorous animals and share similar strategies for finding and exploiting (crop) plant species. Some of them, however, require very different approaches because of their diverse natures, including, most obviously, the differences in size (for example, nematodes are much smaller and vertebrates much bigger) and differences in behaviours (vertebrates are very adroit at learning).

In this chapter we describe organic management approaches to some of these pests. In doing so we have taken a similar approach to that taken in the previous chapter for insect pests, in that we describe the methods that can be taken by organic farmers and growers to manage and control this varied group of pests, placing them in the context of the previous generalist chapters on organic pest and disease management (Chapters 2–6). The same pest tables (Tables 9–17, *see* Chapter 7 pp. 222–269) provide more details on specific pest species attacking crop types, whilst the main text describes the management approaches with a broader brush. As mentioned previously, where more details are required it will be beneficial to consult specialist texts, as a book of this type cannot possibly do justice to the huge amount of information gathered by agricultural researchers over the years.

MANAGING NEMATODES

Background

Nematodes are found in all soils in vast numbers. They are an important component of the detritivore trophic layer in all soil ecologies, and are vital for soil health. They are small (less than 5mm long) and so are difficult to see with the naked eye, although there may be many thousands in a handful of soil. They are unsegmented and wormlike, and are adapted to moving in the film of water that surrounds soil particles; they therefore move more easily in open, lighter soils, provided there is sufficient moisture present. Conversely little movement of nematodes can occur in saturated, densely packed soils. Fortunately only a small proportion of nematode species in the soil are plant parasites, namely those that complete their life cycle on plants, and cause injury and damage to plants. They are amongst the most troublesome pests to diagnose, and once present, are impossible to eradicate with the methods available to organic farmers and growers.

Nematode Pest Groups and Species

Plant parasitic nematodes are generally categorized by their lifestyle and/or their feeding habits, although they all share the characteristic of having a distinctive hollow mouth spear called a stylet. They feed on plant tissue by inserting the stylet through the plant cell walls, injecting digestive enzymes, and then sucking out the resulting liquid contents. The stylet is also used to rasp the surface of plant roots to gain access. Plant parasitic nematodes often have a wide host range, although some species are very host specific.

Those nematodes that move through the soil while feeding are called migratory parasites, whilst those that feed attached to the plant host are sedentary nematodes. Nematodes can also be ectoparasitic – that is, they generally move freely through the soil and feed externally through plant walls – or endoparasitic, that is, feeding within living plant tissues (roots, buds and so on). Crop pests can be broadly categorized into three groups:

Migratory ectoparasites: Normally complete their life cycle in the soil around the plant roots, feeding on the roots, and only causing damage when present in very high numbers.

Migratory endoparasites: Move through the soil seeking out host plants and invade, and feed, inside plant tissue. As they feed within the host, even quite low numbers may cause significant damage.

Sedentary endoparasites: Also move through soil seeking host plants, but once they have invaded the plant they permanently attach themselves to the plant (roots). The plant roots often swell up or change shape, and the nematodes may swell up and become visible with a hand lens.

Migratory Ectoparasitic Nematodes

There are a great many families and species of migratory ectoparasitic root-feeding nematodes. When present in large numbers they can damage a plant's roots and reduce its ability to take up nutrients and water. Important root feeders are more normally known by their Latin names, and include the following:

Dagger nematodes: Can be found on strawberries, raspberries, celery and grasses, and are normally only serious where they transmit viruses (for example *Xiphinema* spp.).

Needle nematodes: Attack a wide range of plants and crops, characteristically attacking the root tips, which may become galled (for example *Longidorus*). They are also vectors for viruses. They have been known to cause 'fanging' in carrots and parsnips (Table 9), but are also known on grasses, sugar beet and tomato, to name a few crops. They are also associated with docking disorder in sugar beet.

Stubby root nematodes: Cause thickened, stubby roots and transmit viruses (for example *Trichodorus, Parartrichodorus*). One of the most serious is tobacco rattle virus (TRV) that causes spraing in potatoes (Table 16). Common weeds such as field pansy, knotgrass, groundsel, shepherd's purse and chickweed are also hosts to the virus.

Stunt nematodes: Not normally serious pests unless present in very high numbers (for example *Tylenchorhynchus*). They are prone to desiccation in dry conditions despite being associated with sandy soils. Some of these nematode species are semi-endoparasites capable of feeding in plant tissue.

Pin nematodes: Not normally serious pests on crops and, like ring (for example *Criconemoides* spp.) and sheath (*Hemicriconemoides* spp.) nematodes, have a more sedentary habit once they find a host plant.

Migratory Endoparasitic Nematodes

Semi-endoparasitic nematodes normally feed with the front part of the body embedded in the host plant whilst truly endoparasitic nematodes feed within plant tissues. Because of this these types of nematode can cause more specific damage to plants and are more often considered economically important. The endoparasites can be migratory or sedentary; the migratory ones include the following:

Spiral nematodes: Migratory nematodes that burrow into the roots of host plants and feed on the cortex (for example *Rotylenchus*). They have been known to damage grasses, tomato, carrots and lettuce when present in high numbers.

Lance nematodes: Have been recorded as attacking peas, carrots and other field crops (for example *Hoplolaimus* spp.) although they are not normally serious pests.

Root lesion nematodes: Also migratory nematodes, they attack a wide range of plants, including crops, and cause internal browning in roots of crops such as lettuce, carrots and peas and in the tubers of potato (for example *Pratylenchus* spp.). These nematodes feed on the root hairs and then burrow into the root, leaving large cavities that are prone to secondary bacterial and fungal infections. Due to extensive root damage above ground the crop plants can appear stunted and yellowing.

Stem and bulb nematodes: One of the few nematodes that regularly feed above ground (for example *Ditylenchus* spp.). The main species are *Ditylenchus dipsaci* (which has many different 'preferred host races') and *D. descructor*, an important pest of potato causing potato tuber rot (Table 16). Apart from attacking a wide range of different types of plant tissue, it also causes damage by secreting enzymes that cause deformation in plant tissue: this can include swollen stems, distorted leaves, buds and flowers as well as necrotic stems and bulbs, from which it gets its name. Secondary infection of bacteria and fungi often follows on from nematode infection. It can multiply rapidly in host plants but is susceptible to desiccation in the top layer of soil where it is found free living. Apart from bulb crops such as narcissus, tulip and onion (Table 9), it is also found on field vegetables, oats, rye and legumes, including white and red clover (Table 14). This nematode can be transmitted on farm-saved bean seed.

Sedentary Endoparasitic Nematodes

The final ecological group of plant parasitic nematodes is the sedentary endoparasites, in which the life cycle of the females is often modified so that they lose mobility and become saccate. This group includes the following:

Nematode injury to clover roots.

Root knot nematodes: (*Meloidogyne* spp.) Form galls on the plant tissue they attack. The young nematodes enter through the root tips and establish feeding sites, where they remain sedentary for their whole life. The females lay brown egg masses on the outside of the roots, which can be visible with a hand lens as brown spherical masses. Roots are often rough, knotted and lumpy and can crack easily, the visible effect of galling. The galling impairs normal plant function and impedes the flow of water and nutrients in the plant, leading to stunting and abnormal growth and eventually yellowing and wilting above ground. These nematodes have a wide range of host plant species, and most field vegetables, crops and ornamentals are susceptible to one or more species of the nematode.

Cyst nematodes: (*Heterodera* spp. and *Globodera* spp.) Cause abnormal plant growth, and the crop often looks small or malnourished with wilting foliage. Roots become thicker and may turn reddish or brownish. When they die the female nematodes form a hard cyst around their eggs which protects the eggs and from which juveniles can hatch over a protracted period, making this a very persistent pest. Cyst nematodes attack a wide range of crops including cereals and field vegetables. The most economically important cyst nematodes in the UK are the potato cyst nematodes *Globodera pallida* and *G. rostochiensis*, which cause significant yield losses in potatoes and other solanaceous crops (Table 16).

Flower and leaf-gall nematodes: *Anguina* spp., as the name implies, also feed above ground; there are several species feeding on a range of crops, principally cereal crops such as wheat and rye.

Life Cycle

Nematodes have a more or less complex life cycle depending on their ecological life strategies (*see* above). Some depend on the presence of a particular host to complete their life cycle, but most are free living at some point. Many have a wide range of host species including many common field weeds, which complicates their management by crop rotation.

Nematodes pass through three life stages: eggs, juveniles or larvae, and adults. Development of the first larval stage takes place within the egg, and the second stage juveniles hatch from the eggs. In many species the second stage juveniles move through the soil and locate host plants, and then attack and feed on or within the host plant. In many of the migratory nematodes some of the larval stages may be free living in the soil and only infect the host plant after further moults. In some cases (for example *Globodera* spp.) a hatching stimulus may be required, normally the presence of a host plant root exudate, before the larvae emerge. This obviously ensures that the juvenile only emerges near to a host plant and increases the likelihood that it will succeed in finding the host plant root. However, in many cases it appears that the juveniles are rather unselective in the types of plant that they attack and often invade unsuitable hosts where they will only develop very slowly or not at all. At whatever stage they infect the plant, the juveniles pass through three moults in total before finally turning into adults.

Reproduction normally requires both male and female nematodes to mate, after which the eggs develop inside the female and are then laid in the soil in or around the host. In some species males are rare, or even unknown, and it is likely that they reproduce by parthenogenesis. In some cases environmental conditions appear to dictate the proportion of males to females produced, and may even cause sex reversal. Females normally lay between 50–100 eggs, although cyst nematodes can produce upwards of 2,000. The entire cycle can take one to two months in optimal soil conditions (around 20°C with sufficient moisture), but can vary with environmental conditions as hatching can be delayed by adverse conditions.

Symptoms and Damage

Symptoms of nematode damage include both above- and below-ground injury. Foliage is often stunted and yellowish, plants may be unthrifty, prone to wilting and appear stressed. Nematode attack should be suspected if such symptoms appear in patches which seem to enlarge from one season to the next. Nematodes may also be spread along the line of cultivation as soil is moved in the direction taken by the cultivation machinery giving rise to elongated patches. However, these symptoms can also be due to other root pests or diseases, or even abiotic factors such as soil compaction, and it is necessary to dig up some crop plants to better assess any nematode presence. Galls or swollen roots are indicative of nematodes. Reddish or brownish roots may also be indicative, but plant diseases or insects can also cause all these symptoms. Insects that cause galls or root damage can generally be seen with the naked eye or a lens, but nematodes and diseases cannot. If nematode damage is suspected it may be necessary to send a soil sample to a diagnostic service for analysis for positive confirmation and to distinguish between nematodes and diseases.

Nematodes cause damage by directly feeding on plant tissues, often the roots. Feeding damage can often impair root and plant tissue function and prevent movement of nutrients and water within the plant, leading to poor plant development and stunting. Some nematodes, such as the root knot and gall nematodes, cause abnormal growth, which has the same effect. Nematodes are therefore more serious when present in larger numbers, when plants are small, or otherwise stressed. The effects of drought, low nutrient availability and soil compaction are all likely to be exacerbated by nematode attack. Small seedlings may die back or fail to develop, whilst larger plants will show yield declines that are generally proportional to the numbers of nematodes present.

Apart from direct damage to the plant tissue, nematode feeding can leave open wounds that allow entry of opportunistic or pathogenic bacteria and fungi that do further damage to the crop. Some nematodes carry plant pathogenic viruses, and transmit these to the plant when they feed. Viruses are picked up while feeding on infected host plants or weeds, and then carried either on the mouth parts or in the gut. Nematodes can harbour viruses and remain infective for some time, although all viruses are shed when the nematode moults. These interactions between nematodes and disease can lead to large yield reductions over and above what might be expected from the nematodes alone.

Risk Assessment

It is difficult to estimate crop losses due to nematodes because their effects are closely linked with a range of other factors, including environmental factors and other pest and disease species. Plant parasitic nematodes, above all other pests, are intimately linked with the complex soil biology around the plant roots, which simultaneously affects not only the pest nematode, but also the host crop plant, and mediates its reaction to the pest attack. So the seriousness of nematode damage can depend on a range of other stress factors apart from the nematode alone. In general there will be a high risk of damage where plant parasitic nematode population is high, where plants are otherwise stressed (for example, low nutrient availability), or soil conditions are poor (low organic matter). Damage is likely to be higher where nematodes transmit plant viruses.

Monitoring

If nematode attack is suspected, it is possible to sample the soil for nematodes. Methods of sampling for nematodes vary depending on the species and habits. Before taking any samples it is best to contact a specialist diagnostic service and ask them what they require, and who will be able to advise on the necessary equipment to do the job. For soil-dwelling or soil-dwelling stages of nematodes, soil sampling will generally be necessary, followed by expert soil extraction to identify the nematodes present. Nematode numbers will generally be higher after a season of growing a susceptible crop. Endoparasitic nematodes can be detected in plant tissues, but once again, this usually requires specialist staining and microscopy for positive identification.

For soil-dwelling nematodes in the UK, it is generally recommended that soil samples are taken on W-shaped transects across fields or plots (of maximum size 4ha), concentrating on any light sandy patches or known hotspots. Soil samples should be taken with a corer or small trowel at a rate of about 70 cores per ha covered to a depth of 20–30cm. The aim is to gather about 1kg of soil per ha. Generally the smaller the area and the larger the number of samples, the higher the levels of confidence in the final answer are likely to be. The soil then needs to be sent to a specialist diagnostic service, which can extract and identify the nematodes. For plant-dwelling nematodes the diagnostic laboratory will normally advise the type and amount of plant material that needs to be collected, as well as how it should be packed and sent.

Management

Nematode populations will tend to increase rapidly in susceptible crops where they are present. The first line of defence in nematode management therefore depends on prevention, as in organic systems they are impossible to kill once they are present in or around the plant. Once present, management of plant pathogenic nematodes relies above all on maintaining a healthy organic soil – one that is high in organic matter with high biological activity and a good open structure. Under these conditions the wide variety of other soil organisms can serve to keep a check on the activity of the plant parasitic nematodes.

Hygiene

It is important to prevent nematodes infesting fields where they are not present. Although they can only move slowly across a field unaided (1m or so per year), it is impossible to eliminate them once they have gained a toehold. Healthy, certified seed and propagating stock is a principal method of avoiding contaminating clean fields and all such material should be free from plant parasitic nematodes. Any plants that appear to be infected with nematodes should be rogued out immediately (although it might be too late in this case).

Any amendments applied to the field crop should be free from nematodes and properly composted to kill any that might be present. Many nematodes survive on weed hosts, and these should be managed in infested fields to prevent the pest bridging from one crop to the next if nematode presence is suspected.

Other basic hygiene measures should include cleaning soil from equipment that has been used in infested fields, especially if it has been in direct contact with the soil. People and animal movement from infested fields should be controlled so as not to spread any nematodes too easily. Irrigation water should be from a clean (uninfested) source, or, if stored in a pond, water should be held to allow soil and nematodes to settle out and water pumped from near the pond surface. Irrigation water flow through fields should be controlled so as not to let masses of water flow from an infested to an uninfected one, carrying with it nematodes and egg masses.

Cultural Management

As already stated, a healthy soil food web is of paramount importance in managing nematodes. Soil amendments and composted materials (for example manures, green manures, compost) are often applied to soil to increase the organic matter content of soils and their biological control potential (*see* below). A wide range of materials has been used that have shown some effect on plant parasitic nematodes, and the effects are usually greater in sandy soils or where plants might otherwise be stressed. It is difficult to put the effects down to any one factor, and reduction in nematodes often follows from improved soil structure and fertility, increased numbers of biological control agents (fungi and bacteria) and nematode antagonists, release of nematocidal chemicals and/or effects on host plant susceptibility to attack (by inducing systemic host resistance). As an illustration of this principle, applying chitinous materials (for example crab shells) as an amendment has been shown to indirectly affect nematode survival by stimulating the population of 'chitin digesting' fungi in the soil to increase, which in turn attack and digest nematode eggs and nematodes themselves, which are covered by chitin. Unfortunately chitin itself is too expensive to use for this purpose in practical farm situations.

Crop rotation is a key component of any nematode management plan. Rotation depends on being able to follow a susceptible crop with a number of non-susceptible crops, normally those in different plant families. This may not always be possible if there are a number of nematode species present that attack different crops. In this case the rotation should be designed to manage the most economically harmful species, or the most difficult to control by any other means. Weed management can also be important if nematodes are bridging the gap between crops on weed hosts, and in this case a weed management plan may also be necessary.

Rotations should include cover crops, and some cover crops have been demonstrated to release nematicidal chemicals when they are incorporated and degrade. For instance French marigolds (*Tagetes patula, T. minuta*) have been shown to be good at controlling root knot and lesion nematodes as they release polythienyls and also have nematode-suppressing rhizobacteria living in association with their roots. Similarly brassicas such as mustard and oilseed rape have been shown to have a suppressive effect on nematodes in following crops when they are ploughed in, due to the release of glucosinolate compounds as they break down in the soil. Whatever cover crops are grown, they will also help to increase soil organic matter and indirectly manage nematode populations.

Cover crops, here caliente mustard, are useful to increase soil organic matter content as part of a nematode management plan.

Trap crops can also be grown as cover crops. They can be used in two ways: either crops that stimulate hatching and allow development of the nematodes but are ploughed in before the nematodes become mature, or those that stimulate hatching but do not allow the nematodes to complete their life cycle. The latter is obviously preferable to the former, as mistiming the destruction of the trap crop could exacerbate a problem. In the former case, potatoes have been grown as a trap crop for potato cyst nematodes. Potatoes are grown for 5–7 weeks and then ploughed in after the nematodes have hatched and started to develop but before they mature. This has been shown to lower populations of cyst nematode by 50–70 per cent. Recently, interest has grown in trap crops that stimulate hatching but do not allow development. In this respect, *Solanum sisymbriifolium* (marketed as Foil-sys™ in the UK), a soleanaceous plant, has been developed as a trap crop that allows potato cyst nematodes to hatch but not develop, and can be used to clean a field before a potato crop is planted. Once again the trap crop can stimulate a 70 per cent hatch, and also produces a significant quantity of green manure that can be ploughed in (before it seeds). The crop does best at warm soil temperatures (above 10°C but preferably 20°C), and therefore needs to be sown after mid May and before the end of July.

In similar vein, plant resistance can be exploited as part of the rotation, and can be expected to be more effective against the sedentary species that feed within the crop plant and are dependent on it to complete their life cycle. Resistant varieties which prevent nematodes feeding and completing their life cycle are preferred to tolerant varieties, which will still allow nematode populations to develop. Tomato and pepper varieties are known which are resistant to nematodes, and in which nematodes fail to develop and reproduce even though they often infect the plant roots. Many newer varieties of potato are also resistant to the cyst nematodes. However, information on nematode resistance of many crops is often unknown.

Fallowing has been shown to reduce nematode numbers in the soil by removing the food source and therefore causing starvation, but the best effects have been demonstrated by keeping the fields weed free and crop free for up to two years. This has reduced, for example, root knot nematodes by up to 90 per cent, and can be reduced to a one-year fallow by regular soil cultivation, which can kill large numbers of nematodes because the soil is disturbed, exposing the nematodes to desiccation in the open air. Fallowing should generally be avoided in organic systems because of its negative impact on soil condition, although in this case it might also help to manage perennial weeds if they are a problem.

Fallowing may be further shortened by using soil solarization methods. Solarization depends on plenty of sunshine (not guaranteed in temperate areas), and works by laying clear polyethylene plastic mulch over moist soil, which then heats the soil to temperatures lethal to nematodes. It can be effective over a period as short as 2–3 months, but may have other negative effects on the soil microfauna and flora. Solarization may also increase the effectiveness of some of the previously mentioned soil amendments, because any volatile nematicidal chemicals will be trapped under the plastic for longer periods of time.

Flooding has been shown to be effective in reducing nematode numbers, especially alternate cycles of flooding and drying, but its usefulness in temperate organic agriculture is likely to be limited, although it has been tried on potato fields in some areas of the Fens. If done badly it also has the potential to spread any nematode infestation even further.

Other management practices that can help mitigate nematode damage include early destruction of crop residues from infected crops to remove a food source and shelter for surviving nematodes, and the use of transplants, which can avoid the period when the plant is most susceptible to infestation (when it is small and establishing).

Biological Control

Some microbial pathogens have been tested as potential treatments against nematodes, and some success has been demonstrated with fungi such as *Trichoderma harzianum* and *Verticillium chlamydosporum*, as well as bacteria such as *Bacillus thruingiensis*, *Pasteuria penetrans* and *Burkholderia cepacia*; however none has become commercially available in the UK or registered for use. The potential exists for such products to be available to organic growers in the future.

Although there are no commercially available biological control agents for use against nematodes, natural soil biological control agents should not be overlooked. A large number of organisms potentially prey on, or compete against, plant parasitic nematodes, and these exert a considerable mortality effect. It is for this reason that organic nematode management relies primarily on maintaining soil organic matter that drives these antagonistic reactions against plant parasitic nematodes. In particular there are a number of fungi that trap and digest nematodes. Techniques that are likely to promote such conservation biological control include the use of green manures, organic mulches and composts.

Direct Control

There are no organically acceptable methods of killing nematodes *in situ*. Most conventional control methods rely on fumigation with environmentally damaging nematacides, or soil sterilization, both of which have drastic negative effects on soil biology, arguably leaving a sterile environment that actually promotes further pathogen attack. Hot water treatment can kill nematodes present on bulbs or other plant parts before sowing if rigorously (and carefully) carried out.

Some plant extracts and essential oils have been shown to be effective in killing nematodes, repelling them or in otherwise disrupting their life cycle. Neem (*Azadirachta indica*) has been demonstrated as having anti-nematode properties, but is not registered for use for this purpose in the UK. Essential oils from plants such as fennel and oregano have also been shown to have effects in laboratory assays, but their effectiveness under field conditions is not known. In practice these types of product are likely to have limited use in organic nematode management programmes.

MANAGEMENT OF MITES

Background

Mites are a large sub-group of arthropods, the overwhelming majority of which play an integral role in terrestrial ecosystems. They occur in all habitats and feed on a wide range of products, normally as detritivores, and hence help to recycle nutrients through food chains. Some are, however, parasitic on plants and animals. Plant parasitic mites colonize and feed on a wide range of plants but tend to be more troublesome where natural controls are absent (for example in protected cropping) and/or biodiversity reduced (for example after overuse of pesticides).

Mite Groups and Species

Although there is a large number of mite families and species, in practice only a restricted number are likely to be encountered in organic field crops or in protected cropping. The main types likely to be encountered are:

Red spider mites (Tetranychidae): The most likely mites to be encountered causing problems in organic systems, especially in protected cropping, and in some circumstances might be present in large numbers. They include the fruit tree red spider mite (*Panonychus ulmi*) and the glasshouse red spider mite (*Tetranychus urtica*).

Gall mites (Eriophydiae): Live within plants (buds, leaves) and cause the formation of galls due to their feeding activities. Blackcurrant bud gall

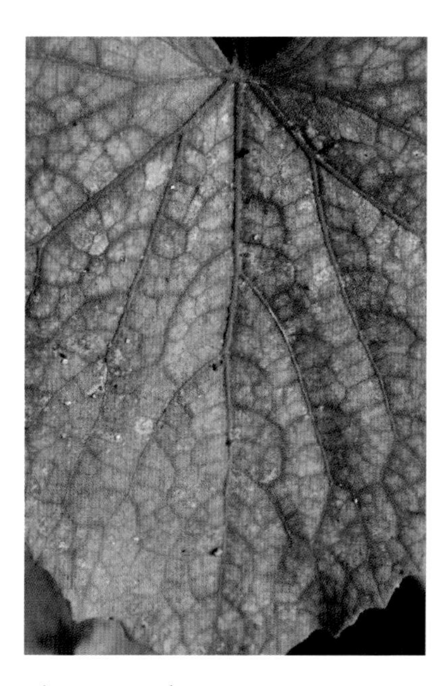

Symptoms of red spider mite on cucumber.

mite (*Cecidophyopsis ribis*) and cereal rust mite (*Abaracus hystrix*) can be found on soft fruit and cereals such as oats, rye grass and meadow grasses respectively.

Life Cycle

Mites reproduce rapidly and have an incomplete metamorphosis. Generally eggs are produced after mating, from each of which a larva hatches. This six-legged larva often lacks a fourth pair of legs and becomes a protonymph (with eight legs) after a resting stage and a moult. The protonymph is followed by a tritonymph, which is similar to the adult but sexually immature. In some conditions an intermediate stage, or hypopus, can be produced, which is a relatively inert dispersal stage, being dispersed by animals, wind or other agents before the sexually mature adults are produced. Different species and groups can have complicated life strategies, dispersal mechanisms and life cycles, including parthenogenesis where females directly produce offspring and males are unknown, rare, or confined to certain periods in the year.

Symptoms, Damage and Risk Factors

Mites cause damage by sucking the contents of plant cells and, especially in the case of galls, by interrupting plant nutrient and water transport. Low numbers are not normally noticeable but large numbers of mites will begin to cause spotting or stippling on leaves, which might later turn bronze or yellow and drop off. Some mites also produce webs that will begin to cover the plant, and other plant parts will become distorted through galling. Mite damage is difficult to diagnose as mite attack is often associated with plants under stress or growing in poor conditions. Plants will often quickly recover from mite attack if stress conditions, particularly lack of water, are removed. However, some mites are vectors of plant viruses and therefore cause damage out of proportion to their feeding; for example the wheat streak mosaic is vectored by the wheat curl mite (*Aceria tosichella*).

Mites are difficult to see with the naked eye and will normally be noticed as the plant symptoms manifest themselves. If mites are suspected it is necessary to use a hand lens to examine the leaves, especially the undersides, where adults and eggs may be visible. They can be shaken off on to a piece of white paper to make identification easier. Webbing can also give away their presence, as can galls. Accurate identification of any particular species will probably require a specialist service, although there are some guides on the internet to the common species.

Management

The mainstay of mite control in organic systems is cultural and biological control, and depends on maintaining a biodiverse system with built-in natural controls. In some cases, as in protected cropping, these elements might need to be reintroduced as commercial biological controls.

Cultural Management

Generally dry conditions and plant water stress are likely to create the best conditions for damaging leaf mite infestations, and so these conditions should be avoided. In protected cropping where mites are more likely to be problematic, taking good care of plants and maintaining and damping down paths and other structures occasionally should help maintain an adverse environment for mites, as can watering the leaves of plants in the polytunnel. If overhead irrigation is used, care should be taken not to encourage disease.

Plant variety resistance has not been well studied for mites. Recent work has shown that, for instance, there are some blackcurrant cultivars that are resistant to infestation by the blackcurrant gall mite (*Cecidophyopsis ribis*), one of the gall-forming mites, indicating that this might be a promising avenue for research in the future. In this case it is also possible to breed the crop for resistance to the virus.

Biological Control

In natural settings mites have a large number of predators, principally other mites, but also small insects. These predators are often easily removed with pesticide treatments, even organically approved ones, and the first line of defence should be not to use these products unless absolutely necessary.

Predatory mites are available commercially to treat spider mite infestations, and this method works well in enclosed spaces. The predatory mite *Phytoseiulus* can control the red spider mite in protected cropping and can bring infestations under control as long as they are not too large. In the latter case it might be necessary to take some direct control measures (*see* below) to reduce the population to a reasonable level before introducing the predator. The predators work best in warm temperatures and when the ratio of predators to prey is about 1:10. In such a system the predators can continue as long as prey is present, but will die off once prey becomes scarce.

When temperatures reach over 30°C *Phytoseiulus* can become ineffective as its reproduction is affected. In these circumstances it can be advantageous to introduce more than one control; for example the predator *Amblyseius californicus* can tolerate high temperatures and low humidities. In complex situations specialist advice may need to be sought in designing complete biological control systems. In all cases, the manufacturer's instructions and advice should be followed for best results.

Direct Control

Strong jets of water directed at, for instance, leaf-dwelling spider mites can dislodge them. Mites generally also prefer dry (and dusty) conditions, so spraying the leaves can help to create conditions that don't favour them as well. For best results it is necessary to get good coverage, including the undersides of leaves. Adding soft soap or oil can also help to 'wet' the leaf and mites, but care is needed with these materials if the plants are water stressed or in hot temperatures (for instance in protected structures) to avoid plant damage. Neem oil has also been tested against mites with some success, but it will also adversely affect natural enemies and may not be registered for this use in many countries.

Gall mites, once present, are difficult to control directly as they are concealed within plant parts, but in small areas they might be controlled by removing the affected parts or destroying the plants. (Lime) sulphur sprays have been used in the past to control this pest, applied during the blossom period.

MISCELLANEOUS ARTHROPOD PESTS

Woodlice Management

Woodlice are rarely problematic in agriculture but recently have been noted as occasional pests under protected cropping. They mainly live on fungi and decaying matter and can only survive in moist places. Where large numbers of woodlice build up they can damage plants at ground level by biting into stalks leaving irregular feeding holes similar to those of leatherjackets and cutworms.

Treatment consists of removing the conditions in which they are likely to thrive. For instance, cultivation of the soil before cropping can expose them, as can removing objects under which they are likely to rest during the day or blocking up any other likely resting places. General glasshouse hygiene should also be practised so remove plant debris, pots and trays, which can all provide suitable habitats for them. Woodlice have few natural predators, although there is some evidence that spiders, centipedes and shrews will eat them.

Millepede Management

Millipedes have similar habits to woodlice and live on decaying vegetable matter. They will occasionally cause damage to germinating crops where they are present in large numbers. As with woodlice, general hygiene should suffice to manage these pests.

Onion with millipedes on roots. (Courtesy Matthew Shepherd)

MANAGEMENT OF SLUGS AND SNAILS

Background

Slugs and snails are land-living molluscs and often particularly troublesome pests in wet temperate climates. Slugs have evolved from snails and have lost the hard protective shell that characterizes the latter. Although the protective shell allows snails to survive desiccation, and possibly predation, it also makes them cumbersome and vulnerable in arable and field vegetable crops. Slugs, with their more streamlined shape that allows them to move easily between cracks and pores in the soil, are the primary mollusc pest in field crops. Slugs are problematic not only in organic but also conventional farming, where the use of molluscides, especially metaldehyde, is likely to become more restricted, necessitating the adoption of organic management methods.

Mollusc Groups and Species

There are three common pest slug species, frequently encountered in organic field production: grey field slugs, round-backed slugs and keeled slugs.

Grey field slugs (*Deroceras reticulatum* and other *Deroceras* spp.): As their name implies, these slugs are small and grey; they can be very numerous in field crops in some situations. They normally pass through one, or at most two, generations per year, with the peak of breeding activity in spring and autumn, although they will in fact breed at any time that conditions are suitable. These slugs are more likely to be seen on the soil surface during the day in damp overcast conditions than other slugs, and have a greater preference for green (leaf) material. They can often be seen climbing on plants in damp conditions.

Grey field slugs.

Round-backed slugs (*Arion* spp.): Also widespread in the UK, although they are usually more noticeable, with some specimens achieving impressive sizes (especially the garden slug *Arion hortensis*). The smaller species are generally the more important pests. These slugs usually have one peak breeding period between spring and autumn, the exact time being dependent on the species.

Keeled slugs (*Milax, Tandonia* and *Boettgerilla* spp.): More localized than the first two slug groups. These slugs are less likely to be seen on the soil surface, and more likely to feed underground on seeds, seedlings and potato tubers.

Snails are not generally field pests although they can be troublesome on smallholdings, in protected structures or certain cropping situations. The most common species include banded, strawberry and garden snails:

Banded snails (*Cepaea* spp.): Common on chalky soils and can be pests in forage legumes.

Strawberry snails (*Trichia striolata*): As their name implies, these are often found in strawberry beds, and can be encountered in other ornamental crops.

Garden snails (*Helix aspersa*): Large familiar snails, widely encountered in gardens, but rarely a pest of field crops, probably because they are large and vulnerable.

Life Cycle

Snails and slugs are hermaphrodite, although most mate and exchange sperm rather than fertilize their own eggs. Some species do, however, self fertilize.

Slugs lay eggs in batches of ten to fifty in soil cavities and under stones or other protection where the soil is moist, but avoid waterlogged sites. When egg batches are dug up they are readily recognized as a mass of spherical translucent white spheres 2–3mm across. Slugs can lay up to 500 eggs in a season, which partially explains their ability to rapidly increase in numbers once conditions are suitable. Small slugs hatch from the eggs after a period of a few weeks in summer, or months in winter, and grow steadily to maturity, at which point they can mate and begin laying eggs to repeat the cycle. The rate of slug growth is highly variable, being dependent on temperature and environmental conditions.

Snails have a similar life cycle but require larger amounts of calcium than slugs in order to secrete their protective coiled shell. Both slugs and snails can live up to four years.

Damage and Risk Assessment

Slugs and snails are more active at dawn and dusk, especially when the soil and plant surfaces are damp or moist, but they can generally be seen at any time of day under damp, overcast conditions. They feed on green plant material (stems, leaves) and also on underground parts such as tubers in potato. Slugs and snails depend on chemoreception to respond to their environment and locate plant food, but they have a poor sense of vision. Some plants contain secondary metabolites that act as repellents or anti-feedants to molluscs. Slugs and snails also use the chemicals in slime trails to interact with other snails and slugs (for example to find mates), and have been shown capable of detecting odours from natural enemies such as predatory beetles. Slime trails also allow homing behaviour, so they tend to return to the same resting places after feeding. Most slugs and snails move slowly and for limited distances (around 5m is typical), so damage is often from slugs already existing in the soil and raises the possibility of slowly clearing areas through good management.

Slugs can cause a great deal of damage to arable crops at or around emergence when they eat off the emerging seedlings or the sprouting seed, and can be one of the factors involved where 'poor emergence' is diagnosed. Vegetable crops can also be affected at germination and emergence, and this can cause serious damage to asparagus where slug feeding can distort the emerging shoots. Slugs can be particularly troublesome to field vegetables at or near harvest when they eat holes in underground parts (carrots, potatoes), damage leaves (lettuce, brassicas) and contaminate produce (all salads) either with slime or their physical presence (see Tables 9–10 and 15).

The following conditions present the main risk factors for slugs:

High moisture: Slugs are most active in wet weather and can often be observed actively feeding after rain during the day.

Mild temperatures: Often linked to overcast conditions and rainfall, and preferred by slugs and snails. They are less active in hot (sunny) conditions and cold conditions, becoming relatively inactive below 5°C, and they will be killed by frost unless they seek protection.

Autumn-sown crops: At greater risk because slug populations are higher at this time of year and more active, although wet springs can also pose a significant risk.

Previous cropping: Can affect subsequent slug populations. Leys, cover crops or crops with dense foliage afford good conditions for slugs to reproduce and survive since they provide a sheltered and moist habitat; furthermore when incorporated into the soil, the organic matter increases the moisture-holding capacity of the soil, also beneficial for slugs. These factors can allow slug numbers to build up, with increased damage to following susceptible crops. Where straw or other crop residues are incorporated or mulches used these can have a similar effect.

Poor agronomic conditions: For example low nutrition and poor drainage may prolong the period at which any crop is at risk to damage from slugs. Conversely irrigation can create favourable conditions for slugs and snails to thrive. Weeds can be beneficial in diverting the slugs from the crop, but can also allow the population to build up quickly, thus storing up future problems.

Heavy soils: Tend to form cloddy, open seed beds through which slugs can move, and which afford plenty of protection.

Minimum cultivation techniques: Can favour the build-up of slug populations. Disturbing the soil with tillage operations can greatly reduce slug numbers as the soil is turned over and slugs exposed. Normally fine consolidated seed beds produced with a number of tillage operations make it more difficult for slugs to move and survive.

Monitoring

Slugs can be monitored to some extent by simply walking crops in wet weather, especially in the evening when slugs are likely to be active. This will give some indication of large populations, as slugs will be seen feeding on plants. However, the majority of slugs will be below ground level and difficult to observe, so using traps can give a better estimation of the likely population levels.

Although many commercial traps are available, these are normally aimed at the amateur gardener (for controlling slugs) and probably too expensive for use on a field scale. Homemade traps are perfectly adequate and can be made from hardboard squares or tiles (of $25cm^2$). Plant-pot saucers are also suitable. A small quantity of bait under the trap will increase its attractiveness to slugs, and chicken layers mash has been found to be good for this purpose (about two teaspoons per trap is fine). Traps should be left out overnight and can simply be lifted to count slugs underneath the next day.

Traps like this work best when the soil is moist, and in sunny weather the traps should be examined before they heat up and the slugs leave them. The optimum number of traps to use depends on the situation. In

Grey slug on brassica showing feeding damage.

arable crops one trap per 2ha spread out over the field (normally on a W transect) should be sufficient to get a good estimate of slug populations, whereas in a mixed vegetable field one to two traps per bed might be sufficient. Traps can be concentrated in areas of higher risk (such as heavy or poorly drained soil). Traps need to be set before any cultivation, as this will reduce slug surface activity.

The interpretation of the trap catches will depend to some extent on the exact situation. It has been suggested that any slugs can be problematic where there is low tolerance to damage (for example potato tubers or salad leaves), whereas only four or more per trap is a problem in winter wheat where more damage can be tolerated or the plant population is able to compensate to some extent.

Management

The main approach to slug management in organic crops is to take an integrated approach to minimizing crop damage and maximizing pest slug mortality. This implies taking a cultural and rotational approach to management, although some direct controls are available for use with permission from a certifying or control body. It should be remembered that a great many slug and snail species are unlikely to do any damage to crops, and increasingly, many are becoming rare or endangered because of habitat destruction and climate change; also the indiscriminate use of molluscides is likely to harm these species.

Cultural Management

There is a large range of cultural practices that have the potential to reduce slug activity. They vary in efficacy, but as part of an integrated approach can be effective in helping to manage slugs. In general they should try and reduce the risk factors cited above, paying particular attention to adequate tillage and avoiding continuous dense foliage and vegetation.

Soil Cultivation and Tillage Adequate cultivation of the soil can significantly reduce slug populations. It is generally agreed that inversion techniques such as ploughing or discing can cause large reductions in slug populations through their physical action in destroying slugs, and that this can be particularly effective if the timing of operations corresponds to the times when the slugs are breeding (although there are many different species of slug that breed at different times).

Soil cultivation should be commensurate with any soil conservation measures or farm conservation plans, but should aim to produce a fine, consolidated seed bed because this will reduce slug numbers (through mechanical action), restrict slug movement below ground and prevent slugs feeding on germinating or emerging crops, and encourage the seedlings to emerge and quickly grow past their vulnerable stage. The worst scenario is when a direct drill has not been covered properly, and the slugs can crawl along the line of the drill and down the open seed slots, destroying each seedling as it emerges. If possible, soil cultivation should be done when temperatures are low because this will expose slug eggs and small slugs to potentially freezing conditions which can kill them.

Mechanical control of weeds has been shown not to control slugs in field vegetable crops, even when done as much as twice a week, possibly because the cultivations are generally quite shallow and the disturbance relatively mild.

Crop Management The timing and placement of crops can help manage slug damage. In cloddy soil, drilling seed deeper might help reduce damage, as the slugs find it more difficult to locate the plants whilst they are germinating and vulnerable. Sowing in dry periods might also help, as slugs will be less active. Sowing crops at least 3m from set aside land or field margins can also prevent large slugs reaching the crop.

Crop management may also be altered to disfavour slugs. Planting out established transplants can avoid the risk of slug grazing on emergence, although crops can still be somewhat vulnerable at this stage. Irrigation in the morning will give soil a chance to dry out before evening when slugs are likely to become active in summer. At the end of the crop cycle it might be possible to harvest crops such as potatoes early before slugs have had a chance to do much damage.

Cover crops and fertility-building crops also pose a risk in allowing slug populations to build up. In general ryegrass should result in less severe slug problems in a subsequent crop as compared to legumes such as red clover or vetch. Legumes such as lucerne have been shown to result in less growth of the slug population than either clover or vetch, and a short overwinter fallow after a fertility crop of vetch has also been shown to prevent a build-up of slug populations, comparable to a fallow alone, but with some benefit of the fertility-building crop.

Temporary makeshift barrier to slug attack to transplant.

Some crops show varietal resistance – for instance, some varieties of potato such as Ambo and Sante seem less susceptible to feeding damage, although this seems to vary with conditions and many contradictory findings have been published. However, for the majority of crop types it has not been studied. It is probable that effectiveness will in any case be limited as most slugs and snails will feed on any plant material when given no choice.

Barrier Methods Barrier methods to slugs are normally only effective and economic on a small scale. They are only effective in preventing slugs immigrating into plots, and any slugs inside the barrier will have to be removed. This can be done by hand, by using nematodes, or by starting with a small area and working outwards to exclude slugs. Many commercial barriers also come painted with a deterrent or impregnated with copper strips that create an unpleasant electrical current across the slug's body, causing them to recoil. Some of the latter can be battery powered, but this does not seem to be necessary. Individual cloches can be fashioned out of various materials (for example, recycled plastic bottles) to enclose plants and prevent slugs and snails gaining access. Bran bait and other materials have been used to form a barrier around plants that slugs and snails prefer to eat or don't like crossing, so protecting the enclosed plant. All these methods are probably only useful to protect high value seedlings or ornamentals in restricted areas.

Biological Control

Natural Predation Slugs have a large number of natural enemies (otherwise we would be knee deep in slugs!). Microscopic parasites such as nematodes, bacteria and fungi are likely to exert a large mortality on (especially) small slugs, whilst predators such as carabid beetles, hedgehogs, frogs and toads can take small to medium-sized slugs. Larger slugs and snails can be eaten by predators such as birds – thrushes, blackbirds, redwings – although many are put off by the slimy mucous secretions of large slugs, which make them very unpalatable. Some birds have learnt to scrape them on the ground to clean off the mucus.

Proactive habitat management will pay dividends in a conservation biological control approach to slug management. Small ponds and shallow water bowls around the holding and in polytunnels can encourage frogs and toads. Piles of brushwood and leaves can provide good shelter for omnivores such as hedgehogs. Birds can be encouraged by nestboxes and winter feeding stations. In all these cases the conservation of a species-rich farm habitat is likely to promote all these natural enemies.

It is also true that they are likely to act more effectively in concert, with any one natural enemy having only a limited effect. For example, slugs are an important prey for carabid beetles (especially *Pterostichus melanarius*), but the beetles are also strongly affected by tillage operations, which will reduce their population in field crops and limit their effectiveness. In addition to

Small ponds and water points can encourage frogs, a natural predator of small to medium-sized slugs.

this, even when artificially introduced in high numbers, beetles have been shown not to have any effect on the contamination of cabbage heads in summer crops, probably because they can switch to other prey if slugs are hard to find, as they are when hidden in cabbage heads. This demonstrates the value in having a many-pronged or multi-predator approach to tackling slugs, as they are likely to be vulnerable to other predators with different searching or hunting habitats or different parasites at different times and under different conditions during the season.

Ducks are reputed to be effective in controlling slugs and snails and have been used for this purpose, especially mallard ducks (*Anas platyrhynchos*), but they also eat crops if given the opportunity and need to be shepherded to some extent.

'Biological' Sprays Slug parasitic nematodes (*Phasmarhabditis hermaphrodita*) and associated bacteria (*Moraxella osloensis*) have been shown to be effective in controlling slugs when applied as a 'biological spray' to the soil surface, although repeated applications may be necessary if conditions are not suitable on initial application. It has the advantage of being able to reach the majority of the slug population that is 'hidden' beneath the soil surface, unlike pellets which need to attract the slugs to feed. The spray is normally applied at sowing or a few days before planting out. It is most effective at killing smaller slugs such as field slugs. The nematode seeks out the slugs in the soil and parasitizes them, causing them to stop feeding after a few days and killing them within two weeks or so. The nematode requires moist soil for at least two weeks after application to be effective, as well as warm soil temperatures (above 5°C). Application dosages and instructions should be rigorously followed for best results,

and the nematodes should be used quickly after purchase (normally sent on request for the treatment).

Such biological control treatment can be expensive, especially when used on a field scale, and may only be justified in high value vegetable crops, although recent work has demonstrated good results with reduced dose treatments when targeted at the vulnerable crop and slug stages. For example, applied in spring the treatment can be used to kill overwintering and recently hatched slugs, and other applications can be targeted on subsequent generations. Thus treatment in lettuce might have to be repeated at intervals as short as two weeks to maintain high crop quality, whereas a single strategic application to Brussels sprouts in late summer when slug populations begin to build can prevent damage to the buttons.

Direct Control

Direct controls are normally those that have an immediate direct effect on slugs, and those that selectively kill pest species are preferred in organic systems. Any products used as molluscides should conform both to any national laws regulating pesticide use and to any organic standards laid down by a control body.

Hand Picking Hand picking is selective, but time consuming, and constrained to larger slugs that can be easily seen. It can, however, work well in restricted areas and environments, for example in polytunnels and seed nurseries. It will work best if picking is carried out at the times the slugs are most likely to be active (after rain, in early evening or at night with a torch), and will need to be repeated often (at least at first). Putting slugs into a bucket of salt water will kill them. In restricted spaces hand picking can be used in conjunction with baited traps, which can increase the effectiveness of this method as the slugs congregate around or under the traps. However some traps – such as beer traps – can also remove predatory soil-dwelling species, which might not be beneficial.

Slug Pellets Commercial molluscidal slug pellets in various forms can be used, depending on the situation and organic standards body involved. Most control bodies would require that the farmer or grower sought permission to be allowed to use these products, and none should be used routinely. It may be possible to target their use – for example, on field boundaries or in traps – to try and avoid any unintended consequences, but on a field scale most are more effective if broadcast on the soil surface. They are normally formulated as brightly coloured pellets with the toxic active ingredient and an edible base to encourage the slugs to feed. They will lose efficiency as they absorb water and rot. Special care should be taken to avoid contamination of any harvested produce. Two products are currently allowed, but restricted, in UK organic systems, but others may become available in the future, as summarized here:

Iron phosphate (Ferramol): Currently the least toxic option to other organisms (although not other species of slugs and snails) allowable in the

UK, but has been shown to affect other organisms including earthworms. It inhibits feeding once ingested. Normally it should be evenly spread on soil at the recommended rate at the time when the crops are the most sensitive to damage.

Metaldehyde: A proprietary molluscide that is allowed in some regions for the control of slugs in restricted circumstances, usually only in traps with a repellent to higher animals, and only when the farm business is severely compromised. In most regions, including the UK, its use would not normally be allowed.

Salt: Has been used on asparagus (a salt-tolerant plant) to deter slugs, but it is questionable whether this is acceptable in organic systems as its effect on soil biology is likely to be adverse over a long period.

Novel products: Various essential oils such as coriander have been trialled as anti-feedants but are not generally effective at a field scale and/or not currently allowed in organic systems. Recent trials with caffeine and garlic have also shown their potential for use against slugs, but it is likely they are indiscriminate in their effects and probably will not be registered for use in the near future.

BIRD MANAGEMENT

Background

Birds are a familiar and much lauded part of the fauna of the UK countryside. The vast majority of bird species in the UK are harmless to crops, and even beneficial in that many species consume vast numbers of insects, especially when rearing their young, not to mention those that consume weed seeds. Some birds, however, are pests and cause real or nuisance damage to crops.

Agricultural crops are good foraging sites for birds as they often provide concentrated quantities of nutritious and readily available food. Birds are also highly mobile and relatively intelligent vertebrate animals, with a more or less developed learning capacity. Consequently, many agricultural crops are subject to some level of bird damage at one time or another. Although bird damage is not a major concern for many farmers, especially in those areas where fields are relatively large, they can become significant pests in small areas – such as allotments and smallholdings – or to high value crops, such as soft fruit or vegetables. Pest status is often accentuated by the fact that human activities are significantly encroaching on wildlife habitats, and managing them, so that foraging space and natural sources of food are scarce.

On the one hand there are large groups of people, and many organizations, dedicated to the preservation and protection of birds, and on the other, groups dedicated to their rearing and shooting. These are not necessarily mutually exclusive. The challenge is to develop methods of bird management that reduce damage to acceptable levels within the constraints of money and time that most farmers and growers face, at the same time as meeting the ethical criteria that bird management demands in a country such as the UK. Any management methods should be taken bearing in mind the sensitive nature of bird control, the strong opinions on all sides, and the necessity of conforming to a range of wildlife laws that protect many birds from deliberate harm. In the UK relevant laws include the various wildlife acts, sport and hunting acts and cruelty prevention acts. Similar laws are likely to be in place in many countries in which temperate organic agriculture is practised.

Common Pest Species

Only a few of the large numbers of species of bird are likely to cause any serious damage to field crops, and these are listed below. Some species such as sparrows (*Passer domesticus)* and starlings (*Sturnus vulgaris*) were commonly regarded as pests in field crops in the past, but their numbers, and the numbers of some other once common species, have shown a dramatic decline in recent decades, and they should no longer be regarded as pests, although it is accepted that their numbers can be locally high.

Pigeons and Relatives

Woodpigeons (*Columba palumbus*), doves (*Stretopelia decaocto*) and feral pigeons (*Columba livea*) can be locally destructive on field crops. Woodpigeons in particular are associated with arable land and feed on clover, cereal and brassica crops (see Tables 14, 11 and 17). Oilseed rape is often subject to high grazing pressure, as are kale and other leafy brassicas, especially in cold and frosty weather. They may also take cereal grain or attack legumes such as peas at and around harvest, but other food sources are usually abundant at this time and they are generally less of a problem. They are usually more of a problem in winter when flocking occurs and feeding is restricted to the shorter daylight hours. Crops such as purple sprouting broccoli can be completely stripped when there is snow on the ground and other sources of food hard to find.

Corvids

Corvids, including rooks (*Corvus frugilegus*) and crows (*Corvus corone*), are gregarious nesters and feeders and can occasionally cause severe, albeit localized, damage. They are common in agricultural landscapes where suitable nesting sites occur, for example rookeries in tall trees. They are generally omnivorous and take a wide range of small animals, insects and carrion (earning them in this latter case a poor reputation with farmers), and searching for such food is probably their main attraction into fields. On occasion they eat grain, especially on lodged cereals, and this is probably where they do most damage. They are also capable of considerable nuisance damage to transplanted crops, where they can pull out large numbers of plants before they are fully established, as they search around and under seedlings and transplants. Brassicas (Table 11, p. 232) and alliums (Table 15, p. 255) seem prone to this type of damage, which necessitates replanting.

Geese

All geese, including Canada geese (*Branta canadensis*), barnacle geese (*B. leucopsis*) and greylag (*Anser anser*), are grazers on grassland and can do damage to grass and cereals where suitable habitats coincide with farmland. Swans and ducks can also do similar damage, but they are not normally responsible for significant damage on farmland. Geese graze grass and cereal crops when they are establishing and growing vegetatively, and damage can be intense along river banks. Later in the season grain can also be taken from cereal crops (Table 17, p. 264). They have been recorded as attacking root vegetables such as swedes and carrots in harsh winter weather, although this is unusual. If present in high numbers they can compact the soil in preferred grazing areas.

Life Cycles and Behaviour

Birds have a familiar life cycle. Depending on the species, courtship between male and female can be more or less elaborate, following which the female lays eggs into a nest prepared beforehand. The adults collaborate to build the nest, hatch the eggs and feed the chicks, though once again the extent of involvement of both parents depends on species. At some point the chicks are fledged and begin to fend for themselves, find mates and repeat the cycle. Large birds normally only breed once per year whilst medium and small birds can raise two, or at most three broods.

Canada goose, a flying rabbit! (Courtesy Sally Cunningham)

Birds are capable of a good deal of elaborate learned behaviour. Two specific behaviour traits linked to learning are likely to be important when considering the management of birds (*see* below): the first is neophobia, or fear of the new, and the second is habituation, or the ability to become tolerant of a stimulus. Behavioural biologists have extensively studied both, which are both to some extent genetically determined and to some extent modified by environmental conditions. Neophobia, both learned and inherited, is likely to be increased in high risk environments and reduced in low risk environments, and it could be argued that the latter includes farmland which has been modified in such a way that there is a low rate of predation and/or risk of disturbance. In this sense birds effectively learn that farmland is a low risk environment for feeding. Coupled to this, they can habituate to deterrent devices, so whilst a loud noise or looming presence might be initially off-putting, with time they will learn that no consequences follow from ignoring the deterrent device. This is far more likely if the stimulus is regular and/or constant in intensity – for instance a regular bang or static scarer device. In both cases, the hungrier the birds and the fewer the alternative sources of food available to them, the more likely they are to quickly overcome their fear of any new deterrent and get used to it.

Damage and Risk

Bird damage to crops is usually very localized and dependent on local bird behaviour and population size. Actual damage is normally the result of consumption of foliage, generally by ripping, which leads to the shredded appearance of leaves, or the consumption of whole grains by pecking.

Pigeon damage is more severe in winter.

Defoliation may lead to loss of vigour and delay in maturation, which can affect marketing. Complete plant loss is rare, although birds occasionally graze off emerging crops so causing poor establishment in parts of a field. Similarly limited areas can be stripped of grain where high concentrations of birds are feeding. In some cases their behaviour can lead to nuisance damage; rooks, for example, commonly pull out transplants leading to a need to replant (sometimes several times). In this case the birds do not generally consume the transplant, but appear to be searching for food in the ground under the transplants.

Damage is usually more severe where birds flock – that is, congregate in large numbers – and where there is a limited range of food available. In the winter months, with limited day length and hence foraging time, crops are likely to present an attractive target. Closer to harvest the higher protein content of the grains is also attractive, although flocking behaviour may be less evident at this time and a wider range of alternative food sources will be available so that damage is less likely to be intensively concentrated in any one crop.

Management

Birds are highly mobile and intelligent animals so it is difficult to effect their management by taking a solely preventative approach, and most reports in scientific literature emphasize that a range of techniques is likely to be more effective when tackling bird pest problems. While total elimination of damage is usually impossible, a combination of techniques over a short time may reduce damage to tolerable levels. Therefore integrating techniques such as bird-resistant cultivars, frightening devices at specific times, and an increase in alternative feeding areas are more likely to be more effective than relying solely on intervention with frightening devices. Different tactics that can be used in bird management are summarized below.

Cultural Management

Because birds are highly mobile there is less scope for cultural manipulation of the crop environment to reduce damage. In a limited number of cases it may be possible to choose crop types and/or varieties that are not attractive to birds. For example, concealed grains need increased time for manipulation and consumption, and/or certain tastes may put birds off (for example saponins in quinoa), but this might also affect harvesting, processing and marketability of crops. Some varieties also mature before or after critical bird damage periods, and so escape periods of heavy attack, typically when many alternative food sources are available to birds. Timing planting and harvesting to miss these critical periods may have a similar effect. Crop rotation may also allow crops vulnerable to bird damage to be planted next to relatively busy roads, footpaths or other locations where background disturbance is likely to be quite high.

As a consequence of the limited possibilities for cultural management, bird management tactics have concentrated on deterrence and trapping to remove them. Deterrence tactics work to scare birds away from the crop by repelling them. For any bird-scaring devices to be effective they need to take the point of view of the bird, and not the human designing them. Damage to crops is generally caused by feeding, and the choices made in feeding are the result of many different behavioural traits and choices on the part of the bird. An overarching explanation is perhaps provided by optimal foraging theory, which posits that the choice to feed will depend on the relative costs of feeding at a particular site compared with feeding at other available sites. From an individual perspective, the bird needs to balance the time and effort taken to feed (intake of energy) with the need to carry out other vital tasks (energy expended) such as looking for food, preening, nest building, mating, keeping out of the way of predators and so forth. In other words, any bird-deterring device should increase the cost of exploiting any particular crop to the level at which exploiting alternative food sources becomes more profitable for the bird. In this sense one of the easiest methods of limiting damage might be to provide birds with a trouble-free source of food. Thus simple measures, such as leaving a cauliflower crop to go on and flower after harvest to provide pigeons with an area to forage in, might induce them to ignore newer and more vulnerable plantings.

Legal deterrents can be divided into four categories: active physical deterrence, physical barriers, visual deterrents, and sound deterrents. The legal use of these deterrents is likely to vary slightly between regions and countries depending on national law. For example, in the UK it is legal to deter birds in most areas providing the deterrent does not trap, injure or kill a bird, that the deterrent or scaring device does not prevent nesting birds access to their active nest site, and that scaring devices should not be used near nest sites of Schedule One (most endangered) species. In the UK the NFU has also produced a bird scarer's code of practice, which aims to minimize disputes between farmers or growers and their neighbours (*see* Further of Information p. 407 for details).

If deterrents are used, then a range of devices or techniques is best, and they are also best changed frequently so that birds cannot habituate to them. It would also be wise not to place too much reliance on any single one, as its effectiveness will always depend on the physiological state of the birds and their hunger. Deterrents that work on innate behaviours linked to survival (for example predator avoidance) are the most likely to endure. Some authorities also recommend starting the control programme before the birds learn that there is a food source available, as it is more difficult to deter them once they have learned that food is available in a certain location.

Active Physical Deterrence In the simplest of cases active physical deterrence is likely to be effective. A highly visible person moving through the crop, performing weeding or other tasks, or just waving their arms, will unsettle most birds. Scarecrows have been used as a substitute for the physical presence of a person, but at best these will only act as a temporary deterrent and will need to be moved frequently to be effective, as birds quickly habituate to them. In some cases more sophisticated scarecrows have been developed that inflate and pop up periodically and even make noises. Spraying water and other such measures are also likely to work, and some devices have radar-activated water streams. Active physical deterrence is likely to be ineffective and too costly on anything but the smallest field scale.

Physical structures in the form of long thin spikes or other uncomfortable shapes have been used, mainly in urban roosting sites, to prevent birds settling on surfaces, and may have some application in farm situations; however they are unlikely to be of use in crops.

Nets and Other Barriers Physical barriers generally work by keeping the birds and the crop apart. Normally such methods would be expected to be completely effective, reducing damage to zero if applied correctly so that birds cannot reach the crop even when sitting on them. A range of barrier methods is available including standard agricultural fleeces and mesh, but also true nets and even webs of fleecy material. In the case of very high value crops exclusion cages made of net can be used. However, the cost of nets and barriers to the grower can be high; they have a high monetary cost to purchase in the quantity required for field crops and are associated with higher labour costs. So for instance weeding is more complicated if nets have to be moved and repositioned. For these reasons they are undoubtedly more practical on smaller areas.

Wires, Tapes or Lines When run over the crop these provide a type of physical barrier and have been observed to prevent bird damage in some cases. How it works is not clear, but it could be that birds have difficulty

Bird netting may be cost effective in small areas.

in flying through the wires without getting tangled up, and/or in the case of tapes, they are perhaps unsettled by the light glinting as these twist in the wind. In both these cases it might be expected that more closely spaced lines would work better, but in fact tape or string suspended 50m apart has been said to deter waterfowl from entering crops. Methods of making the lines more visual, say by attaching CDs, may also serve to increase their effectiveness – although it might also aid the birds in judging where the lines are, and so avoid them. The sound effect of humming line (*see* below) is also said to increase the deterrent effect on birds.

Visual Deterrents These deterrents work via the bird's highly developed visual system. Herbivorous birds generally have lateral and monocular vision (eyes on the side of their heads) and can distinguish the colours important for food and mate recognition, among other things, and so could be expected to react strongly to visual stimuli. A wide range of scare devices have been developed with the aim of deterring birds, including streamers, spinners, flashing lights, mirrors, scare eyes, plastic owl and snake models, balloons, kites and helikites. Some of these can be made from recycled material such as CDs and/or aluminium tins on strings. Most research indicates that birds quickly habituate to most of these devices (over periods ranging from days to a few weeks) and that varying their location, colour and type is likely to be more effective in the medium to long term than relying on one static method. Devices that rely on triggering innate behaviour affecting individual survival would seem to be more effective over longer periods – as, for example, flying devices that imitate predatory birds such as eagles.

Sound Deterrents Sound is also important to birds for mating, communication (distress and alarm calls) and prey location (predatory birds). Noise-generating devices include cannons, exploders and sirens as well as more sophisticated systems that replay bird distress or alarm calls at random intervals. The latter are more likely to be effective when species-specific calls are replayed, which implies they should be deployed on this basis. Research indicates that noise deterrents can be effective under some circumstances, and are best when timed to go off at irregular intervals and moved frequently, but that birds will habituate to them within a period of days to a few weeks. Sound deterrents may be combined with visual deterrents such as pop-up scarecrows.

Biological Control

Aspects of the farm habitat can be managed to attract birds to areas other than the crops, and relatively minor habitat modifications might help reduce damage. For instance trimming and pruning trees and hedges at appropriate times might ensure that alternative food sources are available for birds in hedge lines rather than crops. It might be worth thinking of providing habitat with alternative food sources at critical times (when the crop is susceptible to damage), and this might in any case fit in with, for example, environmental enhancements that provide habitat for other beneficial creatures and/or schemes to promote cover for game birds. In this respect, most authorities point out that if deterrents are to be used it is worthwhile making sure that birds have alternative sites to forage. If this is not the case the deterrent is unlikely to work, as the birds will instead simply become distressed but persist, a theme reiterated across much research into the effect of deterrents on bird pests.

The provision of habitat for top predators, normally birds of prey in the case of birds, may also help deter birds from feeding in crops (which tend to be open and provide birds of prey with a good view). Predators such as falcons can be flown on farms, but their effect is likely to be temporary unless done repeatedly (as for instance in airports). Habitat modifications such as perching platforms can be provided, although their effect is likely to be minor in reducing damage by flocking birds, which can 'mob' such predators. Egg and chick predators such as weasles, foxes and other birds may also be increased by setting aside wild habitat, although the areas needed to achieve this are unlikely to be available to many farmers and growers.

Direct Control

Commercial products have been produced which can be used against birds to repel, trap and/or kill them. However, causing direct mortality of birds for pest control purposes is problematic from a legal and ethical point of view in most countries where temperate and certified organic agriculture is practised. Although it is possible to poison birds, this should be avoided and is often in any case illegal, potentially leading to prosecution. Organic consumers are also likely to be unsympathetic to controlling birds by killing them.

Chemical Repellents Chemical repellents for use against birds have been developed. They are often designed to be applied as sprays to the crop and crop area where the birds are feeding, but in some cases could be applied to roosting or resting sites. They work either as primary repellents, which are painful or irritating upon contact, or secondary repellents, which produce illness or discomfort. Primary repellents cause birds to stop feeding reflexively (without learning), and secondary repellents create a learned aversion response by, for example, inducing nausea, which is then associated with a particular food. They are not generally used in the UK, but have gained acceptance in other countries, and in these cases further information should be sought from the manufacturers of such products. It should be borne in mind that repellents often distribute the problem rather than cure it.

Traps Traps are also available for trapping troublesome birds, which can then be released away from the farm. Such methods are likely to be limited in effectiveness as birds are highly mobile and can return even from large distances; the birds are therefore often killed (but refer to the ethical problems above). Traps are often employed against corvids, in which case sturdy wood and wire funnel cages, sometimes baited, are used and placed in areas where feeding and/or damage is occurring.

Bird traps can be used to remove troublesome birds.

Hunting A final and historic method of control is hunting by shooting or other methods. Fine nets, for example, have been used against smaller birds, but are likely to be non-selective and so should be avoided in organic systems. While shooting is deadly, it is only as good as the hunter is skilled in targeting the birds, and is likely to be more ineffective the more birds there are.

MANAGEMENT OF RODENTS AND OTHER SMALL VERTEBRATES

Background and Pest Species

A range of rodents are potential crop pests, but in practice the damage they are likely to do in the field is very limited compared to other pests and diseases. Some have been introduced and are commensal – that is, survive in numbers only because of people – whilst others have emerged from the natural fauna or are introduced escapees. They are all predated by larger animals such as stoats, weasels, foxes and owls, which might have been removed from the landscape, and programmes to encourage these predators are likely to help manage them. Some of these mammals are considered rare and/or important symbols of the countryside, and so their management should be undertaken with some sensitivity to the organic consumer's perceptions. Common species include rats, mice, voles, moles and the grey squirrel.

Rats (*Rattus rattus* and *R. norvegicus*): Usually pests of stored products rather than field pests, and apart from consuming stored produce, can contaminate it with salmonella as well as transmit diseases such as Weil's disease and plague (but this might be seen as dark propaganda against these fascinating mammals!). Crops such as sweetcorn (Table 17, p. 264) can be attacked and eaten by rats.

House and field mice (*Mus musculus* and *Apodermus sylvaticus*): Mainly problematic in stored produce, but can occasionally cause nuisance damage in digging up and eating germinating seeds, especially peas (Table 14, p. 246), and they can also eat soft fruit on occasion. Field mice have also been implicated in predation of bumble-bee nests.

Vole (*Microtus agrestis* and others): Similar to field mice in appearance and habits, causing similar types of damage. Can cause problems in seed beds and protected crops by tunnelling under plants.

Mole (*Talpa europaea*): An occasional nuisance pest in agriculture as they can burrow under seed beds and cause potholes in pastures and fields. Contamination of grass by soil from molehills can affect the quality of silage, and occasionally there is an increased risk of damaging grass-cutting machinery. They are likely to be more common in organic systems where worms are encouraged.

Grey squirrel (*Sciurus carolinensis*): The predominant squirrel in much of the UK mainland and are normally only problematic in tree and nut plantations, and occasionally orchards. The red squirrel (*S. vulgaris*) is endangered in the UK and should be encouraged.

Life Cycles and Behaviour

Mammals follow similar life-cycle patterns – that is, the adults mate and rear the young in nests or dens. There is often some parental care. Rodents are small creatures characterized by a rapid life cycle, and are capable of rapidly increasing in numbers where conditions are favourable and predators absent. Usually a good deal of ecological and behavioural information exists on rodents, which is specific to the species concerned. This information is available in many good wildlife books and on the internet (*see* Further Information, p. 407) and these should be consulted for the details.

Damage and Risk

Risk of damage from all these mammal pests is very situation specific and likely to be localized. They can cause damage to emerging plants, and this is especially common with peas, which are eaten by mice and voles as they emerge. Growing crops rarely suffer much damage from rodents, but become attractive once they begin to fruit and ripen. Rats and other rodents can cause damage to sweetcorn, for instance, opening the sheaths and chewing the ripening grain. Rats and mice consume cereal grains either on the plant or when they fall to the ground. However, it would be true to say that the main problems and risk are likely to be in storage of grains and fruits, and specialist texts should be consulted on crop protection in stored products.

Rodent (rat) damage to sweetcorn.

Management

Cultural Management

No specific cultural control measures exist for use against rodents over and above the good crop husbandry demanded for the management of other types of pest. As rodents are likely to be encouraged by plentiful food supplies, farm hygiene can be important and destroying grain in stubbles and cleaning up grain spills can be beneficial in denying them these food sources. Removing vegetation from around buildings can be helpful as rodents often don't like to cross open spaces, but this measure is unlikely to be practicable around fields, and in any case will have an adverse effect on other beneficials. Siting susceptible crops away from woodland or other suitable mammal habitat may be possible in some circumstances, but is unlikely to be an option across the whole farm in all years.

Biological Control

Small rodent populations are affected by weather and predators. Cold and wet weather is likely to reduce the survival prospects of rodents, either because of the reduction in food supply or by increasing exposure to harsh conditions. These are probably the main (admittedly uncontrollable) factors in population cycles. Predators also prey on small rodents, which are an important link in the food chain. General habitat management to increase the numbers of generalist predators such as foxes, weasels, stoats and owls will help to reduce and manage populations. Inundative control may be possible with cats, but the number necessary may make this method unfeasible on most farms or holdings, and the effect on bird populations may be unacceptable. It is not known what part natural disease epidemics have in regulating rodent populations on farms, but they are probably important in some circumstances.

Direct Control

Trapping and poisoning have been the traditional methods of dealing with large rodent populations. Whilst this might be necessary in and around farmyards, it is unlikely to be economically viable on a field scale, and in any case the level of economic damage is unlikely to warrant the cost outlay. Trapping can also be effective in protected cropping where there is some control over the influx of rodents from the surrounding environment.

If coagulant or other poisons are used in organic systems they should always be justified (and it may be necessary from legal and public health reasons) and placed in bait stations to which other mammals or animals have no access. Specialist advice should be sought on the range of options available, as many new products are coming on to the market to replace coagulant baits to which some rodent populations are becoming resistant.

MANAGEMENT OF RABBITS

Background

Rabbits (*Oryctolagus cuniculus*) regularly cause a great deal of damage to both vegetable and arable crops; they are a ubiquitous feature of the UK countryside, with the dual identity of 'cuddly' creature to urban consumers and infuriating pest to farmers and growers. It should also be remembered that rabbits are the determining herbivores in some habitats where they graze turf short and prevent the ingress of some plant species and promote the development of others – such as vetches and trefoils – which in turn can promote other biodiversity. They also provide prey for a range of predators, and are thus an important element in the food chain of the countryside. Hares (*Lepus* spp.) also have the capacity to do damage, but they are present in such low numbers that they should be encouraged – indeed they have become iconic of the UK's struggling wildlife in general.

Life Cycle and Behaviour

Rabbits live in warrens, which comprise many burrows and are normally sited in inaccessible areas of hedgerow and other overgrown areas. Many now habitually live close to the surface, as this protects them to some extent from myxomatosis and possibly other epidemic diseases. The main breeding season begins in January and extends to August, although breeding may occur sporadically at other times. Litters are born after about one month's gestation in stops within the burrow; they remain here, largely helpless, for the first fourteen days. The young emerge from the burrow after two to three weeks, and the cycle can be repeated so that a doe can produce six to thirty young per season (typically six to ten). Rabbits are able to respond rapidly to good conditions, and the population can increase dramatically in a short time.

Rabbits usually remain and feed fairly close to the warren, individual home ranges being around 0.5–3ha with some overlap between individuals, so that the individuals in a large warren may range over about 20ha. The warren is dominated by a few bucks, and the does are territorial around the warren, defending a feeding patch from other rabbits. Young male rabbits often leave the warren and routinely disperse up to 4km or more. Apart from structural damage to hedgerows and habitat, large colonies quickly graze vegetation around their warrens and begin to encroach on fields of arable crops and vegetables.

Damage and Risk

Direct damage is by feeding on crops, and indirect damage is due to scrapes and overgrazing which allows weeds to develop. Winter wheat (Table 17, p. 264), for example, can be severely grazed over winter at a time when little other food is available, and can be grazed off completely at field margins and/or severely set back in other areas. Grassland can be similarly grazed, but the damage might not be so apparent. Damage to field vegetable crops depends on the crop, but all green shoots and exposed roots are vulnerable to grazing. Damage can also be done to hedgerows and earth banks as warrens expand.

Risk will be related to the size of the local population of rabbits and their location relative to the field crops in question, as well as the amount of protective cover available. Populations are likely to be largest towards the autumn and before the onset of winter mortality, and crop damage is likely to be a function of the availability of alternative food sources close

Rabbit hole in vegetable bed.

to the warren. Populations increase with increasing temperature and over-winter sunshine (factors that aid survival in lean periods), and it is likely that a warming climate will see a larger population of rabbits.

Management

Cultural Management

Rabbits are large, mobile creatures, but crops closer to warrens are likely to be preferred so when cultivating crops these locations should be avoided if possible. There are some crops that might be less palatable to rabbits, for example potatoes, but their inclusion in a rotation is only likely to be a stop-gap measure, as it will not impact greatly on rabbit populations if alternative food sources are available. There is no apparent varietal resistance to rabbit feeding within crops, although not much work has been done on this.

Rabbit-proof Fencing Barriers in the form of rabbit-proof fences are effective in reducing rabbit damage. Entire fields can be fenced against rabbits, or areas of the field enclosed to protect sensitive crops. In the former case the fence need not be very elaborate and can be made from wire mesh; 1.2–1.4m wide chicken wire of mesh size 25–30mm is adequate. The bottom 30–50cm should be spread along the ground angled outwards from the fence and lightly buried so that grass grows through the mesh to anchor it. Fence posts can be quite flimsy and spaced appropriately to hold the fence up (unless badgers are expected, in which case it may be necessary to put in badger gates to allow them to follow their regular routes). The fence is also best angled slightly outwards to overhang the direction in which the rabbits are coming. All rabbits should be cleared from within the field and any that gain entrance killed as soon as possible so they don't learn that the field represents a source of food. Chicken wire should be placed over gateways as well.

Electric Fencing Electric fences are more normally used around sensitive crops such as carrots or lettuce within field boundaries. They are commercially available and normally made from nylon netting threaded with conducting metal foil. The fences carry a high voltage but low current and rely on shorting to earth through any body parts that come into contact with them. They can be run remotely off car batteries and solar panels, otherwise will need a mains electricity supply. They work more efficiently if they are keep free from weeds, and some growers place mypex strips underneath them to keep them weed free. They are also best angled outwards to prevent the rabbits jumping them and so that their sensitive ears come into contact with the fence. Unfortunately rabbits, with their furry paws, can sometimes get under them without being shocked and if the electricity is off, they may chew holes in the fence.

For larger areas parallel lines of electric wire can be used. Akin to electric fencing used to manage stock, and using the same type of approach, four lines are normally used for cereal crops and six lines for horticultural crops. The wires used are better conductors than the polywire of standard fences and tougher on a rabbit's teeth so should be more likely to deter entrance as the rabbits are more likely to receive a good shock. The wire should be vertically spaced so that rabbits can't squeeze through without touching the wires or jump them.

Mesh and Fleece Mesh and fleece, well pegged down, can also be used as a temporary physical barrier which is likely to deter rabbits, at least in the short term. Effectiveness will be diminished if they learn that food is underneath and/or there are many holes in the material.

Biological Control

The main regulator of rabbit populations in the wild currently seems to be weather and disease. Medium-sized and large predators are now uncommon in the UK and European landscape and these are the predators that might be expected to prey on rabbits. Short of reintroducing them, there is not much that can be done, although habitat management in favour of the predators that still exist, such as stoats, foxes and buzzards, will help. Mink will also prey on rabbits.

Rabbit populations tend to oscillate with the seasons and weather, but they are also affected by the myxomatosis epizootic disease, which is still present in an attenuated form and tends to infect young, non-immune rabbits, killing about 70 per cent of them. Currently the rabbit population is estimated to be some 40 per cent or so of pre-myxomatosis levels, so there is potentially considerable scope for population increases if the disease becomes ineffective. It is incidentally illegal to deliberately introduce a diseased rabbit into a warren to control the population. Another viral disease, rabbit haemorrhage disease, is currently threatening rabbit populations, but to date its impact on the UK mainland seems limited.

Direct Control

An alternative to physical barriers is the use of repellents, which prevent rabbits from feeding or put them off foraging in an area. It would be fair to say that the performance of many of these products seems to be variable. Some are based on natural plant products such as garlic, which would be expected to be suitable for use in organic systems. Lavender and catnip are also reputed to be repellent to some degree to rabbits. However, it would be a mistake to assume that the tastes and sensitivities of rabbits are the same as those of humans, or their reactions the same. Apart from natural products, there is a range of products based on the scat of predators (such as cats that have fed on wild animals), and these are reputed to repel herbivores such as rabbits – but once again, their efficacy is hard to verify. Soaps are also said to put rabbits off feeding.

Although trapping and hunting rabbits has been popular from time immemorial, these measures have largely been abandoned, probably because rabbits are no longer a popular food source, and because of the cost of the labour involved. That said, it is still legal and organically acceptable

Rabbit traps may be relatively inefficient at preventing damage caused by large populations.

to shoot, snare, ferret, dazzle or trap rabbits, although care should be taken to keep within the various wildife laws, and not to be deliberately cruel. Market outlets for rabbits may help recoup some of the costs involved.

All these direct methods are probably relatively inefficient given the size of the rabbit population and the number that can be caught or dispatched in any reasonable time frame. Shooting is probably the most time-consuming and inefficient method of killing rabbits and will probably be of limited effectiveness unless done very regularly. Trapping is arguably more humane and efficient than snaring, although they are both probably also relatively inefficient. Traps and snares should be inspected daily to conform to the law, and any animals found dealt with humanely. Ferreting is normally done in conjunction with trapping or netting the warren entrances.

MANAGEMENT OF LARGE VERTEBRATES

Background

With recent conservation programmes and increased environmental awareness large mammals have been increasing in numbers over recent years and a few have begun to cause some damage to crops, especially as milder winters and lack of predators have allowed numbers to increase due to decreased mortality. The two most problematic are probably deer of various species and badgers, and their management is summarized below. It should be borne in mind that both are iconic species to the large number of wildlife enthusiasts in temperate areas, especially the UK and northern Europe, many of whom are likely to be organic consumers, so that, once again, some sensitivity, and possibly explanation, is likely when managing them on an organic farm.

Pest Species and Life Cycles

Deer

The most widespread species in the UK are muntjac (*Muntiacus reevesi*), fallow deer (*Dama dama*) and roe (*Capreolus capreolus*) deer, although the latter are more predominant in upland areas. Muntjac have no defined breeding season, which partially explains their rapid increase in numbers, whilst the other two species have more or less defined breeding and rutting seasons coincident with the seasons. They prefer habitats with at least some woodland cover present, and are more likely to be problematic in areas with this type of habitat. They are generalist herbivores stripping foliage from plants, including crop plants, and eating it. Damage is most commonly noticed in spring or on new succulent growth and may be especially severe on fruit trees. Damage to trees can be recognized by the rough shredded surface of twigs and stems caused by the lack of upper incisors, which means they strip the plants rather than leaving teeth marks.

Badgers

Badgers (*Meles meles*) are also widespread in the UK but not really serious pests of field crops. They can be locally troublesome where they acquire a taste for sweetcorn and/or carrots, which they are able to dig up causing some degree of secondary damage. They can also be a problem with squashes and pumpkins, in which they make a hole to scoop out seeds and/or roll them around to look for earthworms and other invertebrates underneath. Scratches can lead to callouses and secondary rots. They are also powerful animals and capable of destroying rabbit fencing, which could result in extensive rabbit damage.

Badgers live in setts (underground burrows) commonly found in banks in wooded or partially wooded areas. They actually mainly feed on earthworms and other small invertebrates, but they have become notorious in the UK and Ireland for their disputed part in the transmission of bovine tuberculosis (bTB) to cattle. Paradoxically, while being vilified, they are also subject to protection under law in the UK by the Protection of Badgers Act 1992, which makes it illegal to interfere with a badger sett or to kill them, although exceptions are made for 'pests'.

Damage and Risk

The main damage caused by large vertebrates is feeding in pastures, cereals and horticultural crops (*see* above for details), though in most cases economic damage is actually quite small or negligible. Apart from feeding damage, as large animals they can do considerable damage to fences and other materials (for instance fleeces) on farms. Risk is probably highest on farms with natural woodland cover or other suitable habitat, and is likely to be higher at the times of year when other food sources are scarce.

Management Methods

Deer Management

Management methods for deer, like birds, will largely revolve around deterring them from entering and feeding on the crop. Deterrents are likely to be physical and/or chemical. As with birds, it is likely that a combination of methods will be more effective as many of them will, at best, only be effective for short periods of time.

The main physical deterrents are barriers or fences of various sorts which, given the size of the animals, will need to be strong and/or electrified. They will also need to be tall (over 2m), as deer can leap to a considerable height. In this case solid fence lines can be lined with sharp spikes that might act to deter animals. As with birds, a range of visual or noise-making devices is also available, but again as with birds, might not be effective as animals habituate to them, especially if they are hungry.

A range of chemical deterrents has been promoted, which seems to be dependent on either smelling like people or predators, or in making the crop unpalatable in some way. For instance fragrant soap or human hair hung at head height is said to make deer wary, presumably because it smells as if people are present; but any effect is likely to be short-lived. Cat scat has also been marketed for similar reasons, but as deer in many countries no longer have large feline predators (such as cougars or lions) the effect is likely to be questionable and again short-lived.

Anti-feedants have also been commercialized for use by farmers and growers in different countries. These are often intended to be applied to crops as sprays and make them unpalatable in the short to medium term, but not to taint them for human consumption. Some are based on pepper sauces (chilli), and some are soaps that presumably give a slippery texture to the foliage. Any commercially produced anti-feedants should be passed

Deer fences need to be tall!

by a relevant control body as suitable for use in organic systems, and should be eligible for use under any regional pesticide regulation laws.

Badger Management

Badgers will require a number of approaches to deter them, but most will be based on a combination of stout physical barriers to keep them out. It may also be beneficial to surround susceptible crops (sweetcorn, carrots) with electric fencing to provide an added level of protection within the field. Unlike deer, no commercial deterrents are specifically marketed for use against badgers, but some of the same scaring devices are likely to be as effective as they are with deer. Badger gates in rabbit-proof fencing may reduce incidental damage. It is illegal to interfere with badger setts without a licence, and this includes trapping, snaring, baiting and gassing.

In the UK because of the link with the transmission of bTB between badgers and cattle, badgers are occasionally the subject of government eradication measures by the use of legal poisons. However, most wildlife experts consider that in the long run these will be ineffective in their control and/or detrimental because they will increase the movement of badger populations in disturbed areas. Such programmes are not popular with the public, many of whom are also likely to be organic consumers.

9

Disease Management Strategies in Crops and Systems

Diseases feature prominently in concerns expressed by farmers about converting to organic production, and indeed are among the main problems listed by organic farmers and growers already in production. Diseases are plant disorders caused by pathogenic microorganisms that are not normally visible to the naked eye, and are infectious (that is, they pass from one plant to another). They therefore share some similarities with crop pests, which consume plants and pass from one plant to another. Farmers and growers recognize these pathogens by their effects on crop plants and their symptoms. These include abnormal functioning of the plant, coupled with various physical symptoms such as blotches or lesions, which ultimately lead to reduced yield and/or plant death.

In this book we have restricted our discussion of diseases to infectious microorganisms, although other authors also list nutritional disorders and physiological disorders as diseases. These non-infectious plant 'diseases' are caused by abiotic agents (for example nutrient deficiency, frost or herbicide damage) and we do not discuss these here in any great depth. Alternative texts should be consulted if this type of damage is suspected.

Management of diseases is predominantly by preventative means using many of the strategies and tactics described in Chapters 3 and 4, and these should be read for an overview. Occasionally it may be necessary to have recourse to more direct control measures as discussed in Chapter 5. This chapter aims to synthesize this earlier information and put it in the context of diseases of field crops.

In common with pests, there are many different types of disease-causing pathogens representing a large number of families, species and sub-species. Many, if not most, pathogens are specific to, and only infect, a narrow range of crop species or families, whilst others are able to attack a broader range of plants. These latter diseases are often opportunistic microorganisms that infect plants through wounds or other openings, and which feed saprophytically on the plant if it dies. The former specialist diseases have often evolved complex adaptations to infect and exploit plants and to survive in their absence. By observation over a few seasons farmers and growers can quickly learn to identify the most common diseases in their crops and cropping systems.

In this chapter we discuss many of these key diseases, together with the control measures that are likely to be helpful in managing them. The text discusses diseases and management methods in broad outline at various stages of the crop cycle, whilst the key diseases for crops are listed in the accompanying tables (Tables 18–26). These should be read in conjunction with the text and consulted for specific information on crops and/or diseases. Detailed information on disease-causing organisms and their management is generally more difficult to obtain than that about pests such as insects, and much of it less understandable, but there is still a great deal of information available (*see* Further Information, p. 407). Once again, farmers and growers are well advised to seek out and use these sources to learn about the diseases they encounter; this will help improve and fine tune their management practices over time.

DISEASES OF ORGANIC CROPS

Diseases caused by pathogenic organisms are found on all plants. Being microscopic they are not normally directly observable by farmers and growers, although computer-linked microscopes are available for those interested in exploring these organisms. They are normally recognized by the symptoms they induce in their (crop) plant hosts, and these can be characteristic of specific diseases or groups of diseases. Symptoms can range from leaf spots, to lesions, rots and ultimately plant collapse and death. As with pests, the damage is likely to be more serious when the crops are infected at an early stage of development and/or when the harvestable parts of the plant are infected. With some diseases the capacity to grow through the plant systemically and kill it can make them more threatening than pests, as the ability to check such disease is limited with the organically acceptable control measures available to organic farmers and growers. Consumer tolerance of disease damage is likely to be quite low, and diseases will generally drastically decrease the storage potential of harvested produce. In some cases it may be possible to remove diseased tissue (for instance, outer leaves) but this may increase costs and reduce usable yield.

Disease-Causing Organisms

Infectious crop diseases are caused by pathogens that belong to a number of different groups of microorganisms, namely fungi, protozoa, bacteria, mycoplasmas and viruses. Whilst this might seem clear cut, sometimes it is difficult even for experts to allocate the various diseases unambiguously to the various groups, and even define what the groups are. A certain amount of change in the classification of diseases is therefore to be expected from time to time. Below we simply describe the key characteristics of the main disease types together with some of their symptoms and common names.

Fungi

Fungi comprise a ubiquitous and very large kingdom. They usually have an extensively branched hyphal body or mycelium, whose main structural element is a complex of chitin-like compounds. The large volume-to-surface ratio renders fungi vulnerable to desiccation, and they are normally associated with damp or humid conditions. Fungi do not have the ability to photosynthesize, and by far the largest proportion are saprophytic, playing a vital role in breaking down plant and animal material in soils and in helping to provide nutrients to plants in an assimilable form. Indeed, some go further and form vital symbiotic associations with plant roots, the so-called mychorrizas, which play a pivotal role in making simple nutrients available for plants in exchange for more complex photosynthates. Unfortunately many fungi are also parasitic on crop plants, and invade and kill plant tissue causing an array of symptoms. Those symptoms associated with fungal disease include rots, necrotic or dead patches of tissue, outgrowths of fungal tissue and distortions of plant tissue, to name but a few.

There is a bewildering array of fungal families, groupings and names, and many are difficult to locate in the fungal family tree as their sexual life cycles are unknown (a group sometimes referred to as the 'imperfect' fungi). The most common are listed below according to the way in which they attack and parasitize plants, which is in turn closely linked to the types of symptom they induce in crop plants. Many of the key fungal diseases are also listed in the Tables 18–26 and these should be consulted for specific crops. Although the other plant pathogens listed below (viruses, bacteria) can also be generally categorized in this way, we have excluded them to make the distinction between the various types of plant pathogen clearer. A similar distinction was made between non-obligate, obligate and facultative parasites when discussing the ecology of diseases in Chapter 2.

Necrotrophs These fungi rapidly penetrate the host plant and kill the plant tissues and cells. They rely on breaking down the plant tissue to release the nutrients they need to develop. They are therefore destructive of the plant and are recognizable as damping off diseases, rots, wilts and necrotic spots, all of which indicate that the plant is dying away and being consumed. These diseases are the most rapid acting and likely to be the most damaging. Some may attack a wide range of different hosts. The pathogens responsible cover a wide range of fungal families and species, including familiar diseases such as *Fusarium* spp. (Ascomycota, Hyphomycetes, causing damping off and related diseases), *Phoma* spp. (Ascomycota, Coelomycetes, causing damping off, black leg, cankers and other related diseases), *Septoria* spp. (Ascomycota, Coelomycetes, causing leaf spot diseases) and *Gaeumannomyces graminis* (Magnaporthaceae, causing take-all of cereal).

Biotrophs Once they have invaded a plant, biotrophs divert sugars and other nutrients from the plant to their own use, and are adapted in various ways to do this, both morphologically and biochemically. Obviously this

Fusarium rot on leek.

relationship requires that the plant be alive, and these fungi are often able to maintain the parasitic relationship for some time. Typical biotrophs include powdery mildews (from the family Ascomycota, Erysiphales), downy mildews (Peronosprales), rusts (Basidiomycota, Uredinales) and smuts (Basidiomycota, Ustomycetes). In the former the disease is observable as a white (powdery) mycelium which coats the infected plant's foliage, and from where it penetrates into tissues. In the latter three the mycelia penetrate into the host tissue, which may remain apparently healthy until fungal fruiting bodies erupt through the surface (for example, rust pustules with uredospores). Whilst growing in plant tissues biotrophs can also directly affect plant development processes, and areas of yellowing or distorted tissue might become visible with time, and stems and leaves may take on a twisted and/or distorted appearance.

Hemibiotrophs As the name implies, these fall between the previous two disease types. Invasion of the plant might be a gradual process, taking some time, but eventually the fungal infection begins to rapidly kill the plant tissue, and at this point the parasite switches to breaking down the plant tissues to obtain nutrition. Once again parasites in a range of fungal groups and/or families are hemibiotrophs. Familiar diseases in this category include light leaf spot of brassicas (*Pyrenopeziza brassicae*) (Ascomycota) and anthracnoses (caused by *Colletotrichum* spp. – Ascomycota, Coelomycetes). It might be that many of the necrotrophic fungal pathogens mentioned above belong to this group, but that the biotrophic stage is very short, and as of yet unrecognized.

Downy mildew on onion.

Chromista

These pathogens are also very similar, and often classed together with the fungi. Within the Chromista the Oomycetes contain all the crop plant pathogens, some of which are very familiar to farmers and growers, including late blight of potato (*Phytophthora* spp., Chromista, Oomycota) – *see* Table 25 – and the damping off and root rots caused by *Pythium* spp. (Chromista, Oomycota). This group of fungi is characterized by having both asexual and sexual life cycles, and also motile zoospores that require water to move. They are different from fungi in that the cell walls of the hyphae contain cellulose as the main structural component, rather than chitin.

Protozoa (Protista)

Protozoans are single-celled eukaryotic organisms that are often motile as well. Two classes of protozoa, the Myxomycetes and the Plasmodiophoromycetes, contain plant pathogenic species. The latter includes the familiar clubroot of brassicas, *Plasmodiophora brassicae* (Table 20), which produces motile zoospores and can form thick-walled resting spores that survive many years in the absence of the host plants.

Bacteria

Bacteria are prokaryotic organisms regarded as likely to have evolved before the fungi, plants and animals. They are essential to the functioning of the biosphere, exist in unimaginable numbers, and are central, along with the fungi and other microorganisms, to many soil biological processes. Whilst the vast majority are saprophytic, some have evolved to parasitize plants (and animals). They are generally regarded as less important crop pathogens in temperate regions, probably because they are less well adapted to grow at low temperatures as compared to the fungi. Bacterial infections familiar to farmers and growers include the soft rots (often caused by *Erwinia caratovora* – Table 18) and common scab on potatoes (caused by *Streptomyces scabies* – *see* Table 25).

It is interesting to note that *Rhizobium* spp. bacteria form close associations with members of the legume family, causing them to produce nodules in which the *Rhizobium* fix nitrogen from the atmosphere, a very important relationship for organic farmers and growers who rely on this process to fix nitrogen (N) in their rotations.

Viruses and Mycoplasmas

Viruses are very small 'organisms' that are entirely parasitic and dependent on other organisms to complete their life cycles. Many are pathogenic on plants, where they invade the plant cells and utilize them to produce more viruses. This causes metabolic abnormalities in plants, displayed as stunted growth, patches of necrotic (dead) tissue, yellow or mottled leaves (mosaic viruses), and various degrees of abnormal growth including distortions and tumerous growths. Viruses need to be transported between hosts either in propagating material or through vectors. Common vectors are nematodes, insects, fungi and can even include grafting tools. Propagating material includes cuttings, seeds, pollen bulbs or tubers. Many of the potato viruses (Table 25) are typical in being transmitted either by aphids or in planting material saved from one season to the next.

Mycoplasmas and phytoplasmas are small, primitive bacteria in the form of organisms that have been shown to cause plant diseases such as yellowing, necrosis and stunting, and are known to be spread by insect vectors. Otherwise not a lot is known about these diseases, which are not normally regarded as serious in field crops.

Pathogen Life Cycles

Disease-causing organisms display a bewildering array of different life strategies. As in the case of pests, these strategies have evolved to allow the pathogen to pass from one host to another, to invade and utilize the host, and to survive in the absence of the host when transmission is delayed. Many of the strategies have been alluded to in the previous section (*see* above), and the details will obviously vary not only with the type of organism but also with the specific host-parasite relationship. At the risk of being too general, most of these pathogens display both a sexual and an asexual phase during their life cycle. The asexual phase generally allows a rapid build-up of pathogen inocula (spores and other infective material) during periods favourable to pathogen development. The sexual phase allows genetic mixing of material within a species, and often serves to produce resistant resting spores that can survive long periods in the absence of the host plant. Asexual and sexual phases of the pathogen may occur on different hosts. Spores and other infective inocula may require specific conditions to germinate, or will only germinate in the presence of the host when conditions are favourable to infection. Within the fungi the fruiting bodies producing spores may be quite large and visible to the naked eye (for example rust pustules).

The key stages in the life cycles of typical pathogens have been discussed from an ecological and biological perspective in Chapter 2, which should be consulted for this generalized information on disease dynamics. Some of the key features of the life cycles of important crop diseases are mentioned in the tables (Tables 18–26). Whilst it is not possible to go into the detail of even a fraction of these life cycles here, there are many sources of detailed information which can be consulted (*see* Chapter 10, and Further Information p. 407) when a particular disease is suspected or further knowledge about management is needed.

Symptoms of Disease

Infectious diseases on crops are often recognized at the point when the crop plant begins to function abnormally. Symptoms become visible at this time and include obvious signs such as yellowing or mottling of leaves. Lesions may also become visible on plant parts. Abnormal growths might also be present on the plant (for example, rust pustules) and the plant itself might show twisted or stunted growth. As diseases progress, leaves may become more abnormally coloured, lesions may grow and fuse, and symptoms might spread to all leaves on the plant. The leaves may die and be shed, and the plant may become wilted and/or collapse.

Sometimes the presence of disease is not at all obvious. The plant might appear only slightly stunted or not yield much, without any other obvious signs of distress. In order to confirm the presence of disease in the absence of clear symptoms it might be necessary to resort to sending samples to a laboratory for analysis, but with practice most farmers and growers can become familiar with the symptoms of the most common and serious diseases of their crops.

Plant pathogens can be found on all plant tissues and organs in one form or another. The symptoms can be broadly divided into systemic symptoms that affect all parts of the plant, and local lesions that affect smaller areas or specific plant tissue. Systemic disease symptoms are the most likely to be confused with plant stress caused by other abiotic factors (such as drought or mineral deficiency) and include chlorosis, stunting, wilting and etiolation.

Chlorosis: Yellowing of foliage, which may have a pathogen-specific patternation – for example, mosaics are common with virus infection.

Stunting: Small, unthrifty plants are associated with diseases that disturb plant physiology, such as root rots or viruses.

Symptoms of virus on parsnip.

Wilting: Caused by excessive water loss, or by damage to water-transporting tissue within the plant, including roots and xylem.

Etiolation: Abnormal elongated or other growth, often caused by interference in plant hormone production, or due to hormone analogues or toxins produced by the pathogen.

Local lesion symptoms can be the easiest to diagnose as they are often confined to specific plant tissues and occur in recognizable patterns. They include necrosis, hyperplasia and pathogen growth.

Necrosis: Death of restricted areas of plant tissue, and common with many diseases. Typically areas of brown or black tissue that may be bounded by plant structures such as veins; they may be sunken or may appear as other spots on leaves or other plant parts, and may progress to dieback, rots and blight (sudden blackening and shrivelling). Occasionally other symptoms might include extrusions of gums or fluids from wounds.

Hyperplasia: Distorted growth due to galls or other effects such as leaf curls, blisters and scabs.

Pathogen growth: Includes visible growth over the plant surface, sporulating bodies, sooty moulds and other symptoms.

Disease symptoms also depend to some extent on the host plant's reaction to invasion, the pathogenicity of the invading organism, as well as the environmental conditions under which it is taking place (*see* below). It is interesting to note that some of the symptoms associated with the disease may also be a defensive reaction of the plant to invasion. Necrotic spots, for example, are often a result of the plant cells following a programmed collapse and death as a consequence of pathogen invasion, which often effectively prevents the further spread of (especially) biotrophic pathogens. However, in the end, once the invading organism establishes itself, it also begins to become visible through its effects and through the production of its own structures. Some fungi – for instance mildews and rusts – produce mycelia and spore-producing structures that become visible, while others, such as the wilts, sometimes produce mucilaginous secretions and/or embolisms (gas bubbles) that block up the xylem, responsible for water transport in the plant.

Some of the general symptoms associated with specific types of pathogen have been described in the sections above, and these should be consulted, as should Tables 18–26 that list some of the chief symptoms of the likely key diseases of field crops. A large number of leaflets and web pages also provide detailed information on symptoms, and pictures: *see* previous section and Further Information p. 407).

Disease Monitoring and Risk

Infectious plant disease is often portrayed using the classic 'disease triangle'. In order to manifest itself a disease needs the presence of a virulent pathogen, a susceptible host plant and the right environmental conditions. Each of these can be imagined as representing one side of a triangle, with the length of the side representing the 'amount' of the factor. The area of the triangle, bounded by the three sides, can then be seen as representing the amount of disease. If any one of these three is not present, then there will be no development of disease, because the area of the triangle is zero.

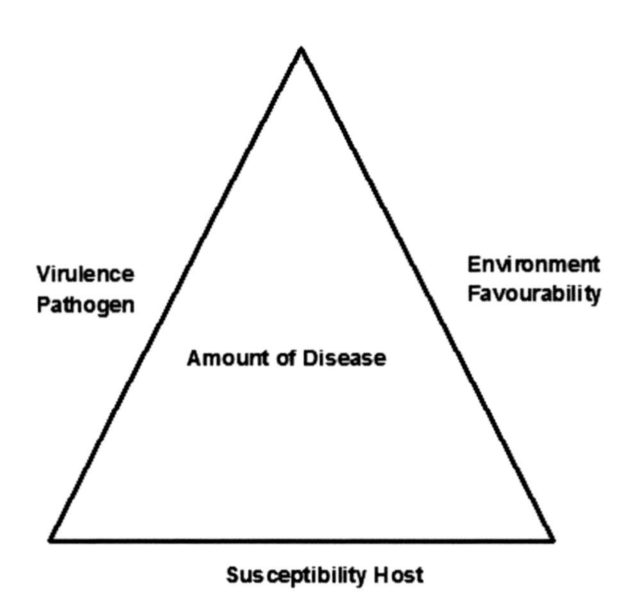

Disease triangle.

Should all the factors be present to any degree, the disease will become established and, given time, develop. If the area of the triangle represents the amount of disease then the longer the sides (or the more of the factor present), the greater the relative area, and hence the more disease will be present. Obviously any disease management strategy should aim at removing one or other of the factors that make up the triangle, and thus prevent the disease establishing. Failing this, the aim of management should be to reduce any of the factors (or the length of the sides of the triangle and hence the area) as much as possible. Organic disease management will therefore be largely a combination of preventing the pathogen reaching the crop, cultivating non-susceptible or resistant hosts, and/or altering the environment around the crop to benefit the crop but not the disease.

Although this may seem straightforward, management decisions will be complicated by the presence of many conflicting demands. For instance, it is unlikely that only one disease will be present, in which case there may be inconsistent requirements in managing two or more diseases. This is still more complicated if pests are also present. The crop itself will also have to be managed according to good husbandry practice, and this might also conflict with the other requirements. In practice the farmer or grower will have to target the key diseases in their system. We have highlighted the main management methods against key diseases in the tables (Tables 18–26), which should be consulted. The main risk factors have also been provided where known or relevant.

The presence of disease in crops should also be monitored, both during the season and between seasons. An analysis of this information will highlight the key diseases on any given farm, and will show any trends (either up or down) between seasons, allowing management techniques to be tweaked or modified. Once again the actual monitoring methods have been discussed in Chapter 6 and are indicated in Tables 18–26, which should be consulted for specific methods.

Over the last decade or so, disease-forecasting systems have also been refined, and this has led to greater understanding of the biological factors that underlie pathogen development. These systems could potentially aid organic farmers in making better risk assessments. Unfortunately many of these models were developed as an aid to better timing of chemical applications, and indicate infection events rather than risks of disease problems *per se*, and are likely to act over too short a timescale to be very useful for organic farmers and growers. For instance, they are unlikely to predict optimum sowing times or other factors over which the organic farmer has some control. Current work (in the UK) is aimed at developing forecasts for the major foliar diseases of brassicas, onions and leeks, and with the availability of cheap computing power and automatic weather data collection, may become useful to farmers and growers in their day-to-day assessment of risk.

Another, perhaps more promising avenue of risk assessment is that of testing for the presence of the pathogen or its resting structures in and around the crop. Rapid diagnostic tests are being developed for many diseases in seeds, plants and soil, and in the future should offer rapid, reliable and economically viable tests for both organic farmers and growers. Diseases for which testing is available include clubroot (*Plasmodiophora brassicae*) in brassicas (Table 20), foot rots in peas (Table 23), and verticillium wilt (*Verticillium dahliae*) in strawberries and potentially other crops. Such tests can be a useful part of general disease-monitoring schemes that enable farmers to build up a picture of diseases on their holding, and which will, over time, enable them to make judgements based on their own and other local experience.

DISEASE MANAGEMENT

Given the large numbers of potential disease-causing organisms in crops, it is difficult to be proscriptive about the management methods that farmers and growers can use. Rotations and cultural control measures are central to disease management in organic systems, and, as previously described, farmers and growers should start from the basis of developing rotations that tackle the weak spots in the life cycle of key pathogens and so prevent their arrival and/or dispersal. Strategic planning of vegetable cropping can also be important to avoid periods of high disease risk. Disease forecasts, where they can be obtained, and which offer guidance during the cropping year, may help in the planning of specific cultural operations such as tillage or harvest, although many will not be particularly sensitive to location or practical farm experience.

Other cultural control measures, including the avoidance of known soil-borne problems and the manipulation of agronomic practices, are also well known tactics for managing diseases. For example, the manipulation of the soil environment with composts, manures, crop residues and other amendments has been shown to be successful in experiments against specific pathogens. Seed-borne infection is one of the most important sources of some major pathogens, and the use of clean, healthy seed is a key requirement for organic farmers and growers. Cultivar resistance to diseases is also well understood, and more use should be made of genetic diversity through species mixtures (intercropping) and cultivar mixtures, at the same time as broadening the range of crops grown in a rotation. Crop covers and sheet mulches can also contribute to disease management in organic systems.

Whilst preventative and cultural control methods are a mainstay of organic disease management, more reactive measures are available, although these are not as well developed or effective as those available for pest species. For instance, some specific biological control agents for disease control are available, but not widely regarded as highly effective (and not approved for use in all places – for instance in the UK). Direct control using biocides has been demonstrated with, for example, plant extracts and foliar feeds, and has been shown to be effective in some conditions – but once again, their use is not widespread.

In the next sections we describe more specific disease-management tactics for farmers and growers in five key areas: seed-borne disease, soil-borne disease, diseases in propagation, diseases in the growing crop (foliar and air-borne diseases) and diseases of grain, fruit and at harvest. Storage is briefly reviewed in the latter section. The text should be read while consulting the tables (Tables 18–26) on key diseases of the major field crop groups or types.

MANAGING SEED-BORNE DISEASE

The principles and advantages of using healthy, vigorous seed are well known and assume great importance to organic farmers and growers because their crops need to get off to a good start and be quickly established. For organic farmers and growers this is doubly important because they have limited means to control an infection once it has taken hold. Apart from seeds, other propagating material such as seed potatoes and onion sets should also be dealt with in the same way as genuine seed, and many of the management principles are the same.

Susceptible Crops

Most crops suffer to some extent from seed-borne diseases. Those perhaps most affected include onions (neck rot, Table 24), cereals (Table 26), brassicas (black rot, Table 20), carrot (Table 18), legumes (Table 23) and potato (Table 25). Seed-borne pathogens can be loosely associated with the seed, in which case they can be easily removed by cleaning, or more tightly bound under the seed coat, in which case cleaning is less likely to be effective. In some cases, for example viruses, the infection may be systemic to the seed, in which case no mere 'cleaning' process is likely to be effective. In such cases it may be necessary to isolate 'clean' meristem cultures of propagating material to culture up disease-free plants. Some of the methods used to clean such plant material might not be acceptable for use in organic growing and propagation systems.

Organic Seed Standards

In some respects seed health is also the area in which farmers and growers have little control. Whilst standards and regulations differ between regions, in Europe organic standards state that organic farmers and growers must use organic seeds and plant materials when a suitable variety is available. If no suitable organically produced varieties are available, then a derogation must be obtained for the use of non-organic seed or planting material. The use of seed treated with pesticides or other chemicals is not permitted under most organic standards, and in the EU was prohibited in 2004. Difficulties lie in defining what varieties are suitable in any particular situation. Varieties listed as resistant to diseases may not be available as organic seed, in which case farmers or growers will need to apply for a derogation to use such seed. Some control bodies may need evidence to back up such requests for derogations, and obtaining untreated seed may

Celery leaf spot, a seed-borne disease.

be problematic in some circumstances as companies may be unwilling to produce untreated batches of seed.

From a farmer's or grower's point of view, any bought-in seed should also be free of pathogens, and this in itself might cause problems. In the UK, for example, there are effectively no official seed health standards for vegetables, and pathogen loading on seed is at the discretion of seed companies. The Vegetable Seed (England) Regulations 2002 simply states that 'diseases and harmful organisms shall be at the lowest possible level that can be achieved'. Most companies aim to achieve this by careful management of organic seed crops (which are often grown in warm, dry climates under irrigation), backed up by the limited range of seed-cleaning technologies acceptable for use in organic systems. There is, however, a concern that derogations are not issued with respect to seed quality (for example disease loading), and that this situation can force growers to accept inferior quality seed.

It therefore has to be accepted that many seed lots will contain some level of pathogen infection. It is unclear what the tolerance levels are for seed loads of pathogens in organic crops, but it is known that even quite high loadings of some diseases – for example, *Alternaria dauci* on carrots (Table 18) – do not appear to have noticeable effects in the field. On the other hand, even quite low levels of neck rot (*Botrytis alli*) on onions could potentially be problematic (Table 24). Whatever the level, farmers and growers would be wise to question their seed supplier if high pathogen loads are suspected on seeds, and in some cases it may be possible (but costly) to get independent testing and advice.

Management Strategies

Cultural Management

Cultural interventions against seed-borne disease should obviously work to prevent the planting of such seed in the first place (*see* above). Once planted there are no interventions that the farmer or grower can take against these pathogens, although maintaining ecologically active soils with a high organic matter content is reputed to help mitigate the effects of diseases in the soil. The mechanisms for this have been discussed previously, and include outcompetition of and predation on the pathogenic organisms in the soil environment. This prevents them becoming established in the crop plant, and is more likely to be effective where the pathogen is loosely associated with the seed coat rather than deep within, or systemic to, the seed.

Ensuring the crop has adequate growing conditions and nutrition is also likely to mitigate the effects of seed-borne disease to some extent by ensuring that the plant is able to resist infection and grow away from disease in favourable periods. Some commercial seed is available as pellets, which enables more accurate placement of individual seeds and may help plants to establish better. The pellets are normally made from clay and other inert materials. Pelleted seed may also be primed to germinate faster and thus grow through susceptible periods more quickly. In both these cases a farmer or grower should check that these processes are acceptable to their certifying or control body.

Recently, seed saving has become an attractive option for farmers and growers. Some organic growers are also looking to develop their own varieties or 'land races', and such schemes normally require the use and preservation of home-saved seed. Farmers and growers should apply the same standards to home-saved seed that are applied by seed companies to commercial seed production. Thus they should only save seed from uninfected crops, and should clean seed where possible and if they have sufficient quantity. Some companies offer the services of portable seed-cleaning rigs for this purpose.

Seed potatoes in particular can be a major source of disease problems. Many of these can be identified on seed, which is why it is important to inspect seed before planting. For example, black scurf *(Rhizoctonia solani)* can manifest itself on organic seed potatoes, especially if the growing season is cold and slow, and can be seen on the seed potatoes (Table 24). Home-saved seed is also likely to be infected with virus, which may only appear in the second year. When bought in it is necessary to take care of the seed after delivery so that problems such as blackleg *(Erwinia caratovora* pv. *atroseptica)* (Table 24) are not aggravated, which can happen if the seed is not kept dry before planting.

Biological Control

Conservation biological control should work to enhance soil ecology, as discussed in previous chapters, as natural microorganisms in the soil are likely to antagonize seed-borne pathogens and may prevent them infecting the plant. Some of these microorganisms could potentially be enhanced in the soil as biological control agents, or even conveniently delivered as seed treatments. However, research in this area has yet to deliver reliable disease control by this means, although some seed treatments have been produced in some countries (not the UK), and are used as part of management for seed-borne (and soil-borne) disease. Previously mentioned biological control agents include *Trichoderma harzianum* (antagonist fungus) and *Bacillus subtilis* (a bacterium), which have been shown to protect root systems against soil-borne pathogens after germination, and in some cases during the cropping period. Other commercial formulations have included the microorganisms *Streptomyces grieseoviridis, Gliocladium virens* and *Streptomyces lydicus.*

Research has also looked at the feasibility of enhancing natural background control by coating the seed in 'nutrients' to encourage the activity of beneficial soil microorganisms, and this is a method that might become available to organic farmers and growers over time. For instance, seed treatment with milk powder and mustard flour has been shown to reduce infection of common bunt (*Tilletia tritici*) (Table 25) in wheat and stem smut (*Urocystis occulta*) in rye. A number of commercial formulations has also been produced, which, it is claimed, enhance germination and establishment, and which contain a mixture of nutrients and/or microorganisms. The use of these in organic farming would be subject to approval by control bodies.

Direct Control

Many pathogen control methods for organic seeds are being investigated by commercial companies, although most have been known about for some time. The most promising research includes work on various heat and cleaning systems for seed. Thorough cleaning can remove pathogens from the seed coat, and heating can kill them without necessarily killing the seed germ as long as the temperature and timing are precisely controlled. Heating systems employ either dry heat or, more commonly, wet heat as steam or water baths. Such heat treatment works well for smaller seeds including brassicas (Table 20), carrots (Table 18), celery (Table 18), tomatoes and peppers, but less well for large seeds.

Some research has also looked at the possibilities of organically acceptable seed treatments that could be used on organic seed to either kill or suppress seed pathogens. Although a range of plant extracts and other materials has shown promise in this respect, it is unclear if they will become acceptable for use in organic systems, or indeed, if they will obtain the necessary permissions under pesticide safety schemes. Bleach (sodium hypochlorite) is also an effective cleaning agent in this respect for removing pathogens from the surface of the seed, but might not be acceptable to all control bodies.

MANAGING SOIL-BORNE DISEASE

Soil-borne diseases can be intractable and frustrating in all farming systems. Soil health is central to organic farming and many of the practices outlined in previous chapters (Chapters 3 and 4) are aimed at maintaining soil health, which depends on more than simply the absence of pathogens.

Experience has shown, and research has tended to support, the idea that pathogens are able to build up in poor soils with low organic matter. This is because pathogens themselves are generally weak competitors with other soil flora and fauna and will tend to be suppressed in biologically active soils. The core principle of managing soil-borne disease in organic systems is to build up soil organic matter while managing soil to provide sufficient nutrition to crop plants at the right time.

Susceptible Crops

Soil-borne diseases can be divided into two groups: those that spend a considerable part, or all, of their life cycle in the soil, and those that overwinter or survive in the absence of the host in soil, normally as some form of resting spore or structure. The former are often diseases of the roots or stems of plants although the infections may spread systemically, whereas the latter are often also spread by airborne means once the disease cycle has been initiated. These are discussed in the section on managing foliar diseases (*see* below), although many of the methods used in managing soil-borne diseases will also be useful in reducing the resting stages of these pathogens in the soil.

Once again, most crops are susceptible to soil-borne diseases when planted into poor seed beds or under poor growing conditions. They include alliums (Table 24), cereals (Table 26), brassicas (Table 20), carrots and relatives (Table 18), legumes (Table 23), lettuce (Table 19) and potatoes (Table 25). There are many persistent soil-borne problems that are difficult to overcome in organic production, even with long and varied rotations, as they are capable of surviving many years in the absence of the crop. Such pathogens include species of *Pythium*, *Fusarium*, *Sclerotinia* and *Rhizoctonia* which are able to infect a wide number of species including weeds, and may even be able to survive saprophytically to some extent.

Sclerotinia, a soil-borne disease, on lettuce.

Management Strategies

Cultural Management

Pathogenic organisms only become important in the soil when the root zone is 'unbalanced' in some sense and crop growth is under stress. In the case of soil pathogens, the physical, chemical and biological reactions in the soil are so complex that it is often difficult to unravel the specific factors or processes responsible for disease control. For the most part, beneficial soil microorganisms hold pathogens in check by competition, by predation and by enhancing resistance of the crop to infection. All cultural operations should be carried out with this in mind, and the following practices are the most likely to aid in the management of soil-borne diseases.

Rotation Rotation of crops is important in the control of soil-borne diseases that are restricted to the soil and where the disease can only survive on the crop host. Classic diseases that need to be managed through rotation include clubroot of brassicas (Table 20) and foot rots of legumes (Table 23). In the absence of the crop host the pathogen is suppressed and/or predated on by other soil microorganisms and will not be able to multiply in the absence of a host. To be fully effective crops of the same crop family and even weeds may need to be controlled in an infected field if they are alternative hosts for the pathogen.

Avoidance Crops should be grown in appropriate locations and conditions, and fields with known soil-borne disease problems should be avoided for as long as possible (*see* 'Rotation' above). Fields that have known drainage and/or compaction problems may also need to be avoided, since these conditions are favoured by diseases such as foot rots. In some cases it might be possible to adjust the sowing or planting dates to avoid periods of high disease susceptibility. For example, sowing potatoes early in the season can lead to greater incidence of stem canker (Table 25) because shoots have more opportunity to be infected as they grow slowly through cold, damp soil; delaying sowing until the soil is warmer can reduce the incidence of the disease.

Incorporation of Green Manures Fertility building and cover crops are regularly incorporated into soils in organic systems, and many studies now show that there can be considerable benefits for disease control to organic growers, either directly through the release of allelochemicals that reduce pathogen numbers, or indirectly through the stimulation of other soil microorganisms that in turn suppress the pathogens. In some cases green manures are undersown beneath crops, and this can also create conditions unfavourable to pathogens. In one trial, for example, carrots undersown with clover showed considerable reductions in cavity spot.

Use of Compost Compost has been shown to suppress pathogen activity in agricultural soils both directly and indirectly. For example, in one research trial the addition of compost increased the marketable yield of potatoes in soil infested with *Verticillium*, and many more examples are

Brassicas undersown with clover. (Courtesy Colin Newsham)

available in the research literature covering crops such as cereals, peas, peppers and cucumbers. The effects are variously put down to competition among pathogens for nutrients in compost, antibiotic production in the compost, parasitism (of pathogens) in the compost, and systematic resistance induced in plants as a result of compost. In addition there are undoubtedly further benefits from the positive effects of compost on the soil environment (for instance improving aeration and soil structure), which promotes pathogen antagonists. Compost produced using waste from onion or allium crops has also been used to manage white rot (Table 24) as it stimulates germination of sclerotia without the host crop being present, thus cleaning the ground to some extent.

Mulching Layers of organic mulch can have similar effects to green manures and composts, and indeed these can be left on the surface as mulches themselves. Mulches can provide a barrier to plant pathogens that need access to the surface to complete their life cycle, but also have positive indirect effects on soil flora and fauna for similar reasons.

Plastic sheeting has also been used as a mulch to prevent spores escaping from contaminated ground, and when clear has also been used to promote soil solarization. As a method of soil disinfection care needs to be exercised in its use in organic systems, and some control bodies might not allow it. It has a complex mode of action comprising physical, chemical and biological components which act to kill many soil micro- and macroorganisms, including pathogens such as *Sclerotium* spp.. In some instances creating a biological 'vacuum' might actually favour soil-borne pathogens which are generally poor competitors in an active soil, but in others it might help reduce soil pathogens.

Incorporation of Crop Residues Incorporating crop residues can serve to reduce soil fungal pathogens. For example, cruciferous residues when incorporated give off glucosinolates as they decompose (the so-called biofumigation effect), and this has been shown to reduce the incidence of soil-borne diseases in subsequent crops. Sealing the incorporated material under plastic mulch or solarization has been used to increase the biofumigant effect.

Cultivations and Tillage Tillage and cultivations can have a direct effect on pathogens in the soil. For example, deep burial of sclerotia of *Sclerotinia* spp. so that they are no longer able to produce apothecia may also be a useful technique. Similar considerations apply for other diseases that germinate at the soil surface to release spores after overwintering. Tillage and cultivations can have both positive and negative effects on soil ecology, as discussed in previous chapters, and this should be borne in mind.

Crop Nutrition A wide range of soil-borne diseases has been shown to be influenced by crop fertilizer application, especially nitrogen, and even though organic farmers and growers don't use such fertilizers, there is no reason to suppose that the relationships don't hold for organic crops. The list of diseases so affected includes potato scab (Table 25), bean root rot (*Fusarium solani* f.sp. *phaseoli*) (Table 23), verticillium wilt, and take-all of wheat (Table 26), all of which are affected positively or negatively by nitrogen application. Apart from the effect of macronutrients and amendments on the management of plant disease, trace elements are also capable of influencing host pathogen relationships by mediating plant responses to pathogens (*see* Chapter 3). Some research has also looked at the potential benefits of soil remineralization through the use of granite dusts, and this has been reported as being beneficial against diseases, partially by supplying rare plant micronutrients; however this effect has not been widely verified.

Management of pH The management of pH is important in managing some soil-borne diseases. For example, raising the pH from 6.5 to 7.5 reduces the incidence of *Fusarium* wilts in tomato and potato (Table 25), lowering the pH with sulfur controls potato scab (*Streptomyces scabies*) (Table 25), and liming to raise the soil pH above 7.0 remains an essential part of clubroot control (Table 20). In some cases it will be necessary to apply for permission to correct nutrient imbalances from a control body, and in most cases they should also be part of a longer-term organic soil management plan for a farm or holding.

Irrigation and Good Water Management Soil moisture is also known to be important in managing soil-borne diseases, including foot rots of peas and legumes (Table 23) which can be reduced by good drainage, and scab on potatoes (Table 25) which can be reduced by irrigation at tuber formation. This may also be the case for scab of red beet, which is susceptible at five to seven weeks after sowing.

Violet root rot on carrot, another soil-borne disease.

Use of Resistant Varieties Some varieties are likely to be more resistant to soil-borne diseases, and these should be used where (admittedly scarce) information is available. Many vegetable crops, for example, have been bred to be resistant to *Verticillium* (often denoted by a V after the variety name), among other diseases.

Biological Control

Most research is agreed on the importance of suppressing plant pathogens in soil. A number of mechanisms are involved, including parasitism and/or predation of dormant and active pathogens by soil microorganisms, competition for resources or nutrients, and production of antibiotics and/or toxins active against pathogens. Given this, conservation biological control should aim to stimulate the natural background biological activity in the soil as much as possible, and many of the cultural techniques mentioned above do this, either directly or indirectly. Various different types of soil micro- and macroorganisms are important in moderating the effect of plant pathogens on crop plants, some of which are described here.

Arbuscular Mycorrhizal Fungi (AM) This and the related vesicular arbuscular mycorrhizal fungi (VAM) are fungi that naturally colonize the roots of most crop plants, although brassicas appear to lack them. They have a mutualistic relationship with their host plants, improving the ability of the host to obtain nutrients (especially P), but also in apparently protecting host plants from disease and in stimulating host plant resistance to disease. In return, the plant supplies the fungi with complex photosynthates. The mechanisms of conferring protection are not fully understood, but one study showed that AM reduced the number of infection loci on roots for

Phytophthora parasitica on tomato. It is also likely that the close relationship between the fungi and the plant will elicit defensive reactions from the plant that might be beneficial when other parasites attack the host. Careful management of crops, including the use of cover crops and sensitive soil management, are required to sustain activity in agricultural soils.

Rhizobacteria and Endophytic Fungi Bacteria which colonize roots (rhizobacteria) have been shown to have an effect in enhancing plant resistance to pathogens, including foliar diseases. The mechanisms are probably similar to those evoked for AM fungi, namely exclusion of pathogens from root infection sites, and stimulation of the host plant immune reaction. Endophytic root fungi probably also have a similar role to play in protecting plants, although their role has not been well explored. For instance, some work has shown that such fungi can protect brassicas against the effects of clubroot (Table 20).

Antagonistic Soil Microorganisms Soil contains a large number of different microorganisms across a range of groups including fungi, algae, actinomycetes, bacteria and protozoa. All of these are vital to the flow of energy and nutrients through the soil environment, and should be conserved as much as possible in an agricultural setting. Whilst it is not possible to detail all their effects, many of them are hyperparasitic on pathogens, and many others simply outcompete them for nutrients and/or other resources in the soil. Active antagonism of pathogens can be due to the production of siderophores (insoluble organic matter), extra cellular enzymes that promote the degradation of the walls of resting spores and mycelium of pathogenic fungi and/or production of antibiotics, amongst other effects. They thus help to reduce and control the numbers of plant pathogens in the rhizosphere, and should be conserved as much as possible. The input of organic matter to soil is central to many crop management programmes in organic agriculture, and this organic matter serves to drive this complex ecology.

It is also from this group that many of the commercially available biological control agents are drawn (*see* below). Typical examples of antagonistic microorganisms include *Pseudomonas* siderophore-producing bacteria, which have been demonstrated to be suppressive to *Fusarium*, *Rhizoctonia*, *Sclerotinia* and *Pythium*, among other soil pathogens. Their function is favoured in neutral or slightly alkaline soil. Similarly *Trichoderma* spp. are soil-dwelling fungal hyperparasites capable of encircling fungal parasitic mycelium and killing them, whilst non-pathogenic strains of *Pythium* spp. are capable of attaching to and penetrating fungal mycelia as well.

Macroorganisms Organisms such as earthworms, springtails and symphylids all live in the soil and constantly ingest and process soil organic matter and soil microflora (including pathogens). They are thus capable of modifying soil conditions and altering the balance between beneficial and parasitic organisms. Whilst their overall contribution is difficult to gauge, they are an important component of healthy soils and should be conserved as much as possible.

Suppressive Soils Some combinations of biological activity, physical conditions and soil types combine to produce naturally 'suppressive soils'. These are defined as soils where, despite favourable conditions for a disease, it cannot establish, or if it establishes, it only does so for a short time or produces no disease symptoms. Such suppressive soils are obviously desirable from an organic farmer's or grower's point of view, but it has proved difficult to recreate them artificially. It is known that diseases are suppressed by a combination of induced plant resistance, hyperparasitism, competition between soil microorganisms and inhibition by secretions of antibiotic chemicals, and in fact they combine many of the elements discussed above in one soil, and this is perhaps what organic farmers and growers should be aiming at.

Commercial Control Agents Failing the production of a suppressive soil, it is possible to target specific plant pathogens with specific antagonists. These are normally applied to the soil to augment the background biological control and are likely to be more effective if the relevant soil management methods discussed above have already been applied. Various microbial species are available in different countries for use against soil-borne diseases and/or propagules of foliar diseases, although it would be true to say that their performance can be variable depending on time and place. Biological control agents that can be obtained commercially include *Trichoderma harzianum, Gliocladium virens, Talaromyces flavus* against *Rhizoctonai solani* and *Pythium ultimum. Coniothyrium minitans* and *Sporodesmium sclerotivorum* are two fungal hyperparasites that have been successfully used against sclerotinia disease. Many of these are commercialized under different trade marks and brand names, which can be obtained on the internet for specific countries and markets.

Sclerotinia, here in celery, potentially susceptible to biological control.

The Use of Amendments Amendments have also been used to stimulate soil biological activity. For example, chitin has been reported to significantly lessen the severity of soil-borne diseases such as root rot of beans caused by *Fusarium solani* and of pea wilt caused by *Fusarium oxysporum* f.sp. *pisi*. by stimulating chitin-digesting microorganisms in the soil. It is more effective when added a month or more before sowing to give beneficial microorganisms time to establish. Similarly, pelleted but spent cultures of *Penicillium* have suppressed *Fusarium* wilt of melons, and have been approved for use by organic growers in some countries. As with chitin, it is thought to work by the stimulation of chitinolytic fungi. Other amendments could potentially have similar effects. It should be remembered that soil volumes are large, and amendments may need to be applied at correspondingly high rates to have a realistic chance of success, and that some amendments may not be compatible with organic farming principles.

Direct Control

Direct control of soil-borne diseases is more problematic in organic systems, and no really effective direct control measures can be applied against soil-borne diseases.

MANAGING DISEASES IN PLANT PROPAGATION

Many commercial vegetables are propagated from seed in module trays and transplanted into the field as young plants. This technique is widely used for brassicas, lettuce and onions, among other vegetable crops. From the point of view of disease management, the use of modules generally ensures better crop establishment and affords some protection from soil-borne diseases such as stem rots and damping off (*see* above). Vigorous, well established crops are also likely to be less vulnerable to pathogen attack. In addition, the crop occupies the land for a shorter period, perhaps giving foliar diseases less chance to establish secondary cycles.

Susceptible Crops

Crops that are commonly grown from transplants, and which can experience problems, include brassicas (Table 20) and salads (Table 19). Seedlings raised under protection are vulnerable to a wide range of diseases, some of which are primarily problems at the seedling stage – for example, damping off and wirestem (*Rhizoctonia solani*) (Table 23) and brassica downy mildew (*Peronospora parasitica*) (Table 20). Other diseases such as lettuce downy mildew (*Bremia lactucae*) (Table 19) may be acquired at the seedling stage and subsequently cause serious problems in the field, perhaps for the rest of the season.

Many commercial plant raisers are now able to provide organically raised transplants produced from organic seed. Growers often buy in commercial transplants, but many small-scale growers raise their own. If bought in, transplants should be subject to checks for vigour and health

Many growers use blocks in plant propagation.

before they are accepted. If raised on the farm, the grower should apply the same hygiene standards as would be expected from a commercial propagator to prevent the infection of planting material (*see* below). Some growers prefer bare-root transplants – that is, plants sown and raised in a specially prepared seed bed for transplanting as bare-rooted plants. In this case, nursery beds should also be subject to hygiene management, and may benefit from the management of soil-borne diseases, especially damping off diseases, as discussed above.

Management Strategies

Cultural Management

High standards of hygiene are of paramount importance when propagating plants, be they in commercial nurseries or on farm. Trays, benches and glasshouses should be scrupulously cleaned, and steps taken to prevent contamination of compost and equipment by soil or other sources of disease inocula. Many diseases are favoured by warm, humid conditions, and control of the glasshouse environment will be a vital part of successful propagation, although there will be less leeway for this in polytunnels and/or nursery beds. In general, good ventilation and moderate temperatures (avoiding extremes) are likely to promote plant health, in part by reducing humidity in and around the foliage, but also by promoting vigorous plant growth by avoiding stress periods (for example excessive transpiration due to overheating). Spacing plants out, depending on raising method and module size, can be important in preventing plant-to-plant spread of disease. Plants should be watered to avoid stress periods that may make them vulnerable to pathogen infection and to avoid high humidity for extended periods of time which also favours pathogen infection of foliage. Any batches of plants with diseases should be removed and disposed of promptly, and ideally the area used for plant raising should not be exposed to air-borne spores from nearby commercial (or other) crops.

When plants are being transported outside to be hardened off, this should be done carefully to avoid physical damage to plants that might allow opportune pathogens to invade the seedlings. Hardening off should be done sensitively to avoid stress by, for instance, gradually increasing the number of hours outside and/or covering the plants to avoid extremes of sunlight and temperature. When being transported to the field, transplants should be packed to avoid physical damage and may need to be watered and shaded if planting is protracted on a hot day. Planting, either mechanical or manual, should be done carefully to avoid damaging plants, and they should be planted at the appropriate depth, avoiding covering them with soil which can also allow opportune pathogens to invade. Plants may need to be watered in to avoid undue stress to the new crop and to encourage it to establish quickly and grow away.

Other cultural control measures include the selection of cultivars with appropriate disease resistance to combat diseases of both seedling and mature plants – as is the case with lettuce downy mildew (Table 20). There is also some evidence that propagating media can be improved by the addition of composts and other organic amendments such as cattle manure, which can help to reduce infections of *Pythium* spp. and *Fusarium* spp.. Some research has even experimented with inoculating plants with mild strains of viruses to elicit natural resistance to other, more damaging, plant viruses and pathogens, but this strategy is not currently commercially practised.

Biological Control

As with pests, the more controlled conditions in plant propagation units has the potential to introduce biological control agents where they have been excluded, or to augment them where they exist. Research has looked at incorporating various microorganisms as biological control agents in compost and/or other plant-raising media. *Trichoderma* and mycorrhizal fungi inoculated into composts, for example, resulted in heavier cabbage transplants, although lettuce plants did not respond in the same way. Currently, however, there is only limited commercial production of such biological control agents due to lack of consistency in effect and difficulty in formulation (*see* above for more detailed discussion of biological control of soil-borne pathogens).

Direct Control

Trays and other equipment should be cleaned between usage, and various disinfectants are allowed in different regions for this purpose and provided there is no chance that they can come into contact with food or plants. Tank dipping involves immersing trays in disinfectant after knocking out or dislodging physical contaminants such as compost, and is suited to smaller operations. Line and stand-alone washing equipment may also be available on a larger commercial scale, which is set up to knock out and brush trays before they are washed with high pressure hot water and/or disinfectant on an automated system. Obviously automated systems have the potential to consume large amounts of water, and water should in any case be collected for recirculation and/or eventual disposal. Care should be taken at all times to keep washing areas clean, and to restrict movement and the potential for reinfecting trays with disease inocula.

There may be some scope for using a limited range of amendments and extracts to control disease at the propagation stage, and various plant extracts and amendments have been tried in this respect (*see* Direct Control of foliar diseases below). There are, however, currently no commercial formulations approved for treating organic transplants in this way. Some research is ongoing looking at a range of organically acceptable seed and transplant treatments for brassicas, and testing their efficacy and cost effectiveness.

MANAGING FOLIAR AND AIR-BORNE DISEASES

Foliar and air-borne pathogens generally cause the most visible disease symptoms on crop plants. Indeed many of these disease epidemics are synonymous with plague and crop failure, as in the classic case of potato blight (Table 25). Many of these pathogens overwinter in crop residues and/or as resting structures in the soil, in which case they are vulnerable to management using some of the methods discussed above for soil-borne diseases. Others overwinter in alternative hosts, and this may make their management more problematic. However, if they survive in the absence of the crop host, they are often capable of prolific spore production in new crop hosts, and of initiating (secondary) disease epidemics when environmental conditions are favourable, which may lead to rapid increases in symptoms on plants, across fields, and even over whole regions.

Susceptible Crops

All crops are susceptible to some degree or other to foliar diseases, and these are a common sight in all field crops. Despite this they are not always damaging, and in many cases (for example brassicas) it will be observed that they are often confined to older leaves, which would in any case be removed before sale. However, their effect is obviously serious when they reduce the leaf area considerably and affect the saleable part of the crop. Crops particularly affected include alliums (Table 24), cereals (Table 26), carrot (Table 18), beans (Table 23), lettuce (Table 19) and potato (Table 25). Diseases that commonly cause damaging epidemics include rusts (*Puccinia* spp. *et al., see* Tables 19, 21, 24 and 26) and downy mildews (*Peronospora* spp. *see* Tables 19, 20).

Potato blight, the 'classic' foliar disease.

Management Strategies

Cultural Management

In keeping with all preventative approaches discussed so far, organic farmers and growers should create as optimal growing conditions for their crops as they reasonably can in order to promote plant vigour and consequently health. Cultural techniques are likely to include having adequate soil and fertility management as part of a rotation, creating appropriate field conditions of planting and growing, and generally using good husbandry techniques in a timely manner. Many of the cultural methods used for managing soil-borne diseases will also have a beneficial effect on reducing foliar diseases and vice versa, and these should be consulted above, especially as concerns soil cultural management techniques, which will have beneficial effects in reducing plant disease in general.

The elements of general hygiene and crop husbandry described below can be used to manage plant diseases.

Rotation of crops: To avoid susceptible crops following on from each other. The longer the time between susceptible crops, the more likely it is that any resting structures or other material on or near the soil surface will be either 'starved out', predated and/or removed in some way. Organic standards in many regions lay down periods that need to be left between crop families. In the UK, Soil Association standards stipulate a minimum of three years between crops of alliums, brassicas and potatoes on the same piece of land, although successional planting within seasons can be allowed. These standards should be considered a minimum, as longer rotations are beneficial in many cases.

Avoid problematic neighbouring crops: These can be a source of air-borne inocula for newly established crops, especially if planted immediately downwind. For instance, conventional oilseed rape fields can be a source of pathogen spores for organic brassica crops, especially at harvest time, when clouds of dust are raised by combine harvesters.

Timely tillage: This is important for establishing the crop at the best time for vigorous development. Apart from providing a good seed bed, different tillage methods may be used for different effects. For instance, deep ploughing may bury pathogens to a depth at which they cannot infect the crop and are likely to be predated on over time. Weed management may also be important in checking the early spread of disease where the crop and weeds are from the same family and share diseases. It should, however, be borne in mind that the strains of diseases on weeds might not be the same as those on crops – for instance white blister (Table 20) on shepherd's purse (*Capsella bursa-pastoris*) is not necessarily infectious to crop brassicas.

Sowing dates: These are likely to be important in reducing disease risk, and high risk periods might be avoided, although this is not likely to be as effective with air-borne diseases as with soil-borne ones, due to the sheer amount of inocula produced and therefore available to land on and infect susceptible plants.

Sowing methods and patterns: These can also strongly influence disease spread. Mixed cropping, varietal mixtures and crop mixes have been investigated, and many systems of mixed cropping are used in all parts of the world (*see also* Plant resistance below). For example leeks undersown with clover develop slightly less rust, and cereal mixes generally show less development of foliar disease. In Europe, the most common mixtures are probably of winter cereals (wheat, rye) and spring cereals (barley, oats), which can also be enriched with legumes (peas or beans). Strip cropping, intercropping and undersowing can also affect disease and are examples of crop mixes that might be tried to reduce disease spread on holdings which have many different crop enterprises.

Plant spacing: How closely plants are sown, or consequent plant stand density, is known to affect disease incidence, and organic growers may be able to manipulate this factor by, for instance, spacing plants further apart in order to reduce humidity in the canopy, thereby reducing disease incidence.

Nutrient balancing and/or adequate pH: Both these factors are important in organic crop production. Liming to raise soil pH above 7.0 to manage clubroot (Table 20) is a well known technique, but it is also known that nitrate availability has an indirect effect on foliar plant pathogens. This

Mixed sowing of lettuce against downy mildew.

has been shown for yellow rust (*Puccinia striiformis*) on wheat (Table 26), where disease severity (and yield loss) increased with increasing rates of nitrogen fertilizer application, and this is likely to be true in organic systems in crops that are subject to flushes of nitrogen in a similar way. Nutritional effects are also likely to influence the natural leaf or soil flora and thereby 'natural' microbial interactions and biological control on foliage.

Good water management: This can help in the management of foliar diseases, the obvious case being the timing of overhead irrigation to avoid periods where humidity can build up in the canopy for long periods, a condition likely to favour infection by foliar pathogens. In some crops such as potatoes or tomatoes, drip or trickle irrigation is not only more efficient but reduces humidity in the canopy and thus reduces the risk of foliane diseases. Irrigation can also help manage and avoid periods of water stress which can render plants susceptible to pathogen infection. Conversely poor drainage and standing water (with consequent anaerobic conditions) are also likely to lead to crop stress and increased disease incidence (for example foot rots).

Removal of diseased plants and volunteers: This policy postpones disease cycles establishing in new crops, and can therefore help reduce or delay disease epidemics, especially on a small scale or where labour is not short. In extreme cases it can be necessary to plough in a crop to prevent a disease establishing itself and passing on to new sowings. On a very small scale removal of diseased foliage may have the same effect. On a larger scale and in the case of potato blight many organic growers flail off the leaves once 10 per cent of foliage is infected in order to prevent blight spores dripping from leaves on to the soil and infecting the tubers.

Residue destruction: This management policy is important to remove overwintering and carryover sites for living (saprophytic) stages of crop pathogens and/or to eliminate resting stages such as spores. Although burning was practised in the past, and one of its effects was to reduce the amount of overwintering disease in residues, it is not generally recommended in organic systems due to its negative effect on the top layer of soil. Ploughing can be used to bury residues, and other cultivation techniques can chop up and distribute crop remains, which should make them more vulnerable to weathering and/or biological action.

Plant Resistance Resistant varieties should be chosen for inclusion in organic cropping systems, and are often seen as the most important defence against plant disease. Varietal resistance against diseases can include either vertical (gene for gene resistance) and/or horizontal (multi-gene) resistance, and there is a feeling that the latter is more suitable to organic farming systems. The corollary to this is that there is a need for dedicated organic breeding programmes to better reflect conditions likely to be encountered on organic farms, including disease pressure. Researchers have, for instance, found that cucumbers bred under organic conditions showed high powdery mildew tolerance, and potato breeding in Switzerland is carried out under natural disease infection with no fungicide application for the same reason. The practice of breeding for organic systems is, however, far from widespread, and growers will generally have to choose those varieties with demonstrated disease resistance produced for conventional farming.

Physical Barriers Soil and crop covers have been used as physical barriers in disease management programmes. For example mulches can help keep soil moist, which might help reduce certain types of mildew – although it might equally create a habitat encouraging damping off diseases unless sufficient space is left between the mulch and the stem. Mulches also create a physical barrier between the crop and the soil, which may help to prevent resting spores germinating and initiating infections in new crops. More recently, coloured plastic mulches have also been used as these seem to have beneficial effects on plant leaf or root growth depending on the colour of the reflected light, and have also been shown to affect, both directly and indirectly, the incidence of plant disease, although the effects are not consistent in this regard.

Biological Control

Conservation biological control is likely to have only limited impact on air-borne diseases, although recent research work has shown that microbial communities on plant foliage can have an important role in suppressing, or at least modifying, pathogen interactions with plants. Farmers and growers should be aware of these largely beneficial populations, which are likely to outcompete pathogens and in all likelihood also produce various antagonistic effects – for example antibiotics or chemical cues that induce resistance mechanisms in plants – thereby providing a level of protection to plants. Some agricultural practices – for example spraying soft soap solution, or overhead irrigation – are likely to induce changes in these leaf communities, which may favour pathogens over the crop plant. The ability to affect microbial populations in such ways also suggests that the positive manipulation of the natural flora might be possible, and in fact natural amendments such as compost teas have been claimed to have this effect, especially when 'natural' fungi and/or bacteria are added during the fermentation process.

Many of the biological conservation techniques aimed at maintaining microorganisms antagonistic to soil-borne pathogens in the soil will also have beneficial effects in reducing foliar disease. They work either directly on the resting stages or spores of pathogens, and also can induce resistance in crop plants (*see* above).

On a farm scale, management of wild (non-crop) plants including habitat enhancement should be beneficial in maintaining antagonists of crop pathogens, although this is likely to be limited and in some cases might even be detrimental. For instance, ergot in cereals in the UK has increased recently, and this might in part be due to wider field margins with many alternative host grasses that are able to carry the disease over between crops.

Direct use of supplemental antagonists or biological control agents in order to augment biological control has had limited success in the field because of the complexity of natural microbial communities. Commercialization of biological control agents is extremely limited in Europe, and only a few control agents are being developed that are likely to be of use for foliar applications. These include the hyperparasitic fungi *Trichoderma harzianum* and *Ampelomyces quisqualis*, which have, for example, been shown to be capable of reducing powdery mildew severity on cucumber, especially at the onset of an epidemic. However, many factors affect control, and there undoubtedly remains a challenge to develop biological control as a practical technique under field conditions.

Direct Control

Organic growers have traditionally applied a range of plant or other extracts with the stated aim of stimulating plant growth and promoting plant and soil health. Some of these extracts have a direct effect on plant foliar pathogens, although in many cases their mode of action is not well understood. Whilst these extracts have been categorized (and their effects summarized below), it should be remembered that their use may be subject to pesticide legislation, and/or their efficacy is likely to be variable depending on the application and environmental conditions. Extracts (many of which are also discussed in Chapter 5) include the following:

Plant extracts: A wide range of plant extracts has been tested and shown to be effective under specific (test) conditions, including dock (*Rumex* spp.), garlic (*Allium sativum*), giant knotweed (*Reynoutria sachalinensis*), horsetail (*Equisetum arvense*), onion and horseradish (*Armoracia rusticana*). It may be appropriate to use purified rather than crude plant extracts to demonstrate high anti-microbial activity, and, for example, lecithin and flavinoid preparations have been tested as disease treatments. Others have been listed in Chapter 5. Plant extracts have various effects on crop plants including structural strengthening (increasing their resistance to hyphal penetration), supply of micronutrients (encouraging vigorous growth to overcome attack), direct suppressant effect on pathogens (preventing germination and/or development), and feeding leaf microbial communities (and hence enhancing biological control). Plant extracts also often seem to have a direct effect on specific pathogens.

All plant extract products applied for disease control must conform to the relevant pesticide laws of the country concerned, and should also be acceptable to the relevant organic control body. It is likely that any application techniques for biological pesticides will need further investigation to optimize application and retention of the active ingredients to improve their effectiveness. It may also be necessary for organic control bodies to clarify the justification for the use of many of these products should they become widely and commercially available.

Seaweed: Widely applied as a foliar stimulant to crop plants. It is now used extensively in the UK and Europe, ostensibly to improve the establishment, growth and development of plants. A secondary effect is in reducing plant disease on crop plants. In the case of seaweed, the mode of action is likely to be complicated and due to the presence of micronutrients and even plant hormone analogues produced by the algal seaweed. In some cases permission to use these products regularly is likely to be required. In the UK a derogation is needed for the use of liquid seaweed on Soil Association certified farms, which is given on an annual basis as long as it is justified at inspection.

Compost teas: Becoming increasingly used in a manner similar to seaweed. These products can be either a result of steeping compost in water or an active fermentation process with added microorganisms. In either case they are claimed to boost plant defences and/or stimulate microbial leaf communities, both indirectly protecting the plant from pathogen attack. Consistency of effect can be difficult to demonstrate with these products.

Manure and straw extracts: Have been shown to be effective against some diseases. For example, *Septoria tritici* blotch and powdery mildew of winter wheat (Table 26) were reduced by straw but not by manure applications to soil, and brown rust was reduced by both straw and manure treatments, and this was attributed to the increased availability of silicon.

Minerals and chemicals: Some minerals and chemicals which are known to have direct effects on either the plant or the pathogen are permitted for use in organic systems. The main products that have been shown to have direct biocidal effects on pathogens are compounds of copper, potassium and mineral sulphur. Copper and sulphur fungicides are available commercially, and can currently be applied in organic systems with permission from a control body. However, copper especially has many problems associated with its use and is currently being phased out of use in EU organic agriculture, although it is likely to be used against potato blight (Table 25) for some time to come. Sulphur has widely been used to decontaminate glasshouses and is also available as a spray to apply to foliage, but may also not be allowable in some regions. Potassium bicarbonate has similarly become available as a spray product and in this case is effective against foliar pathogens but especially powdery mildews. These compounds are discussed in more detail in Chapter 5.

Other products: A range of other products such as milk and oilseed rape (canola) oil have shown activity against diseases such as powdery mildew on courgettes (Table 22) and downy mildew on lettuce (Table 19). Many of these are discussed in more detail in Chapter 5, which should be read for more information – although once again, their use in commercial systems will depend on the specific country and situation.

MANAGING DISEASES OF GRAIN, FRUIT AND AT HARVEST

Diseases that develop late in the crop cycle are more likely to be problematic and affect not only the harvestable yield of the crop but also the ability to store the crop safely and without deterioration. Diseases at this stage of the crop cycle will also be very difficult to control in organic systems, and so every effort should be put into preventing them in the first place.

Susceptible Crops

All crops are susceptible to some degree of disease at fruiting and harvest, and indeed, many of the diseases are likely to be those that have established as foliar or even soil-borne diseases. In this case their management will be as previously discussed. Crops especially susceptible to disease at this stage are those that are likely to be stored for some time and include alliums (Table 24), cereals (Table 26), carrots (Table 18), beans (Table 23), root beets (Table 21) and potatoes (Table 25). Some diseases such as grey mould (*Botrytis cinerea*), bacterial soft rots (*Erwinia* spp.) and sclerotinia rots (*Sclerotinia sclerotiorum*) are particularly serious as they can cause rapid and significant damage to produce post harvest. Other common diseases that might cause problems are *Fusarium* spp., *Penicillium* spp., *Rhizopus* spp., *Cylindrocarpon* spp. and *Geotrichum* spp. These are opportunists which develop as surface moulds or enter tissues already invaded by other pathogens. In all cases, attention to handling and optimum storage conditions should help reduce incidence of these diseases.

Storage diseases are not specifically discussed in this book, but it should be appreciated that many of the diseases that appear in storage have their origin in the contamination of produce either in the field or during harvest

Scab on beetroot.

and processing. These processes should necessarily be carried out in hygienic conditions. It is, however, interesting to note that excessively cleaned produce can sometimes rot more quickly than slightly soiled produce: for example carrots that have not been washed store better, and this effect is partly due to the suppression of pathogenic organisms on the dirty root, whereas even a few surviving pathogens on a clean root may be able to multiply unchecked.

Management Strategies

Cultural Management

The cultural management of disease at this stage is essentially a continuation of previous management tactics – that is, good husbandry practice should aim to avoid crop stress and prevent pathogen establishment. Many of the methods for achieving this are discussed in the previous sections on the management of soil-borne and foliar diseases, which should be consulted for general practice. Other cultural practices that might be more relevant towards the end of the crop cycle include good field practice, appropriate timing of harvest, handling at harvest, post-harvest handling and storage.

Good Field Practice Crop management practices specific to organic systems can be beneficial. For example, the relatively long crop rotations that organic growers practise can reduce the risk of some soil-borne diseases, such as canker (*Phoma betae*) (Table 21) on beetroot and dry rot of potatoes (Table 25). Rotations should in any case help to manage some common diseases that directly affect harvested produce, including grey mould rot

(*Botrytis cinerea*) and scab (*Steptomyces scabies*) (Tables 19 and 25) and (sclerotinia) pink rot (*Sclerotinia sclerotiorum*) (Table 18).

It might also be possible to stagger crops to avoid harvest gluts and storage in the first place. For example, staggered cabbage harvests can be planned using a range of varieties, including those which will stand over winter – such as Savoy, January King and winter hybrids – and those that mature at different times, for example spring and summer cabbage.

Some varieties are much better for storage than others, but information on this aspect is often difficult to come by. Squashes are one of the better known crops in this respect, with different varieties recognized as having different storage characteristics; for example Acorn and butternut varieties store until November/December, and Kabocha, Gem and Crown Prince varieties store until Easter and beyond, with some potentially keeping until the following harvest. This type of evidence suggests that there are some possibilities to acquire varieties more resistant to storage disease, and farmers and growers can acquire this knowledge by keeping records of storage performance.

Other factors to consider in the crop cycle include the irrigation of crops to prevent cracking and internal breakdown, and correcting nutrient deficiencies may also be important to prevent rots and breakdown post harvest. For example, calcium deficiency may be a problem on some soils, leading to physiological storage disorders and consequently secondary rots. Organic crops often have insect damage, which can make them very prone to storage rots.

Cultural recommendations are likely to be crop specific depending on crop and market. For example, the most important disease in organic onions is onion neck rot (Table 23), which can be managed by a wide range of different tactics including crop rotation, use of disease-free seed and sets, choice of less susceptible varieties, avoidance of high populations (550,000/ha is considered optimum), avoidance of overfeeding especially with N, not irrigating within two to three weeks of harvest, allowing crops to dry off naturally in the field if drying conditions are good, or high topping if the conditions are wet and artificial drying is required. It is critical that topped bulbs have at least 10cm of neck so that drying can prevent the spread of neck rot from the cut surface of the neck to the bulb.

In some cases it may be possible to use diagnostic tests to detect storage diseases before attempts are made to store the crop, but as of yet this practice has not become widespread although it may be helpful, in the future, in deciding which parts of a crop to sell and which to store.

Appropriate Timing of Harvest Harvest should be carried out at the right physiological time. Many fruits and grains need to go through a ripening and maturing phase, and if this is complete it is likely to make them more resistant to disease attack, especially by opportune pathogens. In addition, organic crops often have a higher dry matter content and are physiologically more mature at harvest, all factors that might be expected to reduce storage disease: squashes (Table 22), for example, ripen to have a firm hard skin resistant to scratching, although stems should also be left on to prevent rots in storage; and similarly potatoes (Table 25) are often harvested

Defoliation of potatoes to prevent tuber blight.

after blighted foliage has been removed. In this case the tuber skin should be allowed to set in the soil for up to two weeks before harvest to prevent infection with blight spores during the harvest process.

Handling at Harvest How crops are handled at harvest can impact on the expression of crop disease post harvest. Sensitive harvesting which avoids damage is good practice, as are screening lines to remove obviously damaged or diseased produce. Logistical planning for better harvesting lines to prevent damage and contamination with soil may also be necessary, and designing processes for a minimum of handling will help reduce accidental damage. In some larger commercial operations produce may even be packed in rigs in the field. Harvesting equipment, such as cutting knives or collection boxes, should be clean and well maintained to reduce the chance of disease transmission from diseased to unaffected produce. Squash, for example, normally get infections as a result of harvesting damage, and this can be particularly acute if the stalk area is damaged and moisture accumulates in the area. Other produce such as lettuce can be similarly affected.

Post-Harvest Handling Many grains and fruits will need to be handled after harvesting in order to be transported from the field to packing shed or barn and then onwards for processing or sale. Transport from the field should be in appropriate conditions to avoid damage to, especially, fruit, which can encourage rots or other diseases. Temporary or permanent storage may also require that the grains or fruit be cleaned to remove soil or other contaminants, and any obviously diseased material should be removed at this time. Produce might also have to be dried to remove excess moisture, which can also increase disease such as rots. Grain in particular stores better with a low moisture content, whereas vegetables and fruits require some moisture content to remain in a saleable condition (unless deliberately dried).

Storage Conditions are likely to be important in controlling disease development in store. More specialist texts on storage should be consulted for the varied storage techniques that farmers and growers can use and their ability to prevent rot. Suffice to say that many grains and fruits store for longer in cool (3°C) and dry(ish) (RH 70 per cent) conditions without large temperature fluctuations which can encourage condensation and moisture build-up. There are, however, exceptions: for instance pumpkins prefer to be stored at relatively high temperatures (10–12°C).

Hygiene in processing and storage is also important. Water used to cool or wash produce should not contain sterilants not approved for use in organic systems, besides which this process might put harvested produce at greater risk of diseases such as soft rots, for example general bacterial soft rots in carrots (Table 18). Hydrostatic pressure from tank washing potatoes may cause water infiltration of pathogens, predisposing them to soft rot attack.

Some consideration also needs to be given to storage methods – for example, bulk storage in link boxes as opposed to storage in smaller boxes. The former might be more efficient of space, but on the other hand it might be more problematic when produce needs to be inspected and/or picked over to remove diseased produce. It might also be easier to prevent or limit damage to produce stored in smaller quantities than bulk-stored produce. Larger scale commercial organic growers are now using sophisticated stores with environmental control, and these are likely to be required by the larger organic growers to maintain adequate quality. Low cost, low 'tech', and arguably more sustainable methods such as clamping, may be more appropriate for use by smaller scale, box scheme growers. Alternative practices, including field storage with straw and/or polythene mulches, may further reduce energy costs and be more sustainable. Organic carrots, for example, are increasingly 'stored' overwinter in the field under black plastic and straw protection until needed.

Biological Control

Biological controls of crops at this stage of the cycle are limited, although some research work has shown that the application of biological control agents can in principle be used to manage post-harvest, and often opportunistic, infections. Otherwise biological controls are likely to rely on those applied at an earlier stage in the cropping cycle (*see* above), and some of these biological control agents have been shown to be effective against common diseases such as grey mould.

Direct Control

In general the benefits of amendments and extracts on grain, fruit or storage diseases will be derived from treatments applied pre-cropping or to the crop itself. There are no organically approved treatments for produce in store, and any infected produce would generally be discarded to prevent spread.

Table 18: Key Diseases of the Apiacae (umbellifer crops including celery, celeriac, carrot, fennel, parsnip and parsley)

Crop	Disease	Symptoms/Damage/ Period	Life Cycle	Monitoring/Risk	Management
Celery Carrot	Damping off (*Pythium* spp; *Rhizoctonia solani* and *Alternaria radicini*)	Collapsed seedlings in propagation/ trays, damping off of seedlings, stunted growth and root loss in field. *Alternaria* black rot with black lesions around hypocotyls and into cotyledons	Mainly soil-borne diseases but fungi are common on working surfaces and on used equipment which aids spread. *Alternaria* is seed borne	Excess nutrients and/or overwatering are common predisposing factors in damping off diseases	Hygienic raising conditions (where applicable). Avoid overwatering and use clean certified seed of carrots
Carrot	Alternaria (leaf) blight (*Alternaria dauci*)	Dark brown lesions usually appear on tips of leaflets with some yellowing. When severe, leaves become blighted	Seed-borne fungus and can also survive on crop debris	Favoured in warm wet conditions and although yield reduced with early infection, loss of foliage prevents top lifting and bunching. Forecasting systems have been developed but not for practical use	Use healthy certified seed with low disease loading. Some varieties seem less susceptible to the disease

Table 18 (Continued)

Crop	Disease	Symptoms/Damage/ Period	Life Cycle	Monitoring/Risk	Management
Carrot (Occasionally parsnip)	Cavity spot (*Pythium violae* or *P. sulcatum*)	Small sunken lesions that show as sunken pits after washing. Secondary infections lead to more general root rots. On parsnips smaller brown lesions are formed	Soil-borne fungus	Leaving crop in the ground can provoke more severe symptoms. Soil test is available to test for presence of fungus, and such soil can be avoided	Considerable differences in susceptibility between varieties. Short crops and early lifting can reduce symptoms. Good soil health protects against disease
Parsnip Carrot Parsley	Powdery mildew (*Erysiphe heraclei*)	White powdery colonies on older leaves but rarely severe, although common. Occasionally causes leaf drop and affects yield and root size	Spread by conidia in air. Overwinters in crop debris either as mycelia, conidia or ascospores	Not normally important enough to monitor but can be widespread in hot dry weather	No specific measures necessary or effective. Good husbandry should reduce period of infestation and limit damage. Mulching (if appropriate) may help limit infestation and damage
Carrot (Occasionally Parsnip, Celery)	Violet root rot (*Helicobasidium purpureum*)	Dense purple-black mat on roots. Above-ground plants may appear stressed and stunted. Has a wide	Soil-borne fungus with a wide host range. Small black sclerotia form resting bodies that can survive many	Short rotations containing carrots, potatoes and beet can be problematic especially if	Early harvesting can prevent development of disease and crop rotation should leave a minimum of four years between susceptible

		host range including beetroot, potato, swede, turnip and asparagus	years in the soil. Spreads by soil growth during season	weeds such as bindweed are present. Acid soils and moderate temperatures (15°C) favour disease development	crops. Volunteer plants should be removed and care taken not to introduce disease with packhouse waste etc.
Parsnip	Leaf spot (*Ramularia pastinacae*, *Phloeospora heraclei*)	Small brown spots on leaves with white deposit in small patches of plants with dead leaves which can spread quickly to kill leaflets and leaves throughout the crop under certain conditions	Airborne diseases during the season	*P. heraclei* is a new problem that appears to be spreading in the UK occasionally causing severe damage to crops. Otherwise leaf spots not normally serious	Rotation of crops and use of healthy seed
Parsnip	Downy mildew (*Plasmopara nivea*)	Yellowish spots visible on upper leaf surfaces with white mould patches below leaf in same location. Patches turn black	Spread mainly by airborne spores within the crop during the season. Resistant spores remain in crop debris to overwinter	Disease is favoured in humid conditions	Avoid overcrowding crops and destroy diseased material at the end of the season. Managing weedy umbellifers may help to prevent spread of disease early in season

(Continued)

Table 18 (Continued)

Crop	Disease	Symptoms/Damage/ Period	Life Cycle	Monitoring/Risk	Management
		in damp weather, and foliage may shrivel and drop			
Carrot Celery Parsnip	Pink (Sclerotinia) rot (*Sclerotinia sclerotium*)	Rot of leaf stalks which may have pink tinge. Fungus produces familiar white fluffy growth with black resting bodies or sclerotia. Can be a serious problem when develops in stored crops	A soil-borne disease that develops from resistant sclerotia to produce airborne spores that spread to crops. Has a wide host range and can be troublesome in crops such as lettuce	Sclerotia resting bodies persist in soil, and sites that are known to be infected should be avoided for as long as possible	Rogue out infected plants and destroy to prevent spread (composting will not destroy sclerotia). Grow on ridges and/or clip foliage to improve air flow. Biological control with *Coniothyrium minitans*. Harvest early to prevent secondary spread, and reduce number of host crops in the rotation
Celery Celeriac Parsley	Leaf spot (*Septoria apiicola* or *S. petroseleni* on parsley)	Small brown leaf spots with many small black fruiting bodies which progresses to cause browning of foliage in wet and windy weather (blight). Spots spread to stems rendering	Seed-borne disease which develops on the plant to produce spores which are spread by splash. Mainly seed borne but also can survive in crop debris (but not soil) between plantings	Small foci of attack rapidly become large areas of infestation in windy wet conditions	Use healthy seed (treated in hot water) and ensure transplants are healthy. Avoid successional plantings adjacent to growing crops and destroy blighted patches to reduce spread. Harvest early to prevent spread and ensure

		plant unmarketable. Similar disease on parsley			adequate hygiene between fields to prevent spread
Parsnip	Canker (*Itersonilia pastinaceae*; occasionally *Phoma complanata*, *Mycocentrospora acerina*)	Large black lesions on the root crown and sides caused by spores dripping from leaves	Seed-borne disease which develops on leaves and then infects roots when spores washed down on to soil. Infects root through wounds or rootlets	Worst in wet seasons. Large roots show more symptoms than small	Use healthy certified seed. Resistant parsnip varieties are known. Keep crowns covered with soil to minimize damage to root which allows spores to penetrate. Early harvesting if monitoring shows disease is progressing
Parsley Carrot Celery	Virus diseases	Reddening of older leaves and stunted growth with younger leaves being mottled (motley dwarf virus). More severe symptoms include leaf death, root distortion, internal discoloration and black spots and crown rot (yellow fleck and anthriscus virus)	Motley dwarf virus is caused by two viruses (red leaf and mottle virus) spread by aphid vectors. Parsnip yellow fleck and anthriscus yellows are also spread by carrot willow aphid. Celery mosaic virus is the most common virus in this crop; it is transmitted by aphids and can infect other umbellifers	Risk higher when aphid or other vectors present	Manage aphid vectors and umbelliferous weeds

(Continued)

Table 18 (Continued)

Crop	Disease	Symptoms/Damage/ Period	Life Cycle	Monitoring/Risk	Management
Carrots Parsnip Celery Celeriac	Bacterial soft rots (*Erwinina carotovora* and other bacteria)	Common disease often associated with damage to roots or stems by other pests or diseases. Important where crops are left in field or for storage	Soil-transmitted bacteria that opportunistically infect wounds	Poor drainage, potassium deficiency, high manure applications (high N) have all been implicated in increased soft rots	Maintain as long a rotation as possible between susceptible crops. Avoid damaging crops, and/or manage pests and diseases that cause wounds. Maintain balanced, well drained soil

Table 19: Key Diseases of Asteraceae (mainly salad crops including endive, chicory, cardoon, artichoke, lettuce, salsify)

Crop	Disease	Symptoms/Damage/ Period	Life Cycle	Monitoring/Risk	Management
Lettuce	Bottom rot (*Rhizoctonia solani*)	Causes damping off of seedlings and a basal rot of mature plants. Rot starts as leaf infection of small brown spots on midrib of leaf and progresses to form a mat of mycelia joining leaf to soil, and rot of plant tissue	Common soil-borne fungus with many strains	Arises from leaf contact with soil	Active soil biology helps to manage the pathogen but where problems occur plastic mulches can be used to isolate leaves from soil
Lettuce (also Globe artichoke)	Downy mildew (*Bremia lactucae*)	Pale green or yellow blotches on leaves with white fungal spores on lower surface. Often bounded by veins and prominent on older leaves. Affected areas turn brown and may be secondarily infected by rots (e.g. grey mould)	Spread by airborne spores (conidia) but resting spores (oospores, conidia) also produced which enable the pathogen to overwinter in soil	Prominent on seedlings and especially on mature plants towards the end of the season (autumn). Cool, moist conditions favour the disease. Severe disease	Resistant varieties available (single gene resistance) but fungus constantly evolving. Avoid using infected trans-plants. Avoid prolonged periods of wet leaves due to overcrowding, inadequate ventilation or overwatering. Mixtures of varieties can reduce disease spread. Trim infected leaves. Plough in crop debris to prevent spread to successional crops, and isolate

(Continued)

Table 19 (Continued)

Crop	Disease	Symptoms/Damage/ Period	Life Cycle	Monitoring/Risk	Management
				can rapidly be produced in covered crops when foliage wet	spread from mature plants. Destroy surplus transplants plantings to prevent
Lettuce	Grey mould (*Botrytis cinerea*)	Brown blotches and rotting of older leaves and stem rots. Identifiable by the masses of grey spores around the infection	Widespread disease that occurs throughout the year from abundant spores	Stress conditions in the crop (wilting at transplanting) can increase susceptibility to the disease. Common secondary infection on plants	Careful handling of transplants to prevent damage and ensuring transplants don't wilt. Avoid planting deeply or in wet sites. Avoid damage to crop in field operations
Lettuce	Soft rots (*Erwinia carotovora*)	Bacterial rots cause a watery and slimy disintegration of leaves causing yield loss but also reduced market quality and storability	Bacterial disease ubiquitous in soil and infecting plants opportunistically	Favoured by warm to hot humid conditions where there is poor drainage and/or drying conditions. Some evidence that more severe with low K	Good husbandry to provide good growing conditions on well drained land. Avoid mistreatment during harvest and avoid harvesting infected plants. Care when processing and washing for market

Lettuce	Ring spot (*Microdochium panattonianum*)	Small brown circular holes whose centres fall out leaving shot hole symptoms. Sunken brown lesions on midrib (similar to slug damage)	The disease is probably seed-borne but also survives in soil and crop debris. Conidia produced under moist conditions spread disease between plants	Cold and wet conditions favour the disease and infected plants are often found in low-lying or wet areas	Rotations should reduce soil survival. Field hygiene should prevent in and between field transmission. Patches of infection can be rogued or destroyed to reduce spread
Lettuce	Sclerotinia rot (*Sclerotinia sclerotium*; *S. minor*)	Lettuce drop manifests as wilted plants which collapse just before harvest either as scattered plants or larger patches. Fluffy white fungal growth with large black sclerotia distinguishes this disease from other rots	Soil-borne sclerotia produce fruiting bodies that release airborne spores that infect lettuce plants. There may be direct infection and spread by plant to plant contact. Sclerotia can survive for many years in the soil	Fields with a history of sclerotia disease are likely to be high risk. Iceberg types appear to be more susceptible compared to other types	Rogue and destroy infected plants or patches to prevent sclerotia being produced. Polythene mulches may block initial spore production. Crop rotations should be modified to exclude as many host species as possible. Biological control treatments are available to treat sclerotia (*Coniothrium minitans*)

(Continued)

Table 19 (Continued)

Crop	Disease	Symptoms/Damage/ Period	Life Cycle	Monitoring/Risk	Management
Lettuce	Septoria leaf spot (*Septoria lactucae*)	Small chlorotic spots enlarge and turn brown before desiccating. Often delimited by veins in the same way as downy mildew, although a hand lens can be used to distinguish small black fruiting bodies	Seed-borne and capable of surviving on debris between crops. Dispersal in the crop is by splash	Disease is promoted under wet conditions	Use healthy certified seed. Isolate successive plantings and observe field hygiene to prevent spread within fields. Rogue out badly infected patches
Lettuce	Viruses	A wide range of symptoms including inter-vein yellowing or vein clearing, dead leaf margins, blistering or crinkling, stunting, little hearts, and bolting. Common viruses include lettuce mosaic virus, cucumber mosaic, big vein and beet western yellow virus	Usually transmitted by aphids such as *Myzus persicae*, but also other agents and in seed		Buy clean, certified seed and raise plants hygienically so they are protected from vectors. Remove infected plants to prevent field transmission. Avoid planting in areas with susceptible weed species

Table 20: Key Diseases of Brassicaceae (brassicas including cabbages, cauliflower, broccoli, Brussels sprouts, kohlrabi, swedes, turnip, radish, cress)

Crop	Disease	Symptoms/Damage/ Period	Life Cycle	Monitoring/Risk	Management
Brassicas	Black rot (*Xanthomonas campestris*)	Yellow v-shaped blotches on leaf margins and distinct blackening of veins within these lesions. Patchy distribution in crop	Bacterial disease that spreads in propagation, in the field through rain splash, and in plant debris	Seed-borne disease	Use clean, certified seed. Can survive in crop residues so clean machinery of debris. Rogue out plants at early stage of disease. Incorporate debris and follow a long rotation
Brassicas	Virus diseases	Most damaging likely to be cauliflower mosaic virus, turnip mosaic virus, beet western yellow virus. Symptoms include mottling of leaves, vein clearing and vein banding as well as stunting and even leaf blackening. Can reduce storage life. Often difficult to diagnose and distinguish from disorders	Most are transmitted by aphids, especially cabbage aphid (*Brevicoryne brassicae*) and peach potato aphid (*Myzus persicae*)	Presence of aphids can be a warning of virus attack.	Remove volunteer brassicas that may harbour virus and vectors and avoid sequential cropping in same season in same field. Raise plant away from crops and manage aphid vectors. Select cultivars with resistance to virus

(Continued)

Table 20 (Continued)

Crop	Disease	Symptoms/Damage/ Period	Life Cycle	Monitoring/Risk	Management
Brassicas	Wire stem (*Rhizoctonia solani*)	Causes damping off of young seedlings that collapse and die. Older plants show rot symptoms on fleshy tissue. Black spot or canker rot in swedes and turnip	Soil-borne and capable of rapidly growing over soil with high organic matter content	More severe where plants are growing slowly in cool moist conditions or when plants are stressed by drought	Observe hygienic conditions when raising module plants, and don't use transplants that show symptoms of disease as this can spread infection to field
Brassicas	Club root (*Plasmodiophora brassicae*)	Common and widespread. Large swollen roots (galls) form, visible when plant is pulled up. Occasionally above parts of plant may wilt, and if severe, plant will be stunted and unthrifty. Galls may be infected in turn by soft rots, causing unpleasant smell	Soil-borne disease capable of surviving many years between crops. Spores germinate in presence of brassica roots and produce motile spores which gain entry through root hairs which give	Favoured by warm wet conditions for infection but drier conditions after infection causes greater damage. Sites on acid soils and with a history of brassica forage (turnips, swedes) production at most risk. Soil	Hygiene and preventing infection is a priority. Avoid contaminating trays/modules with soil- or water-borne spores. Clean boots and tools when moving from contaminated fields. Avoid manures or composts from animals eating contaminated crops. A long rotation will help (but not prevent) disease. Liming to pH 7 reduces disease. Avoid planting brassicas in heavily infected areas. Resistant varieties (e.g. of swede, cabbage) are available, but this may not be reliable on all sites. Avoid drought

			rise to reinfections	tests can be used to detect infestation	conditions which exacerbate damage. Earthing up plants may stimulate new root growth and ease symptoms. Remove and destroy infected plants if possible
Brassicas	Dark leaf spot (*Alternaria brassicae*, *A. brassicicola*)	Small black spots with yellow halo increase to produce larger brown spots, especially visible on older leaves. Only serious when on harvestable heads, sprouts or leaves. Common on cabbages and Brussels sprouts	A seed-borne disease that also survives on volunteer brassicas and brassica weeds. Also spread by windborne spores	Nearby oilseed rape crops may facilitate spore spread. Disease-forecasting systems are available but of limited use to organic growers	Use healthy seed and resistant varieties. Resistance to the disease is associated with waxiness of leaves, and crops with less wax are more susceptible. Rotation and avoiding overlap of susceptible crops can help
Brassicas	Downy mildew (*Peronospora parasitica*)	Yellow angular leaf lesions common on propagated plants (seedlings) and in the field where lesions turn white and a grey fungal growth is evident on underside of leaves. More prevalent on cauliflower and calabrese. Can cause	Soil-borne, affecting seedlings through roots. Conidia produced on the mould are spread by the wind. Oospores produced for	Encouraged in damp conditions when white mould can be seen beneath leaves	Some varieties are less susceptible. Ensure that propagation plants are not kept wet for prolonged periods with sensible watering and ventilation regimes. Only transplant healthy plants and harden off. Sometimes plants apparently recover

(Continued)

Table 20 (Continued)

Crop	Disease	Symptoms/Damage/ Period	Life Cycle	Monitoring/Risk	Management
		blackening of curd in cauliflower or root in radish	overwintering in soil		
Brassicas	Light leaf spot (*Pyrenopeziza brassicae*)	Black spotting causing thumb-print patterns on Brussels sprouts and spring greens	Spread by airborne spores produced from dead leaves and crop debris and by rain-splashed spores from infected tissue	A forecasting system for leaf-spot diseases of brassicas is available but currently of limited use to organic growers	Some varieties of Brussels sprouts are more resistant. Rotation and a complete break in cropping will help interrupt the disease cycle, as will burying crop residues by ploughing
Brassicas	Ring spot (*Mycosphaerella brassicicola*)	Common in areas of high rainfall. Black or grey spots containing black fruiting bodies. Often on older leaves but can defoliate plants if severe	Airborne spores produced from infected leaves	Can develop very rapidly in wet weather. Overwintered cauliflower, Brussels sprouts and spring greens are at highest risk. Forecasting systems have been	Use resistant varieties. Avoid continual cropping of brassicas during the season, and ensure a break between seasons. Plough in crop residues and isolate blocks of brassicas

				developed but are of limited use to organic growers	
Brassicas Swede Turnip	Powdery mildew (*Erysiphe cruciferarum*)	Leaves covered with a powdery white coating, usually only severe on swede and turnip crops. Badly affected leaves may yellow and die, plants wilt and roots crack	Spread by conidia and survives in crop debris	Prevalence of the disease varies greatly from year to year but generally favoured by hotter, dry weather when plants are small. Also favoured by mild winters	Irrigation may help reduce drought stress and disease development but may not be economically justifiable in low value brassicas. Rotation, burying crop debris and removal of brassica volunteers and weeds can help prevent carryover. Some varieties show resistance under some conditions
Brassicas	White blister (*Albugo candida*) and white spot (*Pseudocerco-sporella capsellae*)	Small white blisters on the underside of leaves which may turn brown and powdery. May cause leaf distortion. White spot more common on turnips and swedes in wetter areas with small white spots on both leaf surfaces which may enlarge to cause leaf death and defoliation	Can be soil-borne but airborne and/ or rain splash spores likely to be most common source of infections. White blister may cause tissue distortion. Resistant	Common from June onwards and some weeds affected by separate strains	Avoid sequential cropping and incorporate crop residues. Some resistant sprout cultivars (white blister) and selective harvesting when infection becoming severe. Remove and destroy diseased plants with light infestations or in small areas

(Continued)

Table 20 (Continued)

Crop	Disease	Symptoms/Damage/Period	Life Cycle	Monitoring/Risk	Management
			oospores or sclerotia remain dormant in soil over the winter		
Brassicas	Canker, black leg and/or dry rot (*Leptosphaeria maculans*)	Ash-grey spots on seedlings may develop into sunken regions on stem at ground level extending down into roots and/or elongated lesions on roots that become cracks. Circular brown spots on leaves which may wilt or turn purple	Seed-borne disease (which may cause damping off) which develops to form airborne or rain-splashed spores. Can survive on plant debris and persist on wild brassicas or relatives (wild radish, charlock etc.)	Nearby oilseed rape crops may increase the chance of disease spread to vegetable brassicas	Use clean certified seed. Remove and destroy infected plant debris, and maintain a long rotation between brassicas. Managing weed relatives may help reduce transmission between crops. Some varieties may be more resistant

Brassicas	Head rots (*Erwinia carotovora, Botrytis cinerea, Pseudomonas* spp.)	Heads with soft and slimy decay, with or without mould growth caused by bacteria or fungi. Spear rot of calabrese. Reduces storability of produce, often because of secondary infections	Causal organisms common in all environments and soil	Common in damper and wet weather and often associated with crop damage, nutrient imbalance and/or deficiencies	As long a rotation as possible between susceptible crops. Avoid mechanical damage to crops and employ good husbandry (good drainage, balanced fertility, adequate air flow etc.), avoiding stressful conditions for crop
Swedes Turnips (also Radish)	Scab (*Streptomyces scabies*)	Angular corky scabs common on roots. Although yield is only marginally affected, marketability may be reduced	*See* scab in potato diseases	*See* scab in potato diseases	*See* scab in potato diseases
Swede Turnip	Violet root rot (*Helicobasidium purpureum*)	Plants appear unthrifty, weak or yellowish while root tissues are affected by purple mycelial growth and secondary bacterial rots	Sclerotia persist for many years in soil, germinating in presence of host, to give rise in infective mycelium. Persists on weed plants	Acid soil and cool conditions favour the disease	Destroy diseased roots and remove crop remains after harvest. Managing susceptible weeds may reduce persistence. Good husbandry and long rotations after infections. Leaf brassicas tend to be non-susceptible, as are other crops

Table 21: Key Diseases of Chenopodiaceae (salad and root beets, chard, spinach)

Crop	Disease	Symptoms/Damage/ Period	Life Cycle	Monitoring/Risk	Management
Beet Chard Spinach	Blackleg (*Pleospora bjoerlingii*) and damping off (*Aphanomyces cochlioides*, *Pythium* spp.)	Damping off and blackleg can cause death of seedlings. Blackleg can cause shrivelling and blackening of stems leading to dry rot in crowns and sunken lesions and internal cavities in (stored) roots	Damping-off diseases are soil-borne. Blackleg is a seed-borne disease which is dispersed by rain-splashed spores during the growing season	Late sowings and poorly drained sites can increase susceptibility to these diseases	Plant clean seed free from *Pleospora*/*Phoma*. Plant in well drained sites and follow good husbandry practice. Very early and late sowings should be avoided. Check root quality before storage
Beet Chard Spinach	Downy mildew (*Peronospora farinosa* (different subspecies))	Dense greyish fungal growth on underside of leaves, especially young ones. Leaves become pale, toughened, distorted and ultimately blackened and necrotic (less so in spinach). In late season the heart becomes blackened and rots	Oospores persist in soil in plant debris or volunteer plants. The disease might be seed-borne in some conditions	Cool, moist conditions favour the disease spread	Use clean seed. Remove volunteer beets and infected plants. Plant vegetable crops away from sugar beet

Beet Chard Spinach	Leaf spots (*Ramularia beticola*, *Cercospora beticola*, *Uromyces betae* and others)	Pale brown or white leaf spots (*Ramularia/Cercospora*) or orange-brown pustules (rust) can appear as the season advances. Important when blemishes reduce marketability of leaves (spinach, chard, bunched beetroot)	Potentially seed borne but also persistent in crop debris	*Ramularia* leaf spot is favoured in moist conditions. Generally worse towards end of season	May spread from sugar beet so site crops appropriately and isolate successive plantings of leaf beet. Reduce late season cropping where problems occur. Remove affected leaves (early in crop development). Destroy remains of crops and volunteers to reduce transmission between seasons. Some varieties may be more resistant
Beet Chard Spinach	Powdery mildew (*Erysiphe betae*)	White growth on leaves normally only important as a blemish on leaf crops although may reduce yields of sugar beet	Overwinters on infected weed hosts, on buried root material and on crop debris	Forecast for powdery mildew on sugar beet has been developed based on winter frosts. Associated with hot weather and plant stress	General crop hygiene and good husbandry
Beet	Violet root rot (*Helicobasidium purpureum*)	Above ground the plant may be stunted or growing weakly. Roots,	Sclerotia may persist for many years in the soil before they germinate and infect	Acid soil and cool temperatures	Follow a rotation avoiding susceptible species (e.g. carrots) for 3–4 years.

(Continued)

Table 21 (Continued)

Crop	Disease	Symptoms/Damage/ Period	Life Cycle	Monitoring/Risk	Management
		crown and stem bases covered in purplish strands with dark spots and felty mass to which soil clings. Root tissue affected by brownish rot associated with soft rots	plants. Probably infects many weed species. Soil spread by mycelial growth between plants	(15°C) favour the disease. More obvious late in the season	Remove diseased plants, unharvested plants and discard. Manage weeds and maintain good soil husbandry practices
Beet	Scab (*Streptomyces* spp.)	Scabby spots and large swollen growths on roots. Small pitted and cracked areas may also be apparent	Caused by an actinomycete, soil-borne and infecting plants through stomata and lenticels	Common in hot dry summers and on light, sandy, alkaline soils. Infection often greatest after old grass has been ploughed in	Irrigation may help reduce susceptibility but is not as effective as with potatoes because infection is near surface. Some varieties appear to be more resistant. Composts and green manures may help suppress disease. Reduce susceptible root crops in the rotation.
Beet Chard Spinach	Viruses (beet yellows, beet mild yellows, cucumber mosaic virus)	Yellowing and thickening of older leaves. Spinach blight can be caused by cucumber mosaic virus. Secondary infections may cause death of plant	Transmitted via aphid vectors	More likely when conditions favour aphid vectors. Forecasts for yellow viruses are available	Remove volunteer plants and plants growing on dumps that can be a reservoir for viruses

Table 22: Key Diseases of Cucurbitaceae (cucumber, courgette, squashes, marrow, pumpkin, melon)

Crop	Disease	Symptoms/Damage/Period	Life Cycle	Monitoring/Risk	Management
Squashes	Grey mould (*Borytis cinerea*) or stem rot (*Didymella bryoniae*)	Soft rot on plants and young fruits easily recognizable by prominent grey fluffy growth	Weak pathogens that gain access to plants through wounds or when plants are stressed. Ubiquitous and common, spread by spores and capable of surviving on plant debris. A wide range of host plants (grey mould)	Associated with periods of wet or humid weather and poor air circulation. Damage during field operations and at harvest can also allow the disease to infect plants	Improve air circulation by removing old leaves when practical, or use wider spacing. Remove rotted fruits or plant parts to prevent spread. Care in field operations to prevent plant damage and disease infection. Sensitive irrigation to avoid splashing soil on to plants
Squashes	Powdery mildew (*Erysiphe cichoracearum, Sphaerotheca fuliginea*)	White powdery colonies on leaf surfaces, especially older leaves. Leaves become tattered and may drop when infection severe. May reduce yield but can also aid ripening of squashes and pumpkins by opening canopy	Overwinter on plant debris and spread by conidia during season	Generally affects older crops and widespread in autumn. Can reduce yield when infection early and limit season. Prefers drier conditions	Some varietal resistance. Rogue badly infected plants to prevent spread and plough in crop remains promptly, especially where younger crops are present. Care needed with sequential plantings

(Continued)

Table 22 (Continued)

Crop	Disease	Symptoms/Damage/Period	Life Cycle	Monitoring/Risk	Management
Cucumber Marrows	Anthracnose (*Colletotrichum lagenarium*)	Pale green transparent spots on leaves, stems and fruits turning reddish and/or dark as disease develops. Small plants may display damping off symptoms	Spread as conidia spores in humid periods by insects or field operations	Humid conditions promote the disease	Good growing conditions, especially adequate ventilation in glasshouses. Hygiene. Remove and destroy infected plants
Squashes	Bacterial leaf spot (*Pseudomonas syringae*)	Brown angular leaf spots which may ooze clear drops of liquid. Leaf spots may fall out and lesions may develop on stem and fruits	Spread by water splash, insects and field operations	Bacterial disease monitored by symptoms in crop	Hygiene in field operations and glasshouses
Squashes	Viruses	A number of viruses including cucumber mosaic virus cause chlorotic mosaics on leaves and sometimes plant stunting and fruit distortion. Cucumber green mottle virus is also found infrequently	Generally spread by aphid vectors or by field operations such as pruning or harvesting. Potentially spread in seed as well	Observation of crop and aphid or other vectors	Resistant varieties are available. Remove infected plants to limit spread and prevent movement of aphid and/or insect vectors. Remove host weeds within the crop (e.g. chickweed). Plough in crop debris. Hygiene at harvesting

Table 23: Key Diseases of Fabaceae (legumes including runner beans, French beans, field beans, peas, leguminous green manures)

Crop	Disease	Symptoms/Damage/Period	Life Cycle	Monitoring/Risk	Management
Peas French beans Broad beans	Damping off and foot rots (*Phoma* spp., *Fusarium* spp., *Aphanomyces euteiches Thielaviopsis basicola*, *Rhizoctonia solani*)	Peas and French beans are susceptible to a range of foot and root rots causing brown or black discoloration of the stem base, foliage turning yellow prematurely, and lodging of plants. Roots may be rotted, stunted and blackened	Soil-borne diseases present as spores or resting bodies. Some may persist saprophytically on plant debris. Infection may be helped by plant wounds	Problems often associated with poor drainage, soil compaction and repeated cropping. The effect of fertility-building legumes is not known. Soil tests can be used to detect the pathogens	Management of soil structure and drainage will reduce foot rot problems, as will promoting a vigorous soil biology. Leave as long an interval as possible between crops if problems persist. Avoid damaging crop during weeding operations
Broad beans Peas	Downy mildew (*Peronospora viciae*)	Infects seedlings and reduces crop emergence and/or causing stunted growth. Early infections with dense purplish growth on leaves produce spores which cause spotting on leaves, shoots and inside pods	Common soil-borne disease with oospores giving rise to new infections during season, which spread by airborne spores (conidia)	Spreads rapidly in cool, moist conditions and wet seasons	Long (five+ years) rotation reduces the risk of soil-borne infection. Resistant varieties of vining and dry peas are available, although disease evolves rapidly to counter resistance
Broad beans	Chocolate spot (*Botrytis fabae*, *B. cinerea*)	Small red-brown spots on leaves which coalesce to form large black lesions which can ultimately lead	Overwinters on volunteers and and crop debris and can also be	Common in wet seasons and on overwintered crops. Becomes	Remove volunteers and incorporate crop debris. Avoid densely sown crops and sheltered sites

(Continued)

Table 23 (Continued)

Crop	Disease	Symptoms/Damage/Period	Life Cycle	Monitoring/Risk	Management
		to defoliation of the plant and rotting of flowers and pods	transmitted in seed	more aggressive in periods of high humidity	to maintain good air flow through crops. Good growing conditions also suppress aggressive phase of disease (good fertility and drainage). Use clean seed. Establish winter crops in late autumn and do not sow spring crops adjacent to winter crops
Broad beans Peas	Rust (*Uromyces fabae*, *U. pisi-sativi*)	Small brown pustules with a chlorotic halo on leaves. Pustules may progress to cover plant giving it a reddish appearance (broad beans)	Survives on crop residues and volunteer plants between seasons	Usually appears late in the season with a limited effect on yield. Lack of P may exacerbate disease	Incorporate or thoroughly compost plant debris at the end of the season. Remove volunteer plants
Legumes	Sclerotinia disease (*Sclerotinia* spp.)	Brown wet rot at the base of the stem accompanied by white fluffy mould. Large black sclerotinia often present in mould. Plants may wilt and topple over	Soil-borne disease with new infections arising from overwintering sclerotinia, which give rise to spores that infect crops	Prefers damp, cool conditions for development	Remove and destroy infected plants and avoid cultivation of susceptible plants in area for 3–4 years. Biological controls may be effective. Remove susceptible weeds

Peas	Wilt (*Verticillium* spp.. *Fusarium* spp.)	Wilting plants, starting with lower older leaves, developing into yellowing and shrivelling of leaves. Brown or black discoloration in stem conducting tissues when cut well above ground level	Persistent soil-borne pathogens that survive as resistant spores. Infection may be aided by nematode or other root damage	Problems persist in infected fields	Rotation with non-susceptible plants to prevent build-up of disease
Broad beans Peas	Leaf and pod spot (*Ascochyta pisi*, *A. fabae*, *Mycosphaerella pinodes*)	Black, brown-yellow or grey spots on leaves as well as lesions on stems and pods which may also bear dark pycnidia. Seedlings may be killed	Seed-borne infection which gives rise to new infections when seed planted. Airborne spores are also produced from crop debris	Usually only severe in wet growing seasons	Use healthy certified seed. Adequate rotation leaving as long a gap as possible between susceptible crops
Peas French beans	Bacterial blight (*Pseudomonas syringae*)	Water-soaked lesions on leaves that enlarge and coalesce but delineated by veins. May cause damping-off symptoms in small plants	Spread by water splash, insects and field operations. Can survive in seed, and this is often the source of new infections	Bacterial disease monitored by symptoms in crop. Frost or otherwise damaged plants susceptible to the disease	Hygiene in field operations and protected cropping

(Continued)

Table 23 (Continued)

Crop	Disease	Symptoms/Damage/Period	Life Cycle	Monitoring/Risk	Management
Peas French beans	Grey mould (*Botrytis cinerea*)	Grey water-soaked lesions on flowers and pods. Rotted pods often have grey fluffy mould growth	Spreads readily by airborne spores (conidia) and persists in soil on crop debris and as resting bodies (sclerotia)	Common when flowering coincides with wet periods and when petals adhere to pods or pods are abraded in windy conditions	Avoid poorly drained or sheltered sites and maintain air flow through the crop to reduce humidity. Avoid damage to the growing crop and/or remove infected plants or plant parts
Peas Lupins	Powdery mildew (*Erysiphe polygoni*)	White powdery colonies that may gradually spread to cover all leaves and pods. Mainly affects marketability of fresh pods and severe attacks will reduce yield	Overwinters on plant debris and spread by conidia during season	Tends to be more severe towards the end of the season, and worse on crops in dry sheltered conditions	Good husbandry and general field hygiene may help delay establishment and spread of the disease
French beans Runner beans	Halo blight (*Pseudomonas savastanoi*)	Small water-soaked spots on cotyledons as they emerge and on new leaves. Leaf lesions turn brown and surrounded by yellowish halos. Inter-veining yellowing on leaves and finally stems. Plants may die (especially small plants) or remain stunted and	Seed-borne bacterial disease spread in splash and gaining entry through stomata. Infected pods produce infected seed	Rain splash rapidly spreads the disease	Use clean certified seed. Do not soak seed to germinate as this infects clean seed (use modules or blotting paper). Some resistant varieties are available but not reliable

		unthrifty. Pods can become infected			
French beans Runner beans	Anthracnose (*Colletotrichum lindemuthianum*)	Brown sunken stripes on stems with reddening on leaf veins which can lead to yellowing and death. Pods can be infected with reddish spots, especially visible in damp weather	Infected pods give rise to infected seeds which transmit the disease between seasons. Spores also spread in rain splash	More prevalent in cool, wet summers	Resistant cultivars should be grown
Peas French beans Broad beans	Viruses (various species depending on crop)	Mottling and mosaic symptoms on leaves that can be difficult to diagnose. Necrotic stains on seed coat are often due to viruses	Generally either seed borne or transmitted by insect vectors and hosts	Manage vectors. *See* Chapter 7, section on managing aphids etc., p. 259	Use clean certified seed. Manage vectors if possible. Rogue out severely infected plants

Table 24: Key Pests of Lilaceae (allium crops including onions, shallots, leeks, garlic, chives, asparagus)

Crop	Disease	Symptoms/Damage	Life Cycle	Monitoring/Risk	Management
Alliums	Leaf rots (*Botrytis squamosa*, *B. cinerea*)	Generally small white spots with halos or water-soaked margins present, turning to yellow or white when dried out. Eventually can cause shrivelled tips and collapse of foliage	Largely spread by airborne conidia (spores)	Difficult to distinguish between diseases and nitrogen deficiency in field. Can increase rapidly in wet conditions. Disease-forecasting models are available but often not user friendly for organic growers	Some varieties less affected. Sheltered sites and densely planted crops most prone to disease outbreaks. Isolate successive crops and don't practise successional year-round production in the same field. Ensure balance fertility avoiding an excess of N or K deficiency
Alliums	Downy mildew (*Peronospora destructor*)	Small purple grey lesions rapidly enlarge to cover foliage and lead to collapse. Spreads rapidly through the crop and becomes unsightly as secondary sooty mould infections take hold. Can cause loss of yield and damage to bulb onions in store.	Airborne spores cause dormant infections in winter onions, which causes new infections from spores produced in spring. Resistant spores (oospores) remain dormant in soil to produce disease when they germinate up to five years later	Mild winters allow fungus to overwinter and cause earlier epidemics. Fungus favoured by periods of humidity (similar to potato blight). Damage likely to be more severe with early infection	Rotation and isolation of organic crops helps to break disease cycle. Synchronous plantings avoid spread to successive crops (especially autumn sown to spring sown). Some varieties appear to be more resistant. Good crop husbandry to maintain air flow in crop

Alliums	White tip (*Phytophthora porri*)	Leaf tips show water-soaked white lesions (*see* also above) often on margins and centre as well. Leaves may hang or appear distorted. Leaf tips turn yellow and then bleached. Patchy distribution in crop	Soil-borne disease which survives as oospores on debris in soil between crops; these germinate to form sporangia (spores) that initiate new infections	Fungus infects onions in periods of heavy rainfall	Observing four-year rotation period should manage disease adequately. Harvest early if disease detected. Don't return allium packhouse waste directly to vegetable growing land without proper composting
Leeks Chives Onions	Rust (*Puccinia allii*)	Orange pustules on leaves, which reduce marketability and eventually cause death of leaves	Complex life cycle with uredospores produced on leeks but other phases of life cycle on other hosts (teliospores)	Occurs from July onwards and builds up through the year into autumn. Can be worse in mild weather as winter checks development	Varietal resistance is recognized. Infected plants should be removed and destroyed early in the season to prevent spread. Avoiding successional leek crops will limit spread but may interfere with marketing period. Balance fertility is important (avoid excess N or low K), as is good husbandry to promote plant vigour
Alliums (especially onions, garlic)	White rot (*Sclerotium cepivorum*)	Bulbs covered with white mould in which tiny black	Persistent soil-borne fungal disease which survives as	Disease active between 10–20°C and more severe in	Spread when soil is transferred or on sets and transplants so these

(Continued)

Table 24 (Continued)

Crop	Disease	Symptoms/Damage	Life Cycle	Monitoring/Risk	Management
		bodies or sclerotia can be seen. Leaves turn yellow and die back. Seedlings or plants may collapse. Often patchy in field but spreads gradually with cultivation and allium cropping	resistant sclerotia in soil which germinate when allium roots are present. Mycelia infect roots and spread from plant to plant, especially in close plantings such as salad onions. New sclerotia produced in rotting tissue	onions. Monitor by crop inspection or soil sampling in previously cropped areas	should be disease free, as should tools and footwear. Germination has been induced in the absence of host using onion waste compost, and this may be effective in managing the disease. Placing onions after brassicas in the rotation may help
Onions	Neck rot (*Botrytis allii*)	Soft necks in field and spreading to bulbs in store. After a period of storage, soft brownish rot becomes apparent at neck. Grey mould often present with spores and sclerotia visible. Can infect wounds on bulb	Seed-borne fungus spread in seeds or sets	Disease difficult to spot in the field, but widespread and present in most seasons	Use healthy, certified seed and sets. Avoid damage to the bulb when weeding mechanically. Leave a longer neck at topping (80mm or more), ensure onions are dried before storage or force air-dried where problems are expected
Onions	Soft rots (mainly *Erwinia carrotovora* or other bacteria)	Poor storage and slimy bulb rotting accompanied by a bad smell of putrification	Soil-borne bacteria present in all soils. Generally infect through wounds	Generally exacerbated by poor drainage, high N and N–K imbalance	Good husbandry to promote vigorous plants combined with rotation

Table 25: Key Diseases of Solanaceae (potatoes, tomatoes, peppers, aubergine)

Crop	Disease	Symptoms/Damage/ Period	Life Cycle	Monitoring/Risk	Management
Potato Other Solanaceae	Black leg (*Erwinia carotovora* var. *atrospetica*) (var. *carotovora* also causes soft rot of tomatoes)	Soft rotting of lower stems which become black and slimy as well as tubers which eventually rot completely. Leaves may curl and lose colour on affected stems followed by wilting. Early infections may cause non-emergence of shoots. Rotting of fruits such as tomatoes (var. *carotovora*)	Transmitted in diseased tubers and in tuber discard piles. Common on seed tubers and contaminates daughter tubers when the seed tuber breaks down. Fruit rot organisms (var. *carotovora*) ubiquitous in soil	Black leg often more severe in wet seasons. Plants stand out when monitoring crop due to leaf roll and pale colour on some stems which can be pulled out easily. Seed testing can give a risk estimate but may be unreliable depending on storage conditions	Use certified seed as free from black leg disease as possible. Minimize damage during field operations and tuber handling and keep (seed) tubers dry. Some varieties show good resistance. Lift and grade tubers before storage to remove diseased tubers and check store regularly. Keep stores well ventilated and dry
Potato Tomato	Early blight (*Alternaria solani*)	Dark brown angular spots with concentric zoning become apparent on leaves, which may go on to shrivel and drop in severe infestations. May infect tomato fruit causing a bottom rot	Persists in plant debris in soil and normally only infects plants that are already stressed	Becoming more common in midsummer but usually not serious enough to warrant action in the UK	General good crop husbandry should be sufficient to manage this disease

(Continued)

Table 25 (Continued)

Crop	Disease	Symptoms/Damage/ Period	Life Cycle	Monitoring/Risk	Management
Potato Tomato	Verticillium wilt (*Verticillium albo-artum, V. dahliae*)	Yellowing lower leaves and (often) leaves dying up one side of the plant	Soil-borne fungus that can also be introduced on seed tubers	Soil tests can be used to identify infested fields	Some varieties seem to be less susceptible. Irrigation early in crop may increase infection but reduce it later in crop cycle
Potato Tomato	Potato blight (*Phytophthora infestans*)	Brown spots on the lower leaves which rapidly enlarge in warm humid conditions into conspicuous black blotches, often surrounded by a green halo with white sporulating fungal growth. Disease spreads rapidly in field from focal points. Infection spreads to shoots and stems and plant may collapse. Spores washed into soil may also infect tubers (in potatoes) which display inwardly spreading brown/black discoloration	Persists over winter in diseased potatoes left in the ground or on discard heaps. These go on to produce diseased shoots which initiate infection by producing airborne spores which disperse most effectively in damp conditions	Most damaging disease in organic potatoes which occurs in all years although timing depends on weather conditions (Smith periods). Regional warnings or forecasts of blight outbreaks based on weather conditions can be obtained from HDC. Disease spreads fastest in warm, humid	Reduce pathogen survival by destroying cull piles and removing volunteer plants as they emerge. Try and lift all potatoes at harvest. Do not sow diseased tubers. Some varieties show at least partial resistance and some newer varieties high resistance. Earth up growing crop to protect against tuber blight. Burn off or remove

		and subsequently rot. Tomato fruit are similarly prone to infection and rot		conditions and stops in dry weather	patches of severely blighted plants when they are detected. Defoliate the crop once blight reaches 5–10 per cent severity and Smith periods are forecast or expected. Allow two weeks after defoliation for skin set and to prevent infection of potatoes whilst lifting and to allow grading out of diseased tubers before storage. Store in dry, cool and dark conditions and check regularly to grade out (and destroy) rotting tubers. Copper sprays can be effective if applied prophylactically

(Continued)

Table 25 (Continued)

Crop	Disease	Symptoms/Damage/ Period	Life Cycle	Monitoring/Risk	Management
Tomato	Leaf mould (*Fulvia fulva*)	Widespread in glasshouses (but not outdoors), causing greyish patches that turn yellow on upper surface with corresponding mould on lower surface	Survives over winter as conidia, which infect new crop. Mould produces copious spores spread on air currents and during crop care	Develops in moist warm conditions but retarded by bright sunlight	Good glasshouse hygiene and crop husbandry. Adequate ventilation, especially at lower levels, and avoidance of high temperatures. Remove lower diseased leaves. Some varieties of tomato are more resistant
Potato	Viruses (including potato leaf roll, potato virus Y, X/A, spraing and others)	Potatoes display a range of symptoms in response to virus infection (including no symptoms, rolling leaves, vein streaking, mottling or yellow streaks on foliage, roughened leaves, leaves hanging by thread, spraing in tubers). Second year symptoms are often more severe than first year symptoms	Viruses are transmitted between crops in infected tubers and also within and between crop plants by a range of vectors including mechanical	Presence of vectors, especially aphids, is likely to lead to virus transmission in crops	Use certified seed to ensure the crop has as small an initial infestation as possible. Test home-saved seed for virus contamination and do not sow where high (more than 5 per cent). Varietal resistance is available against

		and likely to reduce yields more drastically. Interactions between viruses can also lead to more severe symptoms developing	field operations (potato virus X), aphids (leaf roll, potato virus Y and A, paracrinkle viruses), nematodes (spraing (tobacco rattle)) and/or fungus (mop top virus is spread by powdery scab, a soil-borne disease of potato)		some viruses. Vector management may be possible in some cases, for example manage nematodes where spraing is a problem
Tomato and other Solanaceae	Viruses (including tobacco mosaic virus, cucumber mosaic virus)	Variable symptoms depending on conditions and crop, but including leaf mottling, fern leaves, wilting, unthrifty and stunted plants. Fruit may fail to set or drop	As with potatoes, viruses can be transmitted in seeds, mechanically, and/or by vectors	Scout crops for symptoms	Use certified seed and raise plants in hygienic conditions. Don't transplant diseased plants, and transplant early. Remove and destroy infected plants to prevent

(Continued)

Table 25 (Continued)

Crop	Disease	Symptoms/Damage/ Period	Life Cycle	Monitoring/Risk	Management
					spread within crop. Vector management may help to delay spread
Potato	Silver scurf (*Helminthosporium solani*)	Causes silvery blotches visible on washed tubers. Develops during storage, reducing market value. Infection increases water loss from tubers, which may subsequently shrivel	Spreads by airborne spores during storage and spread by infected seed tubers		Early harvest and cool dry storage conditions help to reduce disease severity
Potato	Stem canker/ black scurf (*Rhizoctonia solani*)	Present as hard black patches that can be scraped off tubers causing lesions on sprouts that may girdle stem leading to prolific sprouts, thinner stems and delayed emergence. White mould may be visible around stem. Resulting plants may be stunted, yellow, deformed (aerial tubers) and yield weakly	The pathogen is a common soil organism and can survive saprophytically on plant debris. Hard black sclerotia lie dormant in soil or adhere to tubers to give rise to new infections	Especially troublesome on light sandy soils in cold dry conditions. Early planted potatoes emerging in cold soil are at risk. Pathogen loading on seed can be checked using lab services	Practise a long rotation between potato crops, but be aware that alternative hosts may sustain the pathogen. Plant chitted seed. Shallow planting can aid rapid emergence, bypassing infection. Use certified seed free from infection. Harvest promptly in infested soils

Potato	Black dot (*Colletotrichum coccodes*)	Lesions with tiny black dots on tuber, but also roots, stolons and stems	Survives in the soil and crop debris for long periods and spread on infected seed tubers	Very common during senescence and normally only a problem in stressed crops or in dry warm seasons	Use clean certified seed. Some varieties appear to be more susceptible and should be avoided in problem soils. Early harvesting and storage in cool dry conditions limit impact of disease
Potato	Common scab (*Streptomyces scabies*)	Angular corky scabs common on tubers; serious when extensive on tuber, reducing marketability although yield is only marginally affected. Also found on beets, radish, swedes and turnips	Disease organism is caused by a complex of soil-borne actinomycetes. Spores produced in scabs and released into soil where organisms survive	Most severe in hot, dry summers and on light and alkaline soils	Some varieties more resistant to the disease and should be used in problem soils. Avoid liming before potato crops and maintain soil organic matter. Any waste should be thoroughly composted or discarded. Irrigation during tuber formation can reduce disease severity

(Continued)

Table 25 (Continued)

Crop	Disease	Symptoms/Damage/ Period	Life Cycle	Monitoring/Risk	Management
Potato	Powdery scab (*Sponogospora subterranea*)	Open scabs on tuber producing brown powder spores and cankerous outgrowths. Vector of mop top virus	Spores persist in soil and disease can also be introduced on seed tubers	More prevalent in heavy soils, and tubers more likely to be infected in wet conditions	Varietal resistance if known. Plant clean seed tubers. Avoid waterlogging during irrigation and practise good soil management. Do not feed infected tubers to stock as manure can infect clean fields
Potato	Dry rot (*Fusarium* spp.)	Infection through wounds on tuber causing undefined brown rot spreading through tuber which may also become wrinkled. Tubers rot in ground	Common soil fungi that persist even in the absence of potato host	Tuber susceptibility increases during storage, especially in high temperatures	Avoid damaging seed potatoes, which facilitates pathogen attack. Plant clean, healthy seed and avoid planting rotted tubers even if sprouting
Potato	Gangrene (*Phoma exigua* var *foveata*)	Sunken black areas develop on tubers in storage with more or less large internal cavities which fill with grey or black mould. Well defined edges to the rot. Crops grown from	Tuber or soil-borne disease. Tuber infected during lifting and skin damage during grading	Encouraged by low storage temperatures	Care needs to be taken in grading not to damage tubers and spread disease. Some varieties more resistant. Cure graded seed to reduce disease development

		infected seed may establish poorly			in storage. Remove infected seed before planting
Potato	Pink rot (*Phytophthora erythroseptica*) and other tuber rots, watery wound rot (*Pythium ultimum*), rubbery rot (*Geotichum candidum*))	Pink rot causes tubers to rot from the heel end. Rotted areas turn pink when exposed to air and smell vinegary. Can cause root rot which leads to wilting where infection high. Other rots produce general rot symptoms	Generally soil borne, surviving as resistant spores	Pink rot worse in dry weather with high temperatures and on heavy land. Rots often associated with high temperatures, damage to tubers and soil problems such as compaction or poor drainage	Long rotation between crops and good husbandry and soil management. Grade out infected tubers before storage
Tomato Peppers Aubergines	Grey mould (*Botrytis cinerea*)	Flowers, buds and fruit shrivelled with grey fluffy mould developing. Can also infect stem and other plant parts. Common in glasshouses	A weak but widespread pathogen that infects through wounds and when plants are stressed. Survives on dead or decaying plant material and spreads by airborne conidia		Good husbandry avoiding damp, densely planted stands with high humidity. Increase ventilation in glasshouses and maintain good hygiene

Table 26: Key Diseases of Poaceae (cereals (Gramineae) including wheat, barley, oats, sweetcorn)

Crop	Disease	Symptoms/Damage/Period	Life Cycle	Monitoring/Risk	Management
Cereals	Brown foot rot (*Microdochium nivale* and *Fusarium* spp.)	Very poor plant establishment and a thin crop due to seedling blights. Can also lead to root rotting, brown foot rot and leaf blotching and even glume blotch and ear blights	Seed-borne pathogen. Common in soil and capable of living saprophytically	Wet weather conditions during flowering and grain formation result in ear infection and high levels of seed-borne infection	Plant clean, disease-free seed. General promotion of soil health will help to suppress the disease
Cereals	Eyespot (*Pseudo-cercosp-orella herpitric-hoides*)	Fungal disease of stems causing lesions around the stem from flowering preventing translocation of nutrients to developing grain	Overwinters in crop residues and volunteers, spread by rain-splashed and wind-blown spores	Early sown crops at most risk, as are crops in minimum tillage systems. Diagnostics can identify strains present but not always linked to disease severity at end of season	Resistant varieties (although they may be lower yielding), ploughing in residues, spring crops normally less affected, rotation and break from cereals to reduce disease, control of grass and cereal volunteers, undersowing may help prevent spread by rain splash
Cereals	Barley yellow dwarf virus (BYDV)	Symptoms vary with many different factors but include yellowing or reddening of leaves, stunting, upright posture of stiffened leaves,	Transmitted by aphids that pick up the virus from infected volunteers and	Young plants are most susceptible. Mild winters allow greater aphid survival and higher	Manage volunteer cereals and grass margins to manage aphid vectors. (*See* Chapter 7 for information

		reduced root growth, no heading and yield reduction	grasses. Virus needs live host or vector to survive	risk of early transmission	on management of foliage pests including aphids, p. 211)
Cereals	Septoria (*Septoria tritici* and *S. nodorum*)	Leaf blotch disease causes most damage when it attacks the flag leaf reducing photosynthesis and yield	Spread of spores within canopy. Pycnidia are found on dead overwintering leaves of (winter) wheat	Wet and windy conditions promote the spread of the disease from the soil to lower leaves and progressively higher leaves up the plant. Cereal mixtures may promote disease spread as can short crops or those crops with canopy overlap	Resitant varieties, increased spacing between plants, undersowing can help prevent infection from soil
Cereals	Mildew (*Erysiphe graminis*)	White pustules on ears and leaves turning from white to grey or brown	An obligate parasite of cereals surviving on volunteers and late tillers	Favoured by relatively cool and humid conditions, rapidly growing crops are the most susceptible, especially in nutrient-rich soils	Varietal mixtures have been observed to slow down the disease
Cereals	Rusts (yellow and brown) (*Puccina* spp.)	Orange to yellow pustules on leaves and spreading to stems, glumes and ears in severe cases	Overwinters on volunteers and crops. Many crop-specific strains of the disease	Brown rust favoured by dry windy days which disperse spores, and cool nights with dew favour the build-up	Plant resistant varieties and manage crops to avoid continuous cereal cropping

(Continued)

Table 26 (Continued)

Crop	Disease	Symptoms/Damage/Period	Life Cycle	Monitoring/Risk	Management
				of the disease. Yellow rust favoured by mild winters and cool, moist summers	
Cereals	Bunt or stinking smut (*Tilletia tritici*)	Plants may be slightly stunted but otherwise symptomless. Grain replaced by bunt balls of greasy smelly spores. Infected field may have fishy smell	Transmitted in seed, and spores germinate in seed growing up plant to infect ear and new seed. May also survive in dry soil and spores blown from infected to clean crops	A problem arising from infected seed	Seed needs to be rigorously cleaned and seed from heavily infected fields rejected
Sweetcorn	Smut (*Ustilago maydis*)	Large white or grey swellings on affected cobs bursting to release black spores with smaller galls on leaves and stems. Similar infection may be seen in other cereals	Spores are wind dispersed and can persist in soil for many years. Smut spores can be found on seed but importance is uncertain	Although spectacular, the disease is normally confined to one or a few plants. Favoured in warm to hot weather	Rogue out infected plants and destroy. Rotation and avoidance of fields should any problems become persistent. Some varieties resistant to some strains

10

Diagnosing and Solving Pest and Disease Management Problems

In this book we have presented an overview of pest and disease management practice in temperate organic farming systems, focusing on examples from the UK and northern Europe. However, there is more to managing pest and disease problems than in simply knowing this information. When confronted with a pest or disease it is necessary to follow a series of steps leading to an answer to the question 'Is it worth taking action against this pest or disease?' In this final chapter we attempt to provide the framework that will enable farmers or growers to answer this question for themselves in any specific situation. Hopefully in this book we have provided the raw information that will enable them to begin to do this, and we will, in short, draw on and synthesize the information presented in the previous nine chapters to show how this can be done. This chapter should also be read with an eye on 'Further Information' (p. 407), which provides a good starting point for exploring the large amount of information and knowledge available on organic pest and disease management.

DECISION MAKING

Pest and disease management is really about making a decision on whether to take action against a pest or disease. Once the basic decision to take action has been made then a secondary decision will need to be made about what action to take. Both of these will depend on the context in which these decisions are to be taken, and the information available to the farmer or grower to make them. In short, decision making is likely to be about allocating resources to meet needs within the farm system, of which the resource requirements for pest and disease management is only a part, whilst recognizing that these resources are likely to be limited in some way. An obvious limit is the amount of money available to invest in a solution, but many other limits might also exist, such as space to run an extensive rotation, conflicting demands on time due to weather conditions, and/or material resources available to construct a bespoke solution. Further limitations might be in some sense 'self-imposed' by social or ethical concerns about wider societal issues such as fair trade, biodiversity loss and climate change.

This underlines the point that there are two basic ways of making decisions: positive and normative. In this case positive is used in the economic (or scientific) meaning of the word: that is, a decision based on things that can be measured or quantified. A normative decision is one that incorporates subjective criteria into the decision-making process. Normative decisions are concerned with the goals of the farmer or grower and answering the question 'What ought to be done?', whereas positive decisions are concerned with collecting information and data, evaluating it and answering the question 'What is happening, what is being done, what is the optimal solution?' Needless to say the distinction between these two types of question can be very important to organic farmers and growers who have to balance the conflicts of running a business (that works on the basis of money in the bank and cash flow, a positivistic approach) with organic and/or other ethical principles (that define larger personal and societal goods and goals, a normative approach).

Although both approaches should be used in decision making for managing pests and diseases, it should also be recognized that, in many cases, the two will conflict. Indeed the higher costs of organic produce are, in many ways, a reflection of this basic clash. Thus an economically optimum solution in the short term – for example, apply an insecticide to kill an insect pest – can be contrary to the organic principle of ecology – for example, build biodiversity to encourage natural enemies to control the pest – which is likely to cost more, at least in the short-term time scale of many business decisions.

Below we suggest a conceptual but practical 'model' that farmers and growers can use in their decision-making processes. It is not meant to be proscriptive, but to assist them in following a series of logical steps that work to gather information about a problem, and which might enable them to make better decisions about their pest and disease management practices. At each step the farmer or grower can ask a series of questions of themselves and their observations, which will lead to the decision either to continue to monitor the situation, or to take the next step towards taking some action to change it. Here we merely outline the steps. In the subsequent sections we elaborate on each of them and make more detailed comments on what is required at that specific stage. We would emphasize that this provides a rough framework that farmers and growers can use, but which is really just a common-sense approach of observing the crop, identifying and defining specific problems, researching solutions and then implementing them with the potential of modifying them or reiterating the process if the actions don't result in the desired outcome.

Step 1. Continual monitoring: All crop protection decisions will start with the identification of potential problems in the field. These are likely to be the result of crop walking and other monitoring activities (*see* Chapter 6). If done systematically they can be used to flag up potential problems when, for example, disease or pest presence passes a certain level (Step 2).

Step 2. Problem recognition and definition: The most basic decision to be made is also, and surprisingly, often the most difficult to come to. Crop monitoring programmes are likely to reveal a large number of pest and disease species, but deciding whether they will go on to become a problem can be surprisingly difficult. The most basic piece of information concerns the identification (diagnosis) of the pests or diseases on the crop, and many of the subsequent decisions will flow from this. Once identified, the problem needs to be defined and the risk of damage assessed (*see* Chapters 7–9). Following this, solutions can be proposed (Step 3), or if it is decided that no problem exists, a return to the continual monitoring programme (Step 1).

Step 3. Generation of alternative solutions: The generation of solutions will depend on the definition of the problem (Step 2) and, more importantly, determining what the causes of the problem are. Once the problem and its causes are defined, it is possible to identify potential solutions that treat the causes (*see* Chapters 2–5). This is probably one area in which organic pest management has provided a lead to IPM practice in looking at pests and diseases as not the cause of yield loss, but rather a symptom of deeper causes (such as nutrient imbalance in the host plant). Once potential solutions are generated, and at least some of them are achievable, they need to be evaluated (Step 4); if there are no achievable solutions, return to Step 1.

Alternaria on cauliflower – difficult to recognize and define.

Step 4. Evaluation solutions: Proposed solutions are best evaluated by setting the goals or criteria to be achieved and quantifying the various alternatives as much as possible as to value, cost, risk and/or other relevant factors (*see* Chapter 6). There are many ways of doing this, and an increasing number of (essentially) business decision-making systems and tools are available. However, the important concept is that proposed solutions should be compared on a like-for-like basis. Once a comparison has been made, a decision can be made as to the one that best meets the goals originally defined (Step 5), or it may be necessary to return to Step 3 to evaluate other solutions if no clear-cut decision can be made.

Step 5. Choose solutions: A decision to choose one particular management strategy or tactic should be based on the information gathered in the preceding steps. Sometimes a choice might be obvious – if one solution is better on all criteria, for example. More often this will not be the case. Sometimes logical approaches can help. For example, prioritizing goals or outcomes can sometimes determine which the best solution is. Once chosen, the solution needs to be implemented (Step 6), or occasionally it may be necessary to return to Step 3 or 4 to gather more information.

Step 6. Implement actions: A solution needs to be implemented, and this is best done by making a plan and monitoring the outcomes (return to Step 1).

MONITORING, PROBLEM RECOGNITION AND DEFINITION (STEPS 1 AND 2)

Monitoring

Farmers and growers should constantly monitor their farms or holdings as concerns pest and disease presence. It pays to learn about the common pests and diseases on crops in any region or locality. Pests and diseases will come and go with the seasons and will vary between the seasons. Monitoring will be necessary to capture and evaluate the meaning of this variability. Monitoring methods, including sampling, have been discussed in more detail in Chapter 6, and this should be consulted for more information. Here we reiterate the importance of keeping monitoring records to build up a picture of how pests and diseases are changing over time on the farm.

All this information can be used to ask questions about pests and diseases, and to formulate strategies for dealing with changing circumstances. It is likely that circumstances and the types of pests and diseases will change with time in response not only to specific farm management practices but also to wider factors outside the control of the farmer. Obvious and important factors in this respect include short-term variation in weather and longer-term change due to climate change and reduced availability of resources.

Diagnosis

Diagnosis will necessarily be part of a monitoring and sampling programme (*see* above and Chapter 6). In order to understand the risks posed by pests and diseases, or even to confirm their presence, it is first necessary to identify which of them are present. The diagnostic process is best carried out as a series of systematic observations on the crop and pest, leading to the identification of the pests or pathogen in question. Diagnosis may be difficult at first, but common pests and diseases will quickly become familiar, enabling much quicker identification of any pests and diseases present on a crop plant. Some of the information sources useful for diagnosis are discussed below, and others are presented in the appendix (*see* Sources of Information).

Diagnosis will depend on the observations made during monitoring and sampling (*see* above and Chapter 6). Basic information that will help in a diagnosis, and which should be routinely included on record sheets, includes field information – location, soil type, slope, crop history, cultivation history, neighbouring crops and so on – crop species and variety, as well as crop information – sowing/planting date, weeding dates, amendment or spray application (dates, products and quantities), irrigation dates – and seasonal information (temperature, rainfall, humidity). As concerns pest and disease symptoms, the date of the first appearance should be noted, along with a description of the symptoms and an estimate of incidence or coverage or population (*see* the section on sampling in Chapter 6, p. 168). It should be noted that many aids to recording information now exist (digital cameras, mobile phones, weather stations and suchlike), and that these can make many of these tasks much easier. In fact the problem may become one of too much information rather than scarcity of information.

White blister on brassica – is it a problem?

The actual diagnosis of a problem will follow from the information collected on the symptoms, and will generally run from the general to the specific. Such information will help to build a picture of where the problem is in a field, how it is distributed, and how it is developing over time. It will also help in evaluating how damaging it is likely to be. The steps in doing this will normally include some or all of the following:

Site history: Is there a history of pests and/or disease in the field? What has been done in the field? Are there any topographical features that might lead to problems (for example slope, different soil type, water channels)? Are there any nutrient problems associated with the field?

Rainfall, temperature and humidity: Has there been excessive or a lack of rainfall? Has humidity been high? Have seasonal temperatures been high or low?

Crop history: What stage is the crop at? When was the crop sown? What other field operations have been carried out (for example weeding, or applying amendments, or treatment against pests)?

Distribution: Is the problem confined to one or a few plants? Is there more damage near the edges? Does damage coincide with other obvious factors (for example standing water)? Is the affected area increasing in size with time? Does the increase coincide with other factors (for example weather)?

Plant symptoms: What are the symptoms? Are they local to specific tissue, or systemic in the plant? Are there signs of defoliation? What is the pattern of defoliation? Are pest species present? Is the plant growing abnormally? Are there signs of disease (for example moulds or growths)?

By combining such information, and with the use of identification guides, it should be possible to narrow down, or even definitively identify, the cause of a problem (*see* below and Further Information, p. 407). In the cases where it is not possible to reach a firm conclusion it may be necessary to send samples off to a specialist laboratory or advisory service (*see* Further Information, p. 407) that might be able to identify the problem. These services are normally provided for a fee, or for the cost of membership of the organization concerned. The level of certainty or reliability in the diagnosis is often a reflection of the fee paid!

Problem Definition

Normally the monitoring and diagnostic process will identify the pest and disease problem in a crop. It will also provide much useful information in terms of defining the problem. This has been discussed in Chapter 6 in terms of the probability of the pest causing economic damage to the crop, and in other chapters as the risk to the farmer or grower suffering crop or monetary loss. A final consideration at this point will be a definition of the scope for action that the problem demands.

Evaluation of Risk

Even though a pest or disease has been identified in the crop, it does not mean that it is causing a problem. Evidence gathered as part of the diagnostic process will aid in defining the problem (*see* above). Other sources of information, such as specialist books, leaflets and websites on pests and diseases, will also be helpful in this respect (*see also* Chapter 6 and Further Information p. 407). Farm records from previous seasons will be invaluable in helping to decide whether the presence of the pest or disease is likely to develop into an infestation that will cause economic harm. Records of previous crop damage can be analysed to predict likely damage in the present season, and these may be combined with forecasting or other information that may be obtained to build up a picture of what is likely to happen.

Other supplementary information can also be used to judge risk (*see also* Further Information). For example, long range weather forecasts can help predict likely periods of, say, abnormally cold or warm weather, which will influence the spread of airborne disease, as will periods of high rainfall and humidity. In a similar way, counts of natural enemies during the monitoring process may help in making decisions about whether to treat a pest infestation or not. The types of factor that will influence risk will also change from species to species, and many of these risk factors have been indicated for specific pests and diseases in the tables (Tables 9–26) or in the text in Chapters 7–9.

In the end, the experience of the farmer or grower, supported by the observations and information gathered during monitoring and diagnosis, will determine if the infestation, as identified and defined, presents sufficient risk to production for action to be taken. Once sufficient risk is deemed to be present in the crop, potential solutions to the specific problem need to be proposed and evaluated.

Don't forget to monitor the presence of natural enemies in the crop.

Scope of Action

Once a pest or disease problem has been diagnosed and the risk defined, the timescale over which the problem is likely to be present will also have an influence on the solutions proposed. There is in fact a limited range of likely scenarios at this stage, and these are briefly outlined below, together with an indication of which types of solution are likely to be appropriate in each case.

Immediate action necessary: In some cases immediate action may be necessary. For instance, pest or disease damage may be increasing in the crop and other factors are combining to make economic damage highly likely. Experience might also dictate that damage nearly always arises from a certain level of infestation. In these cases direct control measures may need to be employed, which act quickly to reduce pest or disease multiplication and/or spread. Such adaptive techniques have been discussed in Chapters 4 and 5. In the longer term, for instance in future seasons or when evaluating crop rotation plans, it will be worth taking preventative measures that stop the pest or disease reaching the crop, or which suppress it in other ways. Many techniques for doing this are discussed in Chapters 3 and 4.

Action can be delayed: Risk may be such that it is uncertain that damage is likely to accumulate to the point at which it is worth taking action. Such a situation could arise where the cost of taking action is high, or past experience indicates that actual damage rarely develops to high levels, or that the crop is likely to be harvested before damage becomes serious or affects actual yield. In these cases it may be worth taking preventative measures in the rotation that prevent the pest or disease building up in favourable years or seasons. Many of the preventative measures discussed in Chapters 3 and 4 will be applicable in these circumstances.

No action can be taken: In some cases the damage already done does not justify any preventative action as the crop is effectively destroyed or will only in any case yield poorly. In some circumstances the crop will be at, or close to, the point of harvest. In other cases experience might dictate that damage rarely if ever follows from leaving the pest or disease to develop. Such low key pests or diseases may not warrant any specific pest or disease management over and above the preventative methods discussed in Chapter 3. They should, however, be monitored as it is possible that their status may change over time as the farming system or factors external to the system, such as climate, change with time.

Crop loss: In some cases damage may be so bad that the crop is effectively lost. Although this in itself might not merit any action, the remains of the crop might need to be destroyed to prevent the multiplication and build-up of the pest or disease, and the infection of neighbouring crops or crops following on in the rotation.

Powdery mildew in squash often occurs too late to have much effect.

GENERATION AND EVALUATION OF ANSWERS (STEPS 3 AND 4)

Generating and evaluating solutions to problems are in fact intertwined processes that depend on gathering information, learning and communication. Whilst many of the techniques that can be used in pest and disease management have been discussed during the course of this book, we indicate below some of the ways and methods farmers and growers might use to go about learning more about these techniques, and combining it in new ways to produce new knowledge to use in managing pests and diseases on their farms in ways that suit their farming conditions and farming goals.

Knowledge and Learning Approaches

Generating solutions to problems is fundamentally about learning. There are two aspects to learning: understanding the context in which the problem has arisen, and learning about the pest and/or disease and how it has been managed in other situations. Farmers and growers should use a variety of methods and approaches to learn about pests and diseases, as many sources have different approaches and/or opinions about the information available. All such background information will provide useful knowledge about potential solutions to pest and disease problems as they arise. We would suggest that the following are among the most useful sources of information when learning about a pest and/or disease, some of it generated on the farm, and some obtained from external sources.

Records and notes: The information generated from continual monitoring is invaluable when combined with the 'external' information obtained below. Keeping good records of where, when and what pests and diseases occur around the farm holding is an invaluable skill which might have to be learnt in itself. Linking pest and disease infestations with other records (such as weather and yield) will also help in building up a more complete picture on which to take decisions in future seasons. There are many technological aids to doing this, which can make the process much easier – for example digital cameras, free database software and so on.

Leaflets, literature reviews and books: A large number of leaflets and books have been produced about pests and diseases in commercial horticulture in temperate regions. Much of this information has been generated from intensive research programmes, but a great deal has also been generated by farmers and growers themselves, and much is being added on a day-to-day basis. Although not all of it is useful for organic farmers and growers (for example the large body of information on pesticides), much of it is (for example the detailed information on pest behaviour, comparisons of cultural techniques, biological control information, and so on). Refer to Further Information p. 407 for some basic texts. It is worth finding a few reference texts that are comfortable to use, and learning to use them.

The internet: Many leaflets, and increasingly books, are available, at least in part, on the internet. However, although a great deal of information is accessible through the internet, it should be treated circumspectly and wherever possible confirmed in different ways. Obviously sites that have been subject to some sort of quality control (for example university research sites) will contain better information than sites without such control (but not always). Many research institutes and advisory groups have on-line information, but some of it may not be in a state appropriate for on-farm use (for example, research results that have not been necessarily verified on farm). Commercial companies produce sites with a lot of good information, but this should be tempered with the knowledge that, in the end, they are selling a product. The agricultural trade press puts a lot of information on line, as do scientific journals, but may require a subscription or one-off payment (pay per view) to get all the information.

Trade press: Agricultural magazines and in-house journals produced by certifying bodies or other such interest groups are also a good source of practical information. They often help to bring up-to-date news to farmers and growers, including relevant research results and, perhaps more importantly, case studies of practical experience (*see* below) and the opinions of other growers on specific problems or issues.

Open days, shows and workshops: Many research institutes and certifying bodies organize open days, trade shows and/or learning workshops for farmers and growers. Farmers' groups (*see* below) may also do this. Such forums enable farmers and growers to meet each other and to meet representatives of other stakeholder groups such as researchers, equipment manufacturers and seed company representatives. In short, meetings of this sort can be an invaluable source of a great deal of information in quite a short time period if time (and sometimes money!) can be found to attend.

Field walks: This is one of the most useful ways of learning about practical pest and disease management on the farm. General field walks will often focus in on specific pests and diseases and ways of managing them. The best farm walks will aim to create a discussion between farmers/growers, advisers, researchers and other interested parties. Some field walks concentrate on specific pest and disease topics, but the strength of field walks is that they allow an open discussion of solutions arrived at in many different farming situations, and allow many other factors to be brought in, which often serves to generate new knowledge.

Farmers' groups: Farmers' growers' groups can be formed around specific themes and/or problems, and serve to bring members together to discuss issues and tackle problems together. They are often aimed at business improvement. Focused discussions on business improvements and cost cutting will often focus in on specific pest or disease problems, and such forums can be a good way of putting pest and disease problems in the context of the overall farm business. Such groups will often arrange visits to members' farms and/or other exchange visits that can bring in valuable ideas.

Case studies: These are an effective way of summarizing complex information about farm systems, and also help to put pest and disease problems in perspective. They are often promoted in periodicals and the farming press because they are an effective way of communicating complex solutions to problems, and an invaluable source of acquiring information about farm practice. In some cases they might also help to develop 'benchmark' practices against which farms can judge themselves. Benchmarking is especially popular for costings, enabling farm businesses to identify areas in which they might improve, and is likely to be increasingly used for energy and carbon audits.

Field walks are invaluable for learning new information.

Advisory and research services: These services are geared up to provide information to farmers and growers, although the quality of the information will vary. Advisers are normally contracted to provide specific advice either on a whole farm/business basis or on specific problem areas. Such services can be costly, but the experience of the adviser can often compress the time taken to acquire a range of solutions to a specific problem, and may be invaluable when it comes to evaluating alternatives (*see* below). Research results are often context specific and of less immediate use to farmers and growers because they are often obtained under controlled conditions which may not be relevant to normal farm working practice. However, they can be a good source of innovative ideas which can be developed on farm or with the help of an adviser.

Experimentation: Altering practice on farm and documenting effects and change can be a powerful way of investigating solutions to problems if done systematically, and if records are kept. Most farmers and growers experiment with solutions to problems on their farms, and juggle a range of factors and constraints to arrive at answers. Such basic research is more powerful when shared between farmers and with advisers.

Evaluation of Solutions

Evaluation is a more or less logical process that seeks to choose the best option from a range of options, in this case regarding the management of pests and diseases. Evaluation will require that different management options are spelt out, together with the goals or criteria that need to be attained in order for the management actions to be considered effective or a success. In terms of pest and disease management this will normally be to reduce the level of damage to some tolerable level within a given time frame.

It might therefore be necessary to take action immediately against potato blight that has been detected on 10 per cent of the potato plants in a field, and where it is necessary to prevent the spread of rot to the tubers; whereas the management of wireworm damage in potatoes will require action to be taken over the rotational period if damage to 20 per cent of tubers is only discovered at the point of harvest, but it is desired to reduce it to very low levels the next time potatoes are cropped in that particular field. In both cases any proposed solutions should be evaluated against the stated goals and the practicability of applying them over the time scale defined. Other important factors will be costs (*see* Chapter 6) and risks to the business of not taking action (*see* above). In all cases the devil is likely to be in the details and in the ability to gauge the effectiveness and costs of the various proposed solutions.

A whole branch of management theory has been devoted to designing and producing decision tools that can be used to evaluate courses of action, and much of it is beyond the scope of this book. In general a simple and understandable approach is best which compares proposed solutions on a like-for-like basis. For example, simply designing a table (or a matrix) to include different solutions together with costs, likely effectiveness, the risks of both not doing anything and implementing the solution, as well as any anticipated side effects (such as killing natural enemies), can often go a great deal of the way to helping pick out the best solution in any specific circumstances.

In more complex cases it can be useful to break down the proposed solutions into separate components and to look at these separately in order to build up to the wider picture. By breaking down the solution in this way it might even be possible to combine different elements to provide a more effective overall solution. It may also be beneficial to build up decision trees at this stage to evaluate the various 'what if' alternatives that might arise as a solution is implemented, or to take in a multifaceted approach to managing a problem pest or disease. Researchers have built up complex decision-making and evaluation models for (usually) specific pests and diseases, and these might be useful to farmers and growers, although they are not usually very user-friendly.

CHOOSING ANSWERS AND TAKING ACTION (STEPS 5 AND 6)

Most of the preceding steps will help to narrow down the choice of control method to one or two obvious courses of action. Sometimes the likely most effective solution will not be the most cost effective, and at other times the impact on other farm operations (side effects) may be too great to justify the implementation of a particular solution. Often it may be necessary to take different courses of action. For instance it is quite usual that effective immediate action needs to be taken to save a situation, and then to follow this up with longer term preventative action aimed at preventing a recurrence of the pest or disease at damaging levels. Occasionally the optimum solution will be best in all categories, and no compromise will be necessary. Very occasionally time and money might permit an experimental evaluation of alternative solutions, and this might provide a much clearer choice (*see* above).

In any case, once a solution, or solutions, to a particular problem has been chosen, it should be implemented following a plan of action. Once again whole management books are devoted to this subject, and the point will not be laboured here. Suffice to say that a well thought-out plan, with contingencies for anticipated or even unanticipated events, consistently applied and carried through, stands a better chance of being effective than a poorly planned one that is half-heartedly implemented. Part of the skill of farm management lies in planning pest and disease management programmes and fitting them in with other farm operations without losing the ability to respond flexibly to new situations as they arise.

USING INFORMATION AND KNOWLEDGE

Pest and disease management in organic farming systems is about using information and knowledge to manipulate the farm ecosystem to keep pests and diseases at low and tolerable levels, and ideally at levels below which they are causing economic damage to the business of the farmer or grower. In this book we have taken the approach of exploring the reasons that pests and diseases occur in organic farm systems, the organically acceptable methods that can be used to manage them, and finally those pests and diseases that are likely to be encountered. We hope that a recurring theme running through the book is one of finding out about and attacking the root causes of pest and disease problems, and not the symptoms. As a final summary we advise that farmers and growers ground their pest and disease management practice in sound scientific knowledge and ecological theory.

In particular we would hope that most farmers and growers will now appreciate the importance of finding out information on the most common pests and diseases, and begin to build up a picture of their biology and ecology. This would include ideas about what their preferred conditions are, and their likely natural enemies. As a minimum, farmers and growers should then take steps to create conditions for these natural enemies and beneficials in their farm system, at the same time as learning about management methods that are likely to do the least harm in this respect. All pest and disease management decisions should be made with an idea of what, if any, is the likely impact of doing nothing, and certainly with an understanding of what the costs (and side effects) of any action taken are likely to be. In particular, learning will play a valuable role to designing a farm system that will accommodate pest and disease species without them necessarily becoming a problem.

So we hope that farmers and growers will come to appreciate the value of observing, experimenting and recording their own pest and disease management practices, and in sharing them with other farmers and growers – indeed, to realize that building and developing such a shared resource is as rewarding and beneficial as the farming and growing itself.

Hoverfly on fennel.

Further Information

There is a wide range of good books on identification and control of crop pests and diseases. Some of the more useful for organic farmers and growers are detailed below.

BIBLIOGRAPHY: BOOKS AND KEY REVIEW PAPERS

Agrios, G.N., *Plant Pathology* (Academic Press Inc., 2005).

Altieri, M.A., Nicholls, C., & Fritz, M.A., *Manage Insects on Your Farm: A Guide to Ecological Strategies* (Sustainable Agriculture Network, Handbook Series Book 7, 2005). Available at http://www.sare.org/publications/insect.htm

Barbosa, P. (ed.) *Conservation Biological Control* (Academic Press, 1998).

Briggs, Stephen *Organic Cereal and Pulse Production: A Complete Guide* (Crowood Press, 2008).

Buczacki, S., *Plant Problems, Prevention and Control* (David and Charles, 2000).

Buczacki, S.T. & Harris, K.M., *Pests, Diseases and Disorders of Garden Plants* (3rd edition) Collins Photo Guides (Collins, 2005).

Cavigelli, M.A., Deming, S.R., Probyn, L.K. & Mutch, D.R., *Michigan Field Crop Pest Ecology and Management* (Michigan State University Extension Bulletin E-2074, 2000).

Chaboussou, F., *Healthy Crops. A New Agricultural Revolution* (John Carpenter Publishing 2004). An English translation of a French publication published by Flammarion in 1985.

Cloyd, R.A., Nixon, P.L., & Pataky, N.R., *IPM for Gardeners: A Guide to Integrated Pest Management* (Timber Press, Cambridge, 2004).

Coleman, Eliot *The New Organic Grower: A Master's Manual of Tools and Techniques for the Home and Market Gardener* (A gardener's supply book) (Chelsea Green, 1995).

Cubison, Stella *Organic Fruit Production and Viticulture: A Complete Guide* (The Crowood Press, 2009).

Davies, Gareth (ed.) and Lennartsson, Margi *Organic Vegetable Production: A Complete Guide* (The Crowood Press, 2006).

Daxl, Rainer; von Kayeselingk, Niels; Klein-Koch, Carols; Link, Rolf; and Waibel, Hermann *Integrated Pest Management Guidelines* (GTZ GmbH, Germany, 1994).

Dufour, R., *Farmscaping to Enhance Biological Control* (ATTRA, 2000). Available at www.attra.ncat.org/attra-pub/farmscape.html

Gladders, P., Davies, G., Wolfe, M. & Haward, R., *Diseases of Organic Vegetables* (ADAS, 2001).

Greenwood, P. & Halstead, A., *RHS Pests and Diseases* (Dorling Kindersley, 2009).

Gullan, P.J. & Cranston, P.S., *The Insects. An Outline of Entomology* (2nd edition) (Blackwell Science, 2000).

HDRA *Veg Handbook*

Herrera, C.M. & Pellmyr, O., *Plant Animal Interactions: An Evolutionary Approach* (Blackwell Publishing, 2002).

Hill, D.S., *Agricultural Insect Pests of Temperate Regions and their Control*

(Cambridge University Press, 1987). Sections available in Google books at http://books.google.com/

Holland, J. & Ellis, S., *Beneficials on Farmland: Identification and Management Guidelines* (HGCA Publications, 2008). Available at www.hgca.com

Ingram, David & Robertson, Noel *Plant Disease* (part of the New Naturalist Series) (Harper Collins, 1999).

Jones, F.G.W. & Jones, M.G., *Pests of Field Crops* (3rd edition) (Edward Arnold, 1984).

Koike, S.T., Gladders, P. & Paulus, A.O., *Vegetable Diseases: A Color Handbook* (Academic Press, 2006).

Lampkin, Nicolas *Organic Farming* (Farming Press UK, 1998).

Lampkin, Nic; Measures, Mark; and Padel, Susanne *2009 Organic Farm Management Handbook* (University of Wales, 2008).

Mahr, D.L., Whitaker, P. & Ridgeway, N., *Biological Control of Insects and Mites* An introduction to beneficial natural enemies and their use in pest management (Cooperative Extension Publishing, University of Winsconsin-Extension, 2008). Available at http://learningstore.uwex.edu/pdf/A3842.pdf

O'Toole, C. (ed.) *The New Encyclopedia of Insects and their Allies* (Oxford University Press, 2002).

Price, P.W., *Insect Ecology* (3rd edition) (John Wiley, 1997).

Reijntjes, Coen, Haverkort, Bertus and Waters-Bayer, Ann *Farming for the Future. An Introduction to Low External Input and Sustainable agriculture* (ILEIA, Netherlands, 1992).

Rosenfeld, A. & P. Sumption, P., *Methods of Controlling Pests and Diseases in Organic Field Vegetables* (Institute of Organic Training and Advice: Research Reviews, 2009). Available from IOTA at www.organicadvice.org.uk/papers/Res_review_23_pests_field_vegetables.pdf

Ruberson, J.R., *Handbook of Pest Management* (CRC Press, 1999). Limited sections available in Google Books.

Speight, M.R., Hunter, M.D. & Watt, A.D., *Ecology of Insects. Concepts and Applications* (Wiley-Blackwell, 2008).

Speiser, B., Glen, D., Piggott, S., Ester, A., K. Davies, K., Castillejo, J. & Coupland, J., *Slug Damage and Control of Slugs in Horticultural Crops* (FiBL, Switzerland, 8 pages). Available at http://www.slugcontrol.rothamsted.ac.uk/SlugsBrochure.pdf and associated information at http://www.slugcontrol.rothamsted.ac.uk/

Wilson, Phil & King, Miles *Arable Plants – a field guide* (WildGuides and English Nature, 2003. Also available at http://www.fieldguide.co.uk/?P=home&SHC=2&PSD=1

Zehnder, G., Gurr, G.M., Kuhne, S., Wade, M.R, Wratten, S.D. & Wyss, E., *Arthropod Pest Management in Organic Crops* (published in the *Annual Review of Entomology* 52:57–80, 2007) Available on http://orgprints.org/13126/

LEAFLETS

ATTRA, the National Sustainable Agriculture Information Service. A US-based information service for sustainable and organic agriculture with many online leaflets and sources of information (at http://attra.ncat.org/publication.html). Seasonal information should be treated with caution in areas other than the US, but leaflets include:

- *Deer Control Options*
- *Farmscaping to Enhance Biological Control*
- *Flea Beetle: Organic Control Options*
- *Greenhouse IPM: Sustainable Whitefly Control*
- *Nematodes: Alternative Controls*
- *Organic Alternatives for Late Blight Control in Potatoes*
- *Organic IPM Field Guide*
- *Thrips Management Alternatives in the Field*

BPC (UK British Potato Council) website on pests and diseases with numerous factsheets at http://www.potato.org.uk/department/knowledge_transfer/pests_and_diseases/index.html?menu_pos=knowledge_transfer) including black leg, blight, slugs, wireworms and cutworms.

Fera (Food and Environment Research Agency (UK)) Fera's Plant Clinic provides free pest and disease factsheets at http://www.fera.defra.gov.uk/plants/

plantClinic/plantFactsheets.cfm including (among others):

- *Colorado Beetle*
- *Downy Mildews*
- *Fusarium and Verticillium Wilts*
- *Grey Mould*
- *Phytophthora Diseases*
- *Powdery Mildews*
- *Rusts*

HGCA (Home Grown Cereal Authority (UK levy body)) guides (www.hgca.com) including:

- *Beneficials on farmland: identification and management guidelines*
- *Controlling gout fly on wheat*
- *Insect pests of oilseed rape*
- *Orange wheat blossom midge – assessment and control*
- *Pest management in cereals and oilseed rape – a guide*
- *Predicting and controlling wheat bulb fly*
- *Slug forecasting in cereals*
- *The biology and control of cereal aphids*

HDC (Horticultural Development Council (UK levy body)) Guides and Leaflets (http://www.hdc.org.uk/) including:

- *Brassica Crop Walkers Guide*
- *Carrot Cavity Spot*
- *Carrot and Parsnip Crop Walkers Guide*
- *Control of Bean Seed Fly in Allium Crops*
- *Control of White Rot in Onions*
- *Herb, Pest and Disease Cards*
- *Identification guides: Pests and Diseases*
- *Management of Alternaria Blight on Carrots*
- *Management of Celery Leaf Spot*
- *Managing Rabbit Problems Associated with Horticulture*
- *Slug Control in Field Vegetables*
- *Swede Midge Control in Brassica Crops*
- *Vegetable Diagnostics (illustrated CD Rom)*

Institute of Organic Trainers and Advisers (IOTA) has some reports and links to the organic eprints service. http://www.organicadvice.org.uk/index.htm including:

- *Controlling pests and diseases in organic field vegetables*
- *Organic plant raising*

Natural England (UK government adviser on natural environment) have a wide range of leaflets with advice on wildlife problems that may occur and how these can be resolved (see http://www.naturalengland.gov.uk/ourwork/regulation/wildlife/advice/advisoryleaflets.aspx) including:

- *Badger problems: use of electric fencing to prevent agricultural damage*
- *Electric fence reference manual*
- *Managing deer in the countryside*
- *Moles: options for management and control*
- *Rabbits: management options for preventing damage*
- *Rabbits: use of fencing to prevent damage*
- *Rabbits: use of cage-trapping to prevent agicultural damage*
- *Recommendations for fallow, roe and muntjac fencing*
- *The management of damage by brent geese*
- *The management of problems caused by Canada geese: a guide to best practice*

NIAB produce the *Organic Vegetable Handbook* (latest edition 2007) as well as a number of pocket books (including the *Potato Pocket Book*, the *Veg Finder* and the *Combinable Crops* pocket books) which contain a good deal of information on varietal performance under UK conditions. The information is also available on their website at www.niab.com to members.

Ontario Pest and Disease Control (Canadian Government website at http://www.omafra.gov.on.ca/english/crops/organic/orgpests.htm) with factsheets including:

- *Alternatives to Insecticides for Managing Vegetable Insects*
- *Manage Insects on Your Farm: A Guide to Ecological Strategies*
- *Managing Insects and Diseases in Organic Crops*
- *Managing Late Blight in Organically Produced Tomato and Potato*

Organic Centre Wales Technical Notes (from Institute of Rural Sciences, University of Wales, Aberystwyth, Ceredigion, SY23 3AL) (http://www.organic.aber.ac.uk/publications/technical.shtml) including:

- *Biology and management of soil pests* Technical Note No. 3, 2005
- *Biology and management of slugs* Technical Note No. 4, 2005
- *Biology and management of wireworms* Technical Note No. 5, 2005
- *Biology and management of leatherjackets* Technical Note No. 6, 2005

Scottish Agricultural College produces a range of technical notes (available at http://www.sac.ac.uk/consulting/services/c-e/cropclinic/cropadvice/technotes/) including:

- *TN607 Winter Wheat Disease Control*
- *TN605 Impact of Climate Change in Scotland on Crop Pests, Weeds and Disease*
- *TN603 Soil-Dwelling Free-Living Nematodes as Pests of Crops*
- *TN602 Clubroot Disease of Oilseed Rape and Other Brassica Crops*
- *TN601 Ergot Disease in Cereals*
- *TN582 Managing Set-aside and Fallows for Crop Protection*
- *TN580 Crop Protection in Reduced Tillage Systems*
- *TN552 Wheat Bulb Fly and Other Pests of Cereals*
- *TN551 Pests of Swedes and Turnips: Their Management and Control*
- *TN548 Diseases of Peas*
- *TN492 Aphids and Aphid-Borne Viruses in Potato Crops*

Soil Association Fact Sheets from Producer Services (Soil Association South Plaza, Marlborough Street, Bristol BS1 3NX) including:

- *Materials for Pest and Disease Control in Organic Crops*
- *Slug Control in Organic Systems*

WEBSITES

Websites can be located around the world, and as such seasons, varieties and acceptable treatments legally allowable in organic systems and/or by control and certification bodies may vary. Farmers and growers are asked to bear this in mind when consulting these sites, many of which contain a lot of useful information.

Chemicals Regulations Directorate (UK government body): To ensure the safe use of biocides, industrial chemicals, pesticides and detergents to protect the health of people and the environment (formerly PSD). Information at http://www.pesticides.gov.uk/home.asp

Cooperative extension system (US): Organic agriculture from extension, an interactive learning environment from the land-grant universities across the US. http://www.extension.org/organic%20production

Defra (UK government): Department of Environment, Food and Rural Affairs including many links to important organic information (http://www.defra.gov.uk/farm/organic/index.htm)

Doctor Fungus: A website with descriptions and further information on fungi including agriculturally important pathogens at http://www.doctorfungus.org/mycoses/agri/agri_index.htm

Farming Futures: An umbrella body for a number of UK farming interests and the government that produce a range of factsheets on energy use and climate mitigation strategies at http://www.farmingfutures.org.uk/x360.xml

Farming and Wildlife Advisory Group (FWAG): The UK's leading independent and dedicated provider of environmental and conservation advice to farmers: http://www.fwag.org.uk/

Food Environment Research Agency (FERA) (http://www.fera.defra.gov.uk/): A UK government agency responsible for developing a sustainable food chain. Includes pest and disease information and diagnostic services and leaflets on a range of topics.

Garden Organic (formerly HDRA): A good number of factsheets available free to members on pests and diseases. Vegetable network website. Available at www.gardenorganic.org.uk

IFOAM (International Federation of Organic Agriculture Movements): Principles and contacts for the world wide organic movements at www.ifoam.org

Institute of Organic Trainers and Advisers: Some reports and links to organic e-prints service. http://www.organicadvice.org.uk/index.htm

International Centre for Research in Organic Food Systems (ICROFS): A 'centre without walls', where the research is performed in interdisciplinary collaboration between research groups in different institutions and universities, and whose website provides a lot of research information (*see* organic e-prints as well) at http://www.icrofs.org/

NFU (National Farmers Union): A UK organization that champions and represents farmers' interests and rights. Website has useful information relevant to crop protection practices and other issues at http://www.nfuonline.com/ Leaflet on Bird Scarer's Code of Practice at http://www.nfuonline.com/x5317.xml

Ohioline Farm: Online factsheet service at http://ohioline.osu.edu/lines/farm.html

Online guide to plant disease control: A general guide to disease control with factsheets and pictures but with many references to chemical control. http://ipmnet.org/plant-disease/intro.cfm

Organic e-prints: Online repository of research papers on organic farming including pest and disease management. Free and searchable. http://www.orgprints.org/

Organic Futures: Aims to promote low carbon food from rich soil farming at http://www.organicfutures.org.uk/about_us.html

Organic Growers Alliance: Aims to bring (UK) organic growers together to share knowledge and experience at http://www.organicgrowersalliance.co.uk

Organic Inform (http://www.organicinform.org/): A newsletter and information service for organic farmers and growers.

Organic Research Centre – Elm Farm: Research into organic farming and farming policy in the UK at http://www.efrc.com/including pest and disease research.

OrganicXseeds (UK): A database required under EU Regulation (EC) No. 1452/2003, which regulates the use of seeds and seed potatoes in organic farming. It can be used to search for organically available seed at http://www.organicxseeds.com/oxs/do/Login?paramCountry=188 Some companies might not enter their availability of organic seed on this database and may need to be contacted individually.

People's Trust for Endangered Species (Mammals Trust UK): Concerned with reversing the decline of many of our UK mammals and has factsheets on all UK mammals (including potential pest species) at http://ptes.org/

Pest Spotter from Bayer Crop Sciences (UK): Has a good guide to pests on crop pests and can be searched by crop, season and pest. Available at http://www.pestspotter.co.uk/content.output/292/709/e-Tools/Pestspotter/PestSpotter.mspx along with Slugspotter. Contains a lot of advice on pesticide use which is not applicable in organic systems.

Plant Pathology Internet Guide Book: A resource guide for plant pathology, applied entomology and related fields. http://www.pk.uni-bonn.de/ppigb/ppigb.htm

Radcliffe's IPM World Textbook: General information on IPM and some factsheets at http://ipmworld.umn.edu/

Resource Guide for Organic Insect and Disease Management: An online resource book outlining general principals, with factsheets and appendices on additional practices for organic pest and disease management. http://www.nysaes.cornell.edu/pp/resourceguide/index.php

RHS Help and Advice: Online factsheets on pests and diseases and other information mainly aimed at gardeners but relevant to smallholders and farmers at www.rhs.org.uk/advice/problems_archive.asp

Rodale Institute: An independent organic research organization in the US at http://www.rodaleinstitute.org/home

Rothamstead Research: Cutting-edge research on insect plant ecology, and some information of interest to farmers and growers at http://www.rothamsted.bbsrc.ac.uk/pie/

RSPB (Royal Society Protection Birds): Pages on farming advice for biodiversity: http://www.rspb.org.uk/ourwork/farming/

UC IPM Online: A large number of factsheets on integrated pest management in arable and vegetable crops from the University of California. Most factsheets indicate acceptable organic methods (in the US). Available at www.ipm.ucdavis.edu/index.html

Vegetable MD online: A number of vegetable disease factsheets of relevance to organic management: http://vegetablemdonline.ppath.cornell.edu/

***Which?* gardening factsheets:** Although aimed at gardeners, contain good information on the life histories of pests and diseases and their management at smaller scales. Available at http://www.which.co.uk/advice/gardening-factsheets/pest-andamp-diseases-a-l/index.jsp

PEST AND DISEASE FORECASTING

HDC Pest Bulletin: A pest-forecasting bulletin (for the UK) collated and published by Warwick HRI using information by Rothamsted Research (Richard Harrington & Mark Taylor), Warwick HRI (Rosemary Collier) and their collaborators. Covers brassicas, lettuce leeks and onions, carrot and parsnip, and cutworms. (*available at* www2.warwick.ac.uk/fac/sci/whri/hdcpestbulletin

MORPH Decision Support Software: Allows pest and disease management decisions made by growers every day to be supported by complex research ideas and concepts. Available at http://www2.warwick.ac.uk/fac/sci/whri/research/morph/

Crop monitor: Provided for the HGCA for farmers (in the UK) to provide information sourced from monitoring sites located across the country and reporting up-to-date measurements of crop pest and disease activity in arable crops throughout England. All data gathered are analysed to identify disease and pest risk, seasonal variation in disease development, and the effectiveness of control strategies. Users can be alerted to emerging threats during the growing season, and advised on appropriate courses of action. Available at www.cropmonitor.co.uk/index.cfm

UK Potato Council Blight Watch Service: Contains pages with information on all potato pests and diseases, as well as access to blight warnings during the season at http://www.potato.org.uk/blight (requires registration).

DIAGNOSTIC SERVICES

Fera Plant Health: Diagnostic services (for the UK) and some factsheets (especially on emerging and new diseases) available from CSL. Follow instructions on the website to use the services. Available at http://www.fera.defra.gov.uk/plants/plantClinic/

Plant Health Solutions: Provides independent research and development, consultancy, testing and diagnostic services for horticulture and agriculture. More information available at http://www.planthealth.co.uk/

Index of Pests and Diseases

General Index